Interconnecting Electronic Systems

Interconnecting Electronic Systems

Jerry C. Whitaker

Gene De Santis

C. Robert Paulson

CRC Press
Boca Raton Ann Arbor Boston

Library of Congress Catalog Card Number 92-73269.

This book represents information obtained from authentic and highly regarded sources. Reprinted material is quoted with permission, and sources are indicated. A wide variety of references are listed. Every reasonable effort has been made to give reliable data and information, but the authors and the publisher cannot assume responsibility for the validity of all materials or for the consequences of their use.

Direct all inquiries to CRC Press, Inc., 2000 Corporate Blvd., N.W., Boca Raton, Florida 33431.

International Standard Book Number 0-8493-7407-3.
Printed in the United States of America.

Contents

Preface

From the beginning of electronics, the concepts of system design have evolved and engineering practices have been developed. Signal parameters, connector and cable specifications, and equipment-mounting dimensions have all been standardized. Most of the equipment and hardware used to assemble systems today are available from a number of manufacturers; end-users do not have to custom-build their components. These advancements have helped to reduce significantly the engineering design time required for a given project. Many systems of advanced design with superior performance and improved operating efficiency have resulted. Veteran engineers and technical managers are familiar with these practices and standards. However, this is not necessarily the case for less experienced engineers or new engineers who are just entering the electronics industry.

This handbook has been written to establish a foundation for interconnecting audio, video, data, and radio frequency systems. It describes the steps required to take a project from basic design to installation and completion.

This handbook examines the tasks and functions for which the *system engineer* will generally be responsible. It discusses steps required to complete complex projects. For smaller projects, these steps can be implemented easily and—in some cases—certain steps and documentation can be simplified or eliminated without compromising the success of the project.

Although small projects can be completed by a single engineer, larger projects require the system engineer to work with many other people. The reader will realize that the structure of different organizations within companies varies greatly, as do the responsibilities of the individuals who make up the organization.

Within any company, the function of the engineer will vary. An understanding of the workings of a project organization can help engineers understand their responsibilities and deal with misunderstandings and changes. For this reason, the first chapter of this handbook deals with the important topic of project organization and management. Project tracking, reporting methods, and changes to a project have also been included to show the need for complete documentation, especially on large, costly projects. Many organizations have engineering departments that have established standards for building systems for internal use or, in the case of system integrators who build turnkey systems for their clients, for installation at the client's facility. Either way, this handbook will serve as a valuable reference.

The system engineer is responsible for specifying all of the details of how a facility will be built, and it is that person's responsibility to communicate those details to the contractors, craftsmen, and technicians who will actually build and install the hardware and software. The system engineer is further responsible for installation quality and ultimate performance.

Successful execution of these responsibilities requires an understanding of the underlying technology and the applicable quality standards and methods for achieving them. This handbook is dedicated to that effort.

Contributors

- Ken Ainsworth, Tektronix, Beaverton, OR.
- Michael W. Dahlgren, Novell Service Division, Provo, UT.
- Bradley Dick, *Broadcast Engineering* magazine, Overland Park, KS.
- Marc S. Walker, Broadcast Television Systems, Salt Lake City, UT.

CHAPTER 1

PLANNING FOR FACILITY CONSTRUCTION

1.1 Introduction

When the owner or company executive of a technical facility decides to proceed with new construction or the renovation of a technical plant, this person will probably enlist the help of a qualified architect and general contractor. However, another important team member should be included—the *system engineer*. This person may already be a member of the staff, an outside engineering consultant, or a system integration contractor. Without a system engineer, design and development will not proceed as quickly, and a greater chance exists for serious, costly miscalculations.

The system engineer plays a major role in developing a successful facility plan. A good system engineer has a wealth of technical information. With it, this person can expedite the planning and design process considerably, and can help keep overall facility costs down. Many architectural firms cannot provide technical equipment planning on an appropriate level for technical facilities. And, on an industrial or commercial level, they probably cannot provide anything beyond cursory services, unless they bring in an outside system engineering consultant.

Researching a manufacturer's literature to determine the heat output or power consumption of one piece of equipment, for example, can be a time-consuming task. If an architect has no previous experience with the equipment, this approach can cost time and money. Most owners delegate technical facility planning to their operations and maintenance engineering personnel. This is unfortunate because even with their hands-on experience and knowledge about the equipment, they often find themselves unable to devote the necessary time to the project while they're handling normal duties. And, no matter how knowledgeable these technical people are, they often lack the necessary engineering skills and design experience of a system engineer. There is no substitute for a qualified, experienced system engineer.

1.2 The System Engineer

As an electrical engineering graduate, the system engineer has the training and knowledge required for designing electronic systems. This person has hands-on experience with electronic equipment assembly and testing techniques. With this knowledge, the system engineer can avoid the pitfalls that are often encountered by those with less education and/or experience.

An experienced system engineer is already familiar with design techniques, drawings, and specifications. This person also shares the architect's skills in reading drawings and visualizing real, 3-D environments. A familiarity with facility layout requirements enables the system engineer to recognize and avoid costly problems. It can be heartbreaking, for example, to see the space set aside for future expansion being consumed by the last-minute addition of more equipment racks.

Experienced system engineers, because they are familiar with proper construction and wiring techniques, can also assure quality workmanship on the job. They can catch and correct work that has been done improperly. In other words, an experienced system engineer is management's eyes and ears.

1.2.1 Outside Engineering Contractor

When the size of a project warrants the employment of a system engineer, a company's owner or its executives should approve the expense. The alternative, if frequent changes to a facility are not anticipated, is to go outside of the company and retain the services of a system design engineering consultant or a system integration contractor. There are many circumstances in which consulting system engineers represent the most cost-effective solution to engineering needs. Consultants contribute specialized expertise and experience to projects when these resources are not available within a firm.

Consultants can be used for peak-period or unique projects. They can help equalize the work load of permanent employees. Consultants can provide impartial analysis and can make valuable contributions regarding problems, products, and plans. Because they are independent business people, they know the value of good judgment and unbiased opinion. Consultants can be effective as catalysts for innovation and change when fresh thinking is needed on a project or in a department. Consultants can provide formal and informal training at all company levels. They can be especially useful during expansion, by recommending and training new full-time employees.

When near-term business situations are uncertain, consultants can perform the work without long-term commitments. As independent contractors rather than employees, consultants offer specialized services or skills to clients for a fee. They are reimbursed either on a fixed-fee basis, a fixed fee determined by a percentage of the equipment cost, or for time and expenses. Consultants provide their clients with expertise and specialized knowledge that are usually not available within the client's organization. Because they are exposed to a variety of situations, they have developed experience through the application of many successful approaches to business and technical challenges. Consultants are project- or task-oriented and can devote the necessary time to accomplish an assignment independent of other responsibilities.

Consultants solve problems. They are experts. They are frequently asked to implement solutions—to function as a *system integrator.*

System integration contractors provide turnkey systems design, fabrication, and installation. They can often provide the same skills and services rendered by professional consulting engineers. They can handle all phases of a project, from a project's conception through completion. As the equipment supplier, training and warranty support are also offered as a part of their contract. Typically, system integrators are reimbursed either on a fixed-fee basis or by a percentage of the equipment cost. Although a fixed-fee arrangement is usually best for the owner, because larger system integration contractors are also equipment vendors, they may prefer to charge a commission for larger projects based on equipment costs, rather than charging an engineering fee. These companies will probably specify equipment made by manufacturers with whom they have a dealer agreement. They will also probably specify products that provide them the greatest profit margin.

If competitive bids must be solicited for a project, an independent consultant should usually be retained to prepare the specification documents and a *request for proposal* (RFP). This is done to avoid any advantage or bias that may be built into bid documents prepared by a system integrator. Upon the owner's acceptance of the bid specification, an RFP can be sent to several system integrators. This process also provides checks on the specifications and may introduce some alternative ideas.

Whatever the arrangement, a qualified system engineer can help contain equipment costs, meet construction schedules, and ensure that a technical facility will perform as required.

1.2.2 Design Development

System design is executed in a series of steps that lead to an operational unit. Appropriate research and preliminary design work is completed in the first phase of the project—the *design development* phase. This phase is designed to fully delineate all project requirements and to identify any constraints. Based on initial ideas and information, the design concepts are modified until all parties are satisfied and approval is given for the final design work to begin. The first objective of this phase is to answer the following questions:

- What are the functional requirements of the product?
- What are the physical requirements of the product?
- What are the performance requirements of the product?
- Are there any constraints limiting design decisions?
- Will existing equipment be used, and is it acceptable?
- Will this be a new facility or a renovation?
- Will this be a retrofit or upgrade to an existing system?
- Will this be a stand-alone system?

The equipment and functional requirements of each of the major technical areas are identified by the engineer as this person works closely with the owner's representatives. With facility renovation, the system engineer's first step is to analyze existing equipment. This person will visit the site to gather detailed information about the existing facility. The system engineer, usually confronted with a mixture of acceptable and unacceptable equipment, must sort out the equipment that meets current standards and determine which items should be replaced. After soliciting input from the facility's technical personnel, the system engineer then develops a list of needed equipment.

One of the system engineer's most important contributions is the ability to identify and meet the owner's needs, and to do so within the project's budget. Based on the owner's initial concepts and any subsequent equipment utilization research conducted by the system engineer, the desired capabilities are identified as precisely as possible. Design parameters and objectives are defined and reviewed. Functional efficiency is maximized to allow operation by a minimum number of people. Future needs are also investigated at this time, and future technical systems expansion is considered.

After management approves the equipment list, preliminary system plans are drawn up for review and further development. If architectural drawings of the facility are available, they can be used as a starting point for laying out an equipment floor plan. The system engineer uses this floor plan to be certain adequate space is provided for present and future equipment, and adequate clearance is furnished for maintenance and convenient operation. Equipment identification is then added to the architect's drawings.

Documentation should include, but not be limited to, a list of major equipment:

- Equipment prices
- Technical system functional block diagrams
- Custom item descriptions
- Rack and console elevations
- Equipment floor plans

The preliminary drawings and other supporting documents are prepared to record design decisions and to illustrate the design concepts to the owner and/or facility manager. Renderings, scale models, or full-size mockups may also be needed to better illustrate, clarify, or test design ideas.

Ideas and concepts must be exchanged and understood by all concerned parties. Good communication skills are essential. The bulk of the creative work is carried out in the design development phase. The physical layout—the look and feel—and the functionality of the facility will all have been decided and agreed upon by the completion of this phase. If the design concepts appear feasible, and the cost is within the anticipated budget, management can authorize work to proceed on the final detailed design.

1.2.3 Level of Detail

With the research and preliminary design development completed, the design details must be concluded. The system engineer prepares complete detailed documentation and specifications necessary for the fabrication and installation of the technical systems. These include all major and minor components. Drawings must show the final configuration and the relationship of each component to other elements of the system. The drawings will also show how these components will interface with other building services, such as air conditioning and electrical power. This documentation must communicate the design requirements to the other design professionals, including the construction and installation contractors.

In this phase, the system engineer develops final, detailed flow diagrams that show the interconnection of all equipment. Cable interconnection information for each type of signal is taken from the flow diagrams and recorded on the cable schedule. Cable paths are measured and timing calculations performed. *Timed cable lengths* (used for video and other special services) are entered onto the cable schedule.

Special custom items are defined and designed. Detailed schematics and assembly diagrams are drawn. Parts lists and specifications are finalized, and all necessary details are worked out for these items. Mechanical fabrication drawings are prepared for consoles and other custom-built cabinetry.

The system engineer provides the architect with layouts of cable runs and connections. Such detailed documentation simplifies equipment installation and facilitates future changes in the system. During preparation of final construction documents, the architect and the system engineer can confirm the layout of the technical equipment wire ways, including access to flooring, conduits, trenches, and overhead wire ways.

Dimensioned floor plans and elevation drawings are required to show placement of equipment, lighting, electrical cable ways, ducts, conduits, and HVAC ductwork. Requirements for special construction, electrical, lighting, HVAC, finishes, and acoustical treatments must be prepared and submitted to the architect for inclusion in the architectural drawings and specifications. This type of information, along with cooling and electrical power requirements, also must be provided to the mechanical and electrical engineering consultants (if used on the project) so they can begin their design calculations.

Equipment heat loads are calculated and submitted to the HVAC consultant. Steps are taken when locating equipment to avoid any excessive heat buildup within the equipment enclosures, while maintaining a comfortable environment for the operators.

Electrical power loads are calculated and submitted to the electrical consultant. Also, steps are taken to provide for sufficient power and proper phase balance.

1.2.4 Management Support

The system engineer can assist in ordering equipment and helping to coordinate the move to a new or renovated facility. This is critical if a lot of existing equipment is being relocated. With new equipment, the facility owner will find the system engineer's knowledge of prices, features, and delivery times a valuable asset. A good system engineer will make sure that equipment arrives in ample time to allow for sufficient testing and installation. A good working relationship with equipment manufacturers helps guarantee their support of and speedy response to the owner's needs.

The system engineer can also provide engineering management support during planning, construction, installation, and testing to help qualify and select contractors, resolve problems, explain design requirements, and assure quality workmanship by the contractors and the technical staff.

The procedures described in this section outline an ideal scenario. Management may often try to bypass many of the foregoing steps to save money. This, they reason, will eliminate unnecessary engineering costs and allow construction to begin immediately. By using in-house personnel, a small company may attempt to handle the job without professional help. With inadequate design detail and planning, which can result from using unqualified people, the job of setting technical standards and making the system work then defaults to the construction contractors, in-house technical staff, or the installation contractor. This can result in costly and uncoordinated work-arounds and, of course, delays and added costs during construction, installation, and testing. This also makes the project less manageable and less likely to be completed successfully.

The size of a technical facility can vary from a small, one-room operation to a large, multi-million-dollar plant or network. Management should recruit a qualified system engineer for projects that involve large amounts of money and other resources.

1.3 The Project Team

The persons who plan and carry out a project compose the *project team*. The project team's makeup will vary depending on the size of the company and the complexity of the project. Management is responsible for providing the necessary human resources to complete the project.

1.3.1 Executive Management

The executive manager is the person who can authorize the project's undertaking. This person can allocate funds and delegate authority to others to accomplish this task. Motivation and commitment are important aspects for accomplishing the goals of the organization. The ultimate responsibility for a project's success lies with the executive manager. This person's job is to complete tasks through others by assigning group responsibilities, coordinating activities among groups, and resolving group conflicts. The executive manager establishes policy, provides broad guidelines, approves the project master plan, resolves conflicts, and assures project compliance with commitments.

Executive management delegates the project management functions and assigns authority to qualified professionals, allocates a capital budget for the project, supports the project team, and establishes and maintains a healthy relationship with project team members.

Management is responsible for providing clear information and goals—up front—based upon management's needs and initial research. Before initiating a project, the company executive should be familiar with daily facility operation and should analyze how the company works, how the people do their jobs, and what tools are needed to accomplish the work. An executive should consider certain points before initiating a project:

- What is the current capital budget for equipment?
- Why does the staff currently use specific equipment?
- What function of the equipment is the weakest within the organization?
- What functions are needed, but cannot be accomplished, with current equipment?
- Is the staff satisfied with current hardware?
- Are there any reliability problems or functional weaknesses?
- What is the maintenance budget, and is it expected to remain steady?
- How soon must the changes be implemented?
- What is expected from the project team?

Only after answering the appropriate questions will the executive manager be ready to bring in expert project management and engineering assistance. Unless the manager has made a systematic effort to evaluate all of the obvious points about the facility requirements, the not-so-obvious points may be overlooked. Overall requirements must be divided into their component parts. Do not try to tackle ideas that have too many branches. Keep the planning as basic as possible. If the company executive does not attempt to investigate the needs and problems of a facility thoroughly before consulting experts, the expert advice will be shallow and incomplete, no matter how good the engineer.

Engineers work with the information they are given. They put together plans, recommendations, budgets, schedules, purchases, hardware, and installation specifications based upon the information they receive from interviewing management and staff. If the management and staff have failed to go through the planning, reflection, and refinement cycle before those interviews, the company will probably waste time and money.

1.3.2 Project Manager

Project management is an outgrowth of the need to accomplish large, complex projects in the shortest possible time, within the anticipated cost, and with the required performance and reliability. Project management is based on the realization that modern organizations may be so complex that they preclude effective management using traditional organizational structures and relationships. Project management can be applied to any undertaking that has a specific objective.

The project manager has the authority to carry out a project and has been given the right to direct the efforts of the project team members. The project manager gains power from the acceptance and respect that is provided by superiors and subordinates, and has the power to act and is committed to group goals.

The project manager is responsible for the successful completion of the project, on schedule, and within budget. This person will use whatever resources are necessary to accomplish the goal in the most efficient manner. The project manager provides project schedule, and financial and technical requirement direction. This person also evaluates and reports on project performance. This requires planning, organizing, staffing, directing, and controlling all aspects of the project.

In this leadership role, the project manager is required to perform many tasks:

- Assemble the project organization.
- Develop the project plan.
- Publish the project plan.
- Set measurable and attainable project objectives.
- Set attainable performance standards.
- Determine which scheduling tools (PERT, CPM, and/or GANTT) are right for the project.
- Use the scheduling tools, and develop and coordinate the project plan. This includes the budget, resources, and the project schedule.
- Develop the project schedule.
- Develop the project budget.
- Manage the budget.
- Recruit personnel for the project.
- Select subcontractors.
- Assign work, responsibility, and authority so team members can make maximum use of their abilities.
- Estimate, allocate, coordinate, and control project resources.
- Deal with specifications and resource needs that are unrealistic.
- Determine the right level of administrative and computer support.
- Train project members how to fulfill their duties and responsibilities.
- Supervise project members, giving them day-to-day instructions, guidance, and discipline, as required, to fulfill their duties and responsibilities.

- Design and implement reporting and briefing information systems or documents that respond to project needs.
- Control the project.

Some basic project management practices can improve the chances for success. Consider the following:

- Secure the necessary commitments from top management to make the project a success.
- Establish an action plan that will be easily adopted by management.
- Use a work breakdown structure that is comprehensive and easy-to-use.
- Establish accounting practices that help, not hinder, successful project completion.
- Prepare project team job descriptions properly up front to eliminate conflict later.
- Select project team members appropriately the first time.

After the project is underway, follow these steps:

- Manage the project, but make the oversight reasonable and predictable.
- Persuade team members to accept and participate in the plans.
- Motivate project team members for best performance.
- Coordinate activities so they are carried out in relation to their importance, with a minimum of conflict.
- Monitor and minimize interdepartmental conflicts.
- Get the most out of project meetings without wasting the team's productive time. Develop an agenda for each meeting, and start on time. Conduct one piece of business at a time. Assign responsibilities where appropriate. Agree on follow-up and accountability dates. Indicate the next step for the group. Set the time and place for the next meeting. Then, end on time.
- Spot problems and take corrective action before it is too late.
- Discover the strengths and weaknesses in project team members, and manage them to obtain desired results.
- Help team members solve their own problems.
- Exchange information with subordinates, associates, superiors, and others about plans, progress, and problems.
- Make the best of available resources.
- Measure project performance.
- Determine, through formal and informal reports, the degree to which progress is being made.
- Determine causes of and possible ways to act upon significant deviations from planned performance.
- Take action to correct an unfavorable trend, or to take advantage of an unusually favorable trend.
- Look for areas where improvements can be made.
- Develop more effective and economical methods of managing.
- Remain flexible.
- Avoid "activity traps."
- Practice effective time management.

When dealing with subordinates, employees should:

- Know what they are supposed to do, preferably in terms of an end product.
- Have a clear understanding of what their authority is, and its limits.
- Know what their relationship with other people is.
- Know what constitutes a job well done in terms of specific results.
- Know when and what they are doing exceptionally well.
- Be shown concrete evidence that there are rewards for work well done and *exceptionally* well done.
- Know where and when they are falling short of expectations.
- Be informed of what can and should be done to correct unsatisfactory results.
- Feel that their superior has an interest in them individually.
- Feel that their superior believes in them and is enthusiastic for them to succeed and progress.

By fostering a good relationship with associates, managers will have less difficulty communicating with them. The fastest, most effective communication takes place among people with common viewpoints.

1.3.3 Engineering Manager

The engineering manager in a technical facility usually manages the technical staff, which may be made up of graduate engineers and technicians. The engineering manager is committed to technical quality and the functional integrity of the facility.

If a company has no project manager, the engineering manager may assume this role.

1.3.4 System Engineer

The term *system engineer* means different things to different people. The system engineer provides the employer with the experience gained from many successful approaches to technical problems developed through hands-on exposure to a variety of situations. This person is a professional with knowledge and experience, possessing skills in one or more specialized and learned fields. The system engineer is an expert in a given field, highly trained in analyzing problems and developing solutions that satisfy management objectives.

Education in electronics theory is a prerequisite for designing systems that employ electronic components. As a graduate engineer, the system engineer has the education required to design electronic facilities correctly. Knowledge of testing techniques and theory enables this person to specify system components and performance, and to measure the results. Drafting and writing skills permit efficient preparation of the necessary documentation needed to communicate the design to the technicians and contractors who will have to build and install the system.

Training in personnel relations, a part of the engineering curriculum, helps the system engineer deal with subordinates and management. A good system engineer has a wealth of technical information that can be used to speed up the design process and help in making cost-effective decisions. If the system engineer does not have the needed information, this person knows where to find it.

The system engineer performs the following functions:

- Receives input from management and staff.
- Researches the project and develops a workable design.
- Solves technical problems related to the design and integration of the system into a facility.
- Concentrates on results and focuses work according to the employer's objectives.

The degree to which these objectives are achieved is an important measure of the system engineer's contribution. In some cases, the system engineer may have to assume the responsibilities of planning and managing a project.

The system engineer's duties will vary, depending on the size of the project and the management organization. Aside from designing the system, this person has to answer questions and solve problems that may arise during hardware fabrication and installation. The system engineer must also monitor installation quality and workmanship. Hardware and software will have to be tested and calibrated upon completion, which is also a concern of the system engineer. Depending on the complexity of the new installation, the system engineer also may have to provide orientation and operating instructions to the users.

Other key members of the project team include the following:

- Architect—responsible for design of the structure.
- Mechanical engineer—responsible for HVAC and other mechanical designs.
- Structural engineer—responsible for concrete and steel structures.
- Construction contractors—responsible for executing the plans developed by the architect, mechanical engineer, and structural engineer.

Small in-house projects can be completed on an informal basis. This is probably the normal routine for uncomplicated projects. In a large facility project, however, the system engineer's involvement usually begins with preliminary planning and continues through fabrication, installation, and testing. A project's scope will determine the work that will be required of the system engineer. The scope is an outline of the endeavors to which pursuits, activities, and interests will be confined. Consequently, the scope of the project must be formulated and agreed upon by the project participants. The extent and the limits of the work also must be determined. In this case, the intent of the scope is to fully delineate the work to be carried out by the project's system engineer. Subjects to be considered include:

- What work is to be done by the system engineer.
- What the results of the work will be.
- What the end product of the work will be.

1.4 Budget Requirements Analysis

The need for a project may originate with management, operations staff, technicians, or engineers. In any case, some sort of logical reasoning or a specific production requirement will justify the need. On small projects, such as the addition of one piece of equipment, money must only be available for the purchase and installation costs. When the need justifies a large project, the final cost is not always immediately apparent. The project must be analyzed by dividing it into its constituent parts or elements:

- Equipment and parts
- Materials
- Resources, including money and time needed for project completion

An *executive summary* or *capital project budget request*, which contains a detailed breakdown of these elements, can provide the information management needs to determine the return on investment, and to make an informed decision on whether or not to authorize the project.

A capital project budget request, which contains the minimum information, may consist of the following items:

- *Project name*—a name that describes the result of the project, such as "control room upgrade."
- *Project number* (if required). A large organization that does many projects will use some kind of project numbering system, or it may use a budget code assigned by the accounting department.
- *Project description*—a brief description of what the project will accomplish, such as "design the technical system upgrade for the renovation of production control room 2."
- *Initiation date*—the date the request will be submitted.
- *Completion date*—the date the project will be completed.
- *Justification*—the reason the project is needed.
- *Material cost breakdown*—a list of equipment, parts, and materials required for construction, fabrication, and equipment installation.
- *Total material cost.*
- *Labor cost breakdown*—a list of personnel required to complete the project, their hourly pay rates, the number of hours they will spend on the project, and the total cost for each person.
- *Total project cost*—the sum of material and labor costs.
- *Payment schedule*—an estimation of individual amounts that will be paid out during the course of the project, and the approximate dates that each will be payable.
- Preparer's name and the date prepared.
- Approval signature(s) and date(s) approved.

More detailed analysis, such as return on investment, can be carried out by an engineer. Financial analysis, however, should be left to the accountants, who have access to company financial data.

1.4.1 Feasibility Study and Technology Assessment

In cases where an attempt must be made to implement new technology, and where a determination must be made as to whether certain equipment can perform a desired function, a feasibility study should be conducted. The system engineer may be called upon to assess the state of the art in order to develop a new application. In addition to a capital project budget request, an executive summary or a more detailed report of evaluation test results may be required to help management make its decision.

1.4.2 Project Tracking and Control

A project team member may be selected by the project manager to report the status of work during the course of the project. A standardized *project status report* form can provide consistent and complete information to the project manager. The purpose is to supply information to the project manager regarding work completed and money spent on resources and materials.

A project status report containing minimum information should contain the following items:

- Project number (if required)
- Date prepared
- Project name
- Project description
- Start date
- Completion date (the date this part of the project was completed)
- Total material cost
- Labor cost breakdown
- Preparer's name

1.4.3 Change Order

After all or part of a project design has been approved and money has been allocated, any changes may increase or decrease the cost. Several factors can affect the cost:

- Material
- Resources, such as labor and special tools or construction equipment
- Costs incurred because of manufacturing or construction delays

Management should know about such changes, and will want to control them. For this reason, a method of reporting changes to management and soliciting its approval should be instituted. The best way to do this is with a *change order request* or *change order.* A change order includes a brief description and reason for the change and a summary of the effect it will have on costs and the project schedule.

Management will exercise its authority to approve or disapprove each change, based upon its understanding of the cost and benefits and the perceived need for the modification of the original plan. Therefore, the system engineer should provide as much information and explanation as may be necessary to make the change clear and understandable to management.

A change order form, containing the minimum information, should contain the following items:

- Project number
- Date prepared
- Project name
- Labor cost breakdown
- Preparer's name
- Description of the change
- Reason for the change

- Equipment and materials to be added or deleted
- Material costs or savings
- Labor costs or savings
- Total cost of this change (increase or decrease)

1.5 Electronic System Design

Performance standards and specifications must be established in advance for a technical facility project. This will set the performance level of equipment that is acceptable for the system and affect the size of the budget. Signal quality, stability, reliability, and accuracy are examples of the kinds of parameters that must be specified. Access and processor speeds are important parameters when dealing with computer-driven products. The system engineer must confirm whether selected equipment conforms to the standards.

At this point, it must be determined what functions each component in the system will be required to fulfill, and how each will perform with other components in the system. The management and operation staff usually know what they would like the system to do, and how they can best accomplish the task. They should select equipment that they think will do the job. With a familiarity of the capabilities of different equipment, the system engineer should be able to contribute to this function/definition stage. Following is a list of questions that must be answered:

- What functions must be available to the operators?
- What functions are secondary and, therefore, not necessary?
- What level of automation should be required to perform a function?
- How accessible should the controls be?

Over-engineering or over-design must be avoided. Such serious and costly mistakes are often made by engineers and company staff when planning technical system requirements. A staff member may, for example, ask for a feature or capability without fully understanding its complexity or the additional cost it may impose. Other portions of the system may have to be compromised to implement the additional feature. An experienced system engineer will be able to spot this and determine whether the trade-offs and added engineering time and cost are really justified.

When existing equipment is used, an inventory list should be made. This is the preliminary part of a final equipment list. Normally, when confronted with a mixture of acceptable and unacceptable equipment, the system engineer must determine what meets current standards and what should be replaced. Then, after soliciting input from facility technical personnel, the system engineer develops a summary of equipment needs, including future acquisitions. One of the system engineer's most important contributions is the ability to identify and meet these needs within the facility budget.

A list of major equipment is then prepared. The system engineer selects equipment based on experience with the products and on owner preferences. Existing equipment is often reused. A number of considerations are discussed with the facility owner to determine the best product selection. Some major points include:

- Budget restrictions
- Space limitations
- Performance requirements

- Ease of operation
- Flexibility
- Functions and features
- Past performance history
- Manufacturer support

The system engineer's goal is to choose and install equipment that will meet the project's functional requirements efficiently and economically. Simplified block diagrams of the video, audio, control, data, RF, and communication systems are drawn and then discussed with the owner and presented for approval.

1.5.1 Developing a Flow Diagram

The flow diagram is a schematic drawing used to show the interconnections among all equipment that will be installed. It differs from a block diagram because it contains much more detail. Every wire and cable must be included on these drawings. See Figure 1.1 for a typical flow diagram of a video production facility.

The starting point for preparing a flow diagram can vary depending on the information available from the design development phase of the project, and on the similarity of the project to previous projects. If a similar system has been designed previously, the diagrams from that project can be modified to include the equipment and functionality required for the new system. New equipment models can be shown on the diagram in place of their counterparts, and minor wiring changes can be made to reflect the new equipment connections and changes in functional requirements. This method is efficient and easy to complete.

If the facility requirements do not fit any previously completed design, the block diagram and equipment list are used as a starting point. Essentially, the block diagram is expanded and details are added to show all of the equipment and interconnections, and to show any details necessary to describe the installation and wiring completely.

An additional design feature that is desirable for specific applications is the capability to disconnect a rack assembly easily from the system and relocate it. This would be used if a system was pre-built at a system integration facility and later moved and installed at the client's site. With this type of situation, the interconnecting cable harnessing scheme must be well planned and identified on the drawings and cable schedules.

1.5.2 Estimating Cable Length

Cable lengths are calculated using dimensions taken from the floor plans and rack elevations and should be included on the cable schedule. The quantity of each cable type can then be estimated for pricing and purchasing. A typical cable schedule database printout is shown in Figure 1.2.

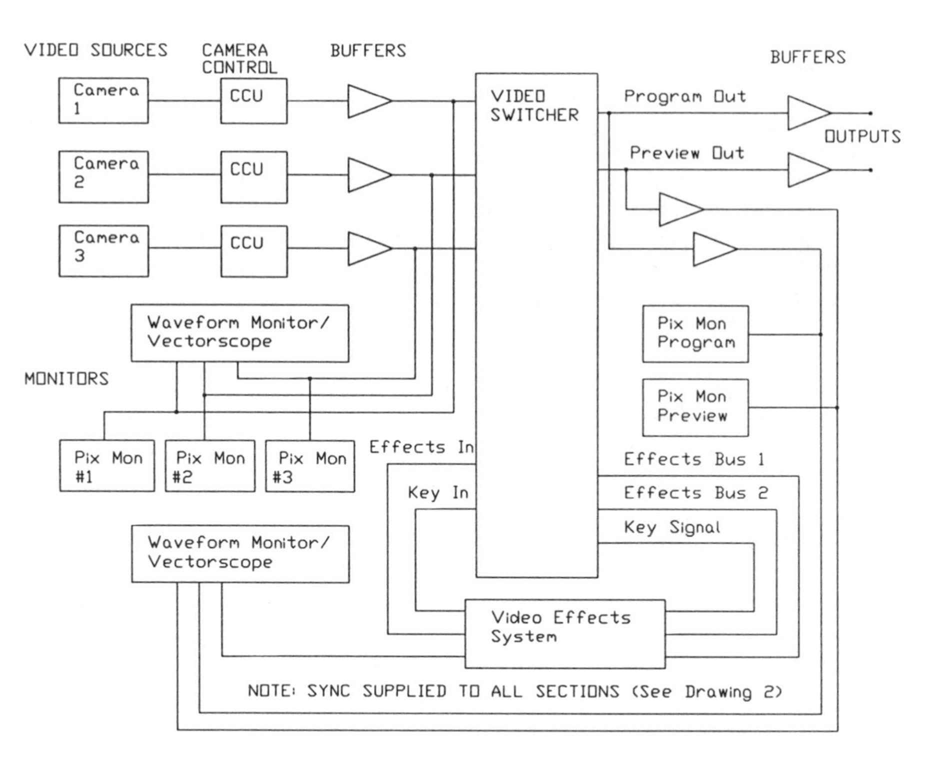

Figure 1.1 Example flow diagram of a video production facility.

1.5.3 Signal Timing Considerations

For certain signal paths, the length of the cable connecting two pieces of equipment may critically affect the timing relationship of that signal as it relates to others in the system. Calculate these critical cable lengths and include them on the cable schedule.

Electrical signals travel through cable at a velocity determined by the physical properties of the cable. Using the published value for the velocity of propagation, calculate the amount of delay in a given length of cable. The velocity for video cables may vary from 66 percent to 78 percent, depending on the manufacturer. The delay may also be determined experimentally by simply measuring the amount of delay produced in a given length of cable being used.

Most modern video equipment incorporates adjustments for the timing relationships of the output signal with respect to the input or reference signal. Some equipment with multiple inputs includes internal adjustable delay lines at the inputs to allow time phase adjustment of several untimed sources.

When building a facility that has critical timing requirements, keep cables as short as possible to minimize signal attenuation and crosstalk. This requires keeping interconnected equipment as close together as possible. It is best to locate all of the distribution equipment in the same or adjacent racks. Because most video cabling among distribution elements must be timed or of matching lengths, short cables make the job manageable, and, at the same time, cable costs are kept low.

1.5.3.A *Cable Loss and Equalization*

A cable's frequency response decreases with increasing frequency. The loss can be compensated for by using an equalizing amplifier with a response curve that complements the cable loss. For video applications, a typical distribution amplifier (DA) has six outputs isolated from one another by fan-out resistors. Because the equalization is adjusted to produce a flat response at the end of a length of a specific type of cable, all of the cables being driven by the amplifier must be the same type and length.

1.6 Facility Design

The best way to design a facility is to begin with the architectural drawings of the existing building or planned construction. If architectural drawings are not available, it is necessary to have the architect prepare them. For small renovation projects, the system engineer may prepare the needed drawings to plan equipment layout.

Before any details are confirmed, a site visit should be made to record and confirm building space dimensions, clearances, and access to building services. Also, existing rack and console dimensions and locations should be measured. If the site is a distance away, photograph important elements, such as existing construction details or current equipment configurations, to reduce the need to travel back to the site.

1.6.1 Preliminary Space Planning

Whether the project involves new construction or renovation of an existing building, current facilities and equipment are reviewed to determine a starting point for the planning process. Building and room layouts are determined by studying each function and its relationship to all others. Functional requirements of each operational department are assessed to determine the gross space requirements of areas to be expanded or renovated. Key facility personnel are interviewed to determine past experiences, future trends, operational requirements for immediate use, and future needs of the facility. This should include the number of present employees and those anticipated in the future.

Environmental factors, such as noise, vibration, RF interference, power line interference, temperature, and humidity also must be considered. Accessibility to utilities, such as communications, power, air supply, fuel, and water, must be calculated. Air conditioning is a major concern in all large facilities that employ a lot of equipment or lighting.

After management approves the equipment list, a rough schematic layout is prepared in conjunction with the architect's preliminary drawings. The system engineer examines this layout to be certain that it provides adequate space for present and future equipment and for maintenance and operation. Equipment identification is then added to the architect's schematic, and the procedure continues to the design-development phase. An example of an architectural floor plan is shown in Figure 1.3. Equipment placement in rack assemblies is illustrated in Figure 1.4.

Design renderings (drawings or paintings created by an artist or drafter to show a realistic flat or perspective view of a design) are then produced. Full-color 3-D models

WIRE NO.	DESCRIPTION	CONN	SOURCE	DESTINATION	CONN	CABLE TYPE	TIME	CABLE LENGTH	NO.OF WIRES	COLOR CODE
1031	BB1	BNC	DA01 OUT 1	CCU1 GENLOCK IN	BNC	RK7560		26'	1	YEL
1032	BB1	BNC	DA01 OUT 2	CCU2 GENLOCK IN	BNC	RK7560		26'	1	YEL
1033	BB1	BNC	DA01 OUT 3	CCU3 GENLOCK IN	BNC	RK7560		26'	1	YEL
1034	BB1	BNC	DA01 OUT 4	CCU4 GENLOCK IN	BNC	RK7560		26'	1	YEL
1035	BB	BNC	DA02 IN LOOP	DA03 IN	BNC	RK7560		1'	1	YEL
1036	BB1	BNC	DA02 OUT 1	P1 REF VID IN	BNC	RK7560		21'	1	YEL
1037	BB1	BNC	DA02 OUT 2	P2 REF VID IN	BNC	RK7560		25'	1	YEL
1038	BB1	BNC	DA02 OUT 3	P3 REF VID IN	BNC	RK7560		27'	1	YEL
1039	BB1	BNC	DA02 OUT 4	P4 REF VID IN	BNC	RK7560		29'	1	YEL
1040	BB	BNC	DA03 IN LOOP	DA04 IN	BNC	RK7560		1'	1	YEL
1041	BB1	BNC	DA03 OUT 1	P5 EXT REF IN	BNC	RK7560		29'	1	YEL
1042	BB1	BNC	DA03 OUT 2	P6 REF VID IN (FUTURE)	BNC	RK7560				
1043	BB1	BNC	DA03 OUT 3	R1 REF VID IN	BNC	RK7560		29'	1	YEL
1044	BB1	BNC	DA03 OUT 4	VTR H TBC COMP VID IN	BNC	RK7560		34'	1	YEL
1045	BB	BNC	DA04 IN LOOP	DA05 IN	BNC	RK7560		1'	1	YEL
1046	BB1	BNC	DA04 OUT 1	VTR I TBC GENLOCK IN	BNC	RK7560		34'	1	YEL
1047	BB1	BNC	DA04 OUT 2	CG1 GEN LOCK VID IN	BNC	MINI		47'	10	1
1048	BB1	BNC	DA04 OUT 3	CG2 ENC PGM IN	BNC	MINI		39'	10	1
1049	BB1	BNC	DA04 OUT 4	CG3 GEN LOCK VID IN	BNC	RK7560		10'6"	1	YEL
1050	BB	BNC	DA05 IN LOOP	DA51 IN	BNC	RK7560		6'	1	YEL
1051	BB1	BNC	DA05 OUT 1	DVE VBS/S IN	BNC	RK7560		8'9"	1	YEL
1052	BB1	BNC	DA05 OUT 2	TBC GENLOCK IN	BNC	RK7560		5'6"	1	YEL
1053	BB1	BNC	DA05 OUT 3	TCG 2 VID IN	BNC	RK7560		5'5"	1	YEL
1054	BB1	BNC	DA05 OUT 4	CG3 CC SYNC IN	BNC	RK7560				WHT
1055	TBC ADV SYNC OUT	BNC	VTR H TBC ADV SYNC OUT	VTR H SYNC IN	BNC	RK7560		4'6"	1	WHT
1056	TBC ADV SYNC OUT	BNC	VTR I TBC ADV SYNC OUT	VTR I VID IN 1	BNC	RK7560		4'6"	1	WHT
1057	SC TO VCR	BNC	TBC VTR SC OUT	VTR I SC IN	BNC	VP618PE		4'6"	1	BLK
1058	BB1	BNC	P5 REF VID IN LOOP OUT	P5 TBC COMP VID IN	BNC	RK7560		6'	1	YEL
1059										
1060	BB2	BNC	DA51 OUT 1	TEST SRCE SWR IN 1	BNC	RK7560	E	7'6"	1	YEL
1061	BB2	BNC	DA51 OUT 2	PS EXT REF IN	BNC	RK7560	G	25'7"	1	YEL

Figure 1.2 Wiring database printout for a portion of the facility illustrated in Figure 1.1.

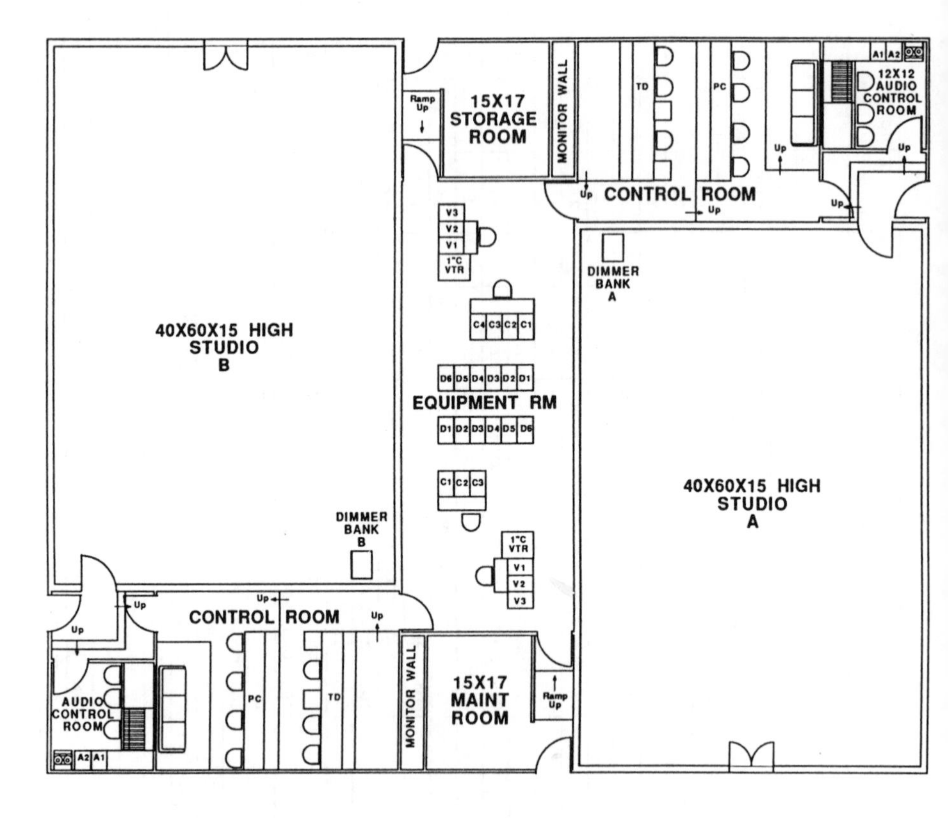

Figure 1.3 Architectural floor plan of a new facility.

can be generated by a computer for viewing from different perspectives. The printout of any view can be used as the rendering.

A color and materials presentation board is usually prepared for review by decision makers. The presentation may include the following:

- Artist or computer renderings
- Color chips
- Wood types
- Work surface laminates
- Metal samples
- Samples of carpeting, furniture fabrics, and wall coverings

Several different combinations may be prepared. The samples and renderings are attached to a board or heavy paper stock for easy presentation.

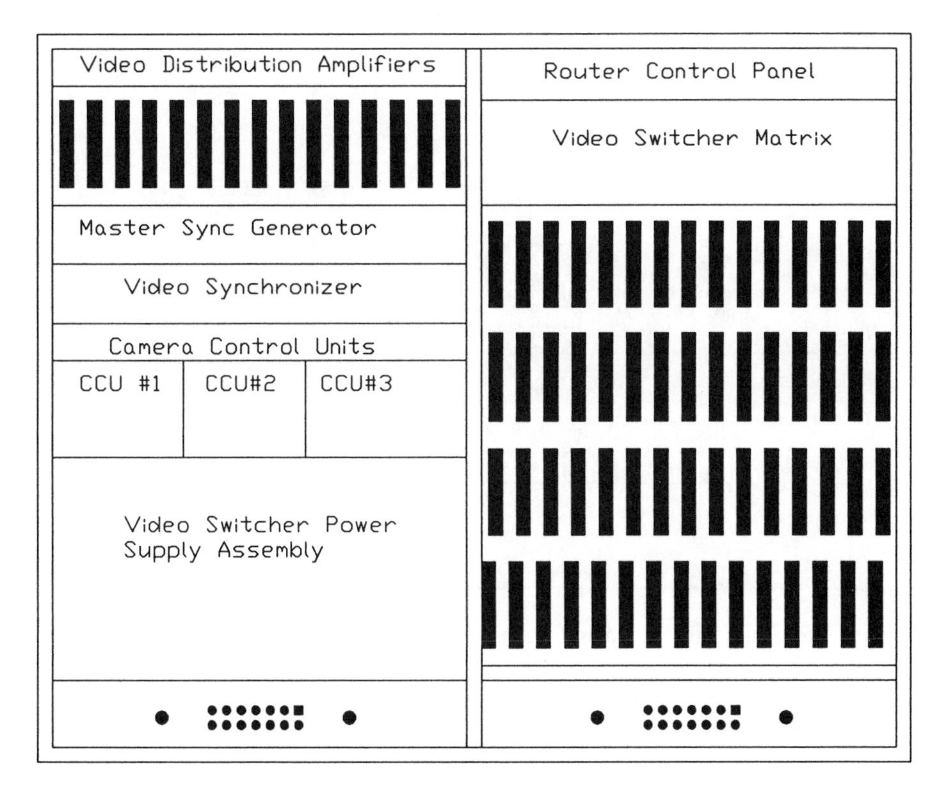

Figure 1.4 Equipment placement in a rack assembly.

1.6.2 Design Models and Mockups

When a drawing cannot be interpreted easily by the owner and/or staff, a scale model or full-size mockup of the facility (or portions of it) can be constructed. This will help familiarize them with the design, allowing them to make decisions and changes. Models may also be used to present a design concept to company executives. Models can provide a cost-effective way to evaluate new ideas. Inexpensive materials can be formed to represent racks, consoles, or equipment. For example, the top and four side views of an enclosure can be drawn or plotted at a reduced scale on stiff paper so that the drawings touch at adjoining surfaces. When cut out, they are glued together to form a 3-D model. Flaps added to the drawing make it easier to join the surfaces. The more detail provided in the mockup, the better.

Blocks of wood can also be cut to the shape of the equipment being modeled. Cut-out drawings of the equipment features are pasted on the block's surfaces to add realism. Plastic scale models of structural components, piping, furniture, and other elements are available from model manufacturers and can be used to enhance the presentation.

Full-scale mockups, like models, can be built using any combination of construction materials. Stiff foam board is a relatively easy and inexpensive material to use to prepare full-size models. Pieces can be cut to any shape and joined to form 3-D models

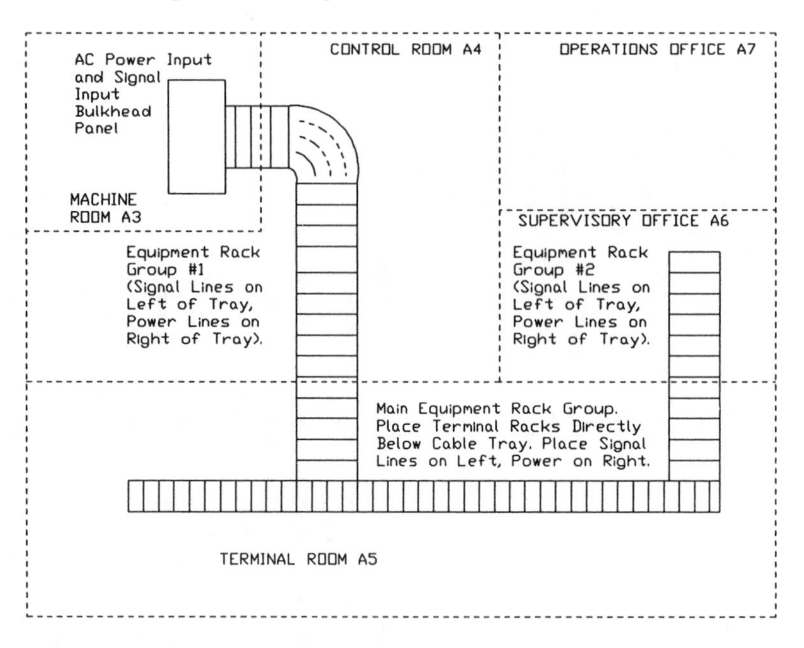

Figure 1.5 Cable wire way plan for a new facility.

of racks, consoles, and equipment. Actual-size drawings of equipment outlines, or more detailed representations, can then be pasted in place on the surfaces of the mockup.

1.6.3 Construction Considerations

Demolition and construction of existing structures may have to be specified by the system engineer. Electrical power, lighting, and air conditioning requirements must be identified and layout drawings prepared for use by the electrical and mechanical engineering consultants and the architect.

During preparation of final construction documents, the architect and the system engineer can confirm the layout of technical equipment wire ways, including access to flooring, conduits, trenches, and overhead raceways. At this point, the system engineer also provides layouts of cable runs and connections. This makes equipment installation and future changes much easier. An overhead cable routing plan is shown in Figure 1.5.

When it is necessary to install coaxial cables in conduit, follow National Electrical Code (NEC) requirements for conduit fill and the number of pull boxes. More pull boxes or larger conduit is required in conduit runs that have many bends. Specify direct-burial-type cable when the conduit or cable trays are underground, and where there is a possibility of standing water. Conduit and cable trays should be designed to accommodate the minimum bend radius requirements for the cables being used. The recommended minimum bend radius for coaxial cable, for a single permanent bend, is 10 times the cable diameter.

Debur and remove all sharp edges and splinters from installed conduit. Remove construction debris from inside the lines to prevent damage to the cable jacket during pulling. Cover openings to the conduit to prevent contamination or damage from other construction activities. If the cable is damaged during pulling, moisture could enter the cable.

1.6.4 Component Selection and Installation

Equipment selection is normally based on the function it will perform. User input about operational ease and flexibility of certain models is also important. However, to ensure that the most cost-effective choice is made, consider certain technical issues before making a decision. The system engineer should research, test (when required), and provide the technical input needed for selecting hardware and software.

Equipment features and functional capabilities are probably the main concerns of the users and management. Technical performance data and specifications are important considerations that should be contributed by the system engineer during the selection process. For a piece of equipment to qualify, technical specifications must be checked to ensure that they meet set standards for the overall system. Newly introduced products should be tested and compared before a decision is made. The experienced system engineer can test and measure the equipment to evaluate its performance.

A simple visual inspection inside a piece of equipment by an experienced technician or engineer can uncover possible weaknesses, design flaws, and problem areas that may affect reliability or make maintenance difficult. It is therefore advisable to request a sample of the equipment for evaluation before committing to its use.

The availability of replacement parts is another important consideration when specifying products. A business that depends on its equipment functioning to specifications requires that the service technician be able to make repairs when needed in a timely manner. Learn about manufacturer replacement parts policies and their reliability. When possible, select equipment that uses standard off-the-shelf components that are available from multiple sources. Avoid equipment that incorporates custom components that are available only from the equipment manufacturer. This will make it easier to acquire replacements, and the cost of the parts will—most likely—be less.

When possible, specify products manufactured by the same company. Avoid mixing brands. Maintenance technicians will more easily become familiar with equipment maintenance, and experience gained while repairing one piece of hardware can be directly applied to another of the same model. Service manuals published by the same manufacturer will be similar and therefore easier to understand and use to locate the information or diagram needed for a repair.

Commonality of replacement parts will keep the parts inventory requirements and the inventory cost low. Because the technical staff will be dealing with the manufacturer on a regular basis, familiarity with the company's representatives makes it easier to get technical support quickly.

Sometimes components are selected that are not really compatible, such as differing signal levels or impedances. The responsibility then falls on the system engineer to devise a fix to make that component compatible with the rest of the system. The component may have been originally selected because of its low price, but additional components, engineering, and labor costs often offset the expected savings. Extra wiring and components can also clutter the equipment enclosure, hampering access to the equipment inside. Nonstandard mounting facilities on equipment can add unnecessary cost and can result in a less than elegant solution.

1.7 Technical Documentation

Engineering documentation describes the practices and procedures used within the industry to specify a design and communicate the design requirements to technicians and contractors. Documentation preparation should include, but not be limited to, the generation of technical system flow diagrams, material and parts lists, custom item fabrication drawings, and rack and console elevations. The required documents include the following:

- Documentation schedule
- Signal flow diagram
- Equipment schedule
- Cable schedule
- Patch panel assignment schedule
- Rack elevation drawing
- Construction detail drawing
- Console fabrication mechanical drawing
- Duct and conduit layout drawing
- Single-line electrical flow diagram

1.7.1 Documentation Tracking

The documentation schedule provides a means of keeping track of the project's paperwork. During engineering design, drawings are reviewed, and changes are made. A system for efficiently handling changes is essential, especially on big projects that require a large amount of documentation.

Completed drawings are submitted for management approval. A set of originals is signed by the engineers and managers who are authorized to check the drawings for correctness and to approve the plans.

1.7.2 Symbols

Because there are only limited informal industry standards for the design of electronic component symbols to represent equipment and other elements in a system, custom symbols are usually created by the designer. Each organization develops its own symbols. The symbols that exist apply to component-level devices, such as integrated circuits, resistors, and diodes. Some common symbols apply to system-level components, such as amplifiers and speakers. Figure 1.6 shows some of the more common component-level symbols currently used in electronics.

The proliferation of manufacturers and equipment types makes it impractical to develop a complete library, but, by following basic rules for symbol design, new component symbols can be produced easily as they are added to the system.

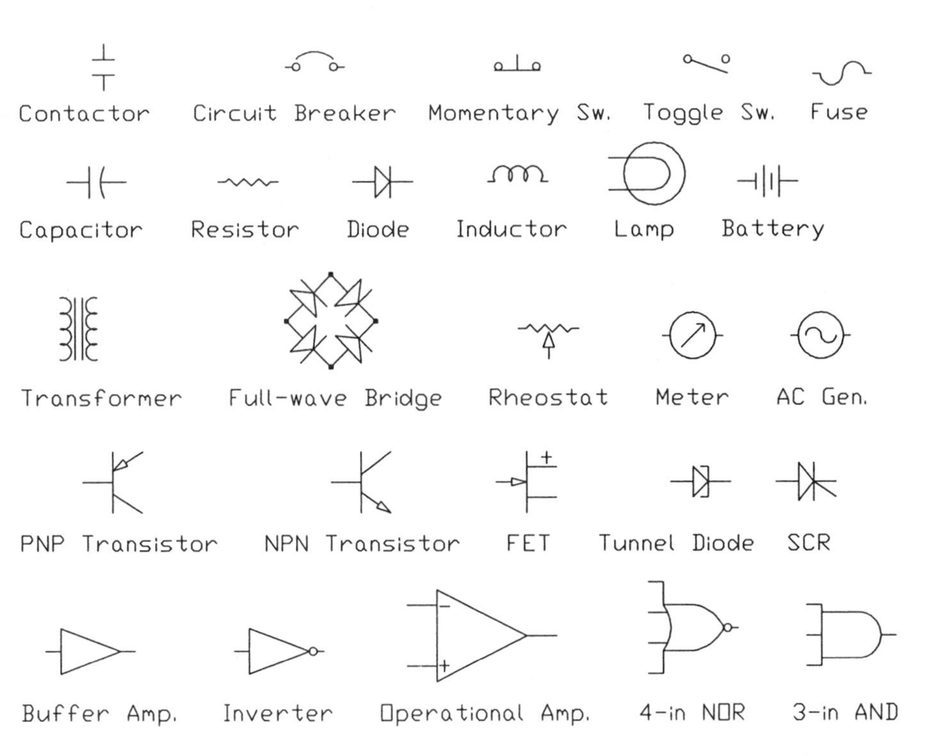

Figure 1.6 Schematic representations of common electrical components and devices.

For small systems built with a few simple components, all of the input and output signals can be included on one symbol. However, when the system uses complex equipment with many inputs and outputs with different types of signals, it is usually necessary to draw different diagrams for each type of signal. For this reason, each component requires a set of symbols, with a separate symbol assigned for each signal type, showing its inputs and outputs only. For example, a videotape recorder (VTR) will require a set of symbols for audio, video, sync, time code, and control signals, as illustrated in Figure 1.7.

If abbreviations are used, be consistent from one drawing to the next, and develop a dictionary of abbreviations for the drawing set. Include the dictionary with the documentation.

1.7.3 Cross-Referencing Documentation

In order to tie all of the documentation together and to enable fabricators and installers to understand the relationships between the drawings, the documents should include reference designations common to the project. That way, items on one type of document can be located on another type. For example, the location of a piece of

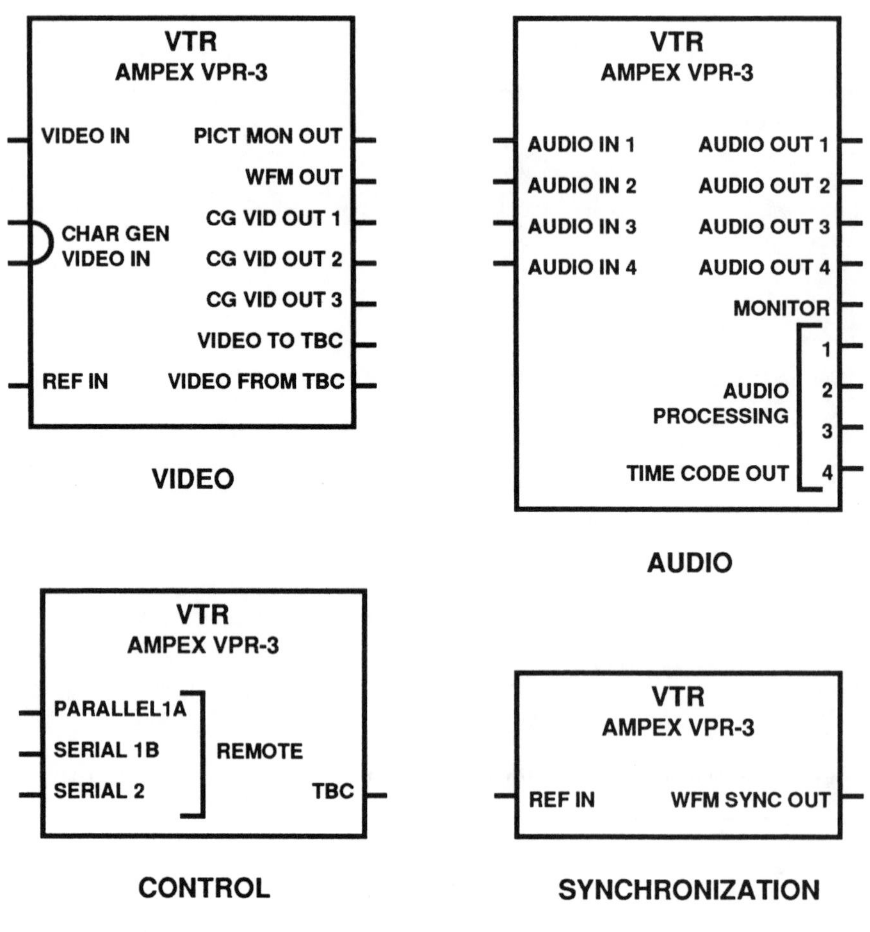

Figure 1.7 Symbol set for a VTR.

equipment can be indicated at its symbol on the flow diagram so that the technician can identify it on the rack elevation drawing and in the actual rack.

A flow diagram is used by the installation technicians to assemble and wire the system components together. All necessary information must be included to avoid confusion and delays. When designing a symbol to represent a component in a flow diagram, include all of the necessary information to identify, locate, and wire that component into the system. The information should include the following:

- Generic description of the component or its abbreviation. When no abbreviation exists, create one. Include it in the project manual reference section and in the notes on the drawing.
- Model number of the component.
- Manufacturer of the component.
- All input and output connections with their respective name and/or number.

1.7.4 Specifications

Specifications are a compilation of knowledge about how something should be done. An engineer condenses years of personal experience, and that of others, into the specification. The more detailed the specification, the higher the probability that the job will be done right.

The *project manual* is the document where specifications and other printed project documentation is compiled.

1.7.5 Working with the Contractors

The system engineer must provide support and guidance to contractors during the procurement, construction, installation, testing, and acceptance phases of a project. The system engineer can assist in ordering equipment and can help coordinate the move to a new or renovated facility. This can be critical if a great deal of existing equipment is being relocated. In the case of new equipment, the system engineer's knowledge of prices, features, and delivery times is invaluable to the facility owner.

The steps to assure quality workmanship from contractors on a job include the following:

- Clarify details.
- Clarify misunderstandings about the requirements.
- Resolve problems that may arise.
- Educate contractors about the requirements of the project.
- Assure that the work conforms to the specifications.
- Evaluate and approve change requests.
- Provide technical support to the contractors when needed.

1.7.6 Computer Tools

Technology is evolving so rapidly that it takes a lot of time just to keep up with the changes. Competition forces change and improvements that would otherwise not take place at such a rapid pace. In this environment, engineering skills must be augmented with tools to speed the design process. Computers can also help. Many of the tasks required of the system engineer can be accelerated and the resulting documentation enhanced with the aid of computers. Computer aided design (CAD) tools include application software from simple word processing and spreadsheet programs to complex simulation, 3-D graphic modeling, and artificial intelligence. Computers are commonly used in new construction and renovations to perform the following tasks:

- Document tracking
- Documentation preparation
- Correspondence
- Report generation
- Technical manual publication
- List management

- Mechanical design
- Electrical design
- Schematic capture

CHAPTER 2

USING EQUIPMENT RACKS

2.1 Introduction

In a professional facility, most equipment will have to be rack-mountable. To assemble the equipment in racks, the installer needs to know the exact physical location of each piece of hardware, and all information necessary to assemble and wire the equipment. This includes the placement of terminal blocks, power wiring, cooling devices, and all signal cables within the rack. Equipment locations can be shown on a rack elevation form. Other forms and drawings can specify terminal block wiring, ac power connections, patch panel assignments, and signal cable connections. An example of an equipment location drawing is shown in Figure 2.1.

Drawings showing the details of assembly, mounting hardware, and power wiring generally will not change from rack to rack. Therefore, they can be standardized for all racks to avoid having to repeat this part of the design process. Exceptions can be shown on a separate detailed drawing. This approach is illustrated in Figure 2.2.

When more than one rack is to be assembled side by side, it is normal practice to show the entire row on one drawing. The relationship of all of the equipment in adjacent racks can then be easily seen on the drawing (see Figure 2.3).

2.2 Industry Standard Equipment Enclosures

The modular equipment enclosure, frame, or equipment rack is one of the most convenient and commonly used methods for assembling the equipment and components that make up a technical facility. The ANSI/EIAJ RS-310-C standard for racks provides the dimensions and specifications for racks, panels, and associated hardware. Other specifications, such as the European International Electrotechnical Commission (IEC) Publication Number 297-1 and 297-2 and West German Industrial Standard DINJ 41494 Part 1, have matching dimensions and specifications.

Applicable standards for equipment racks include the following:

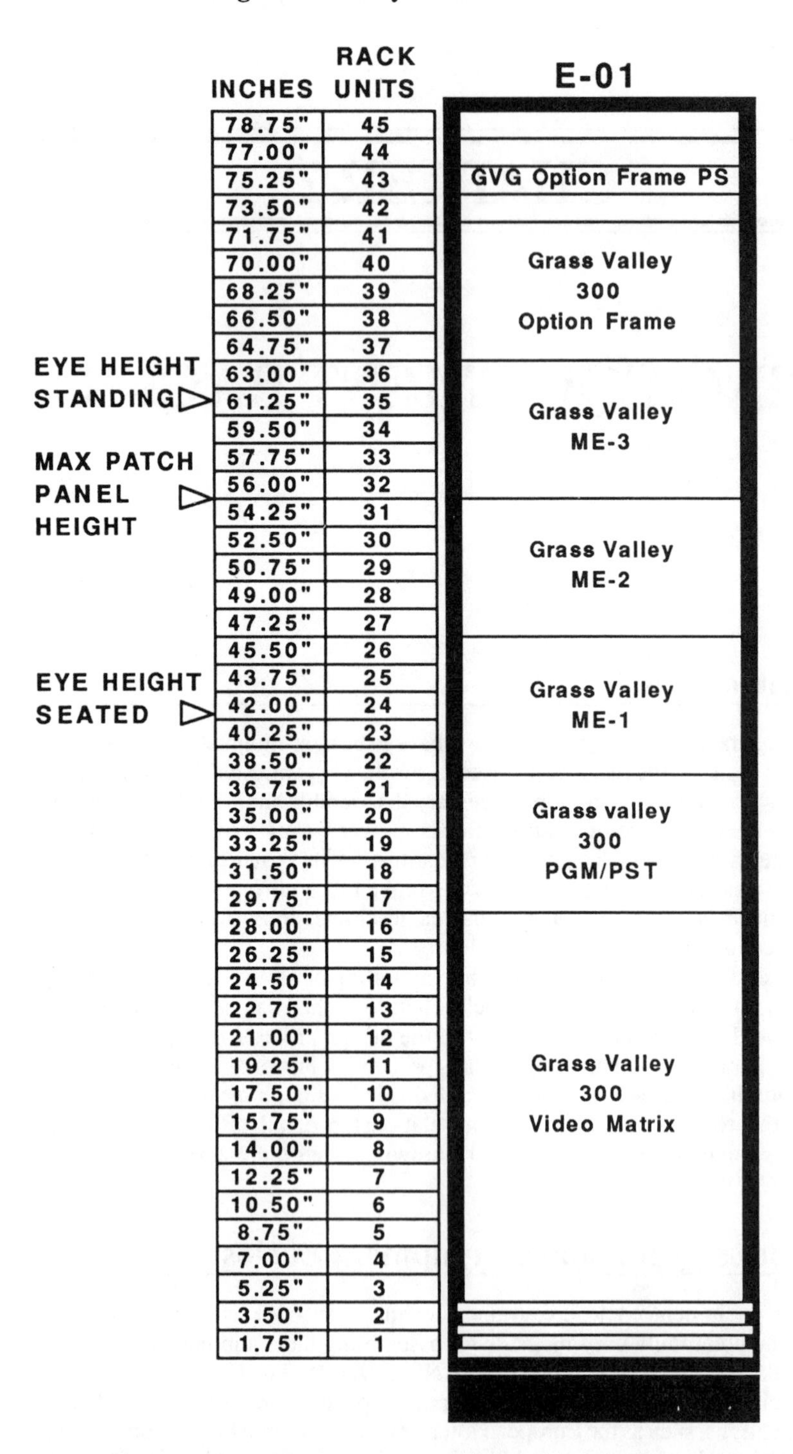

Figure 2.1 Equipment location drawing for a rack enclosure.

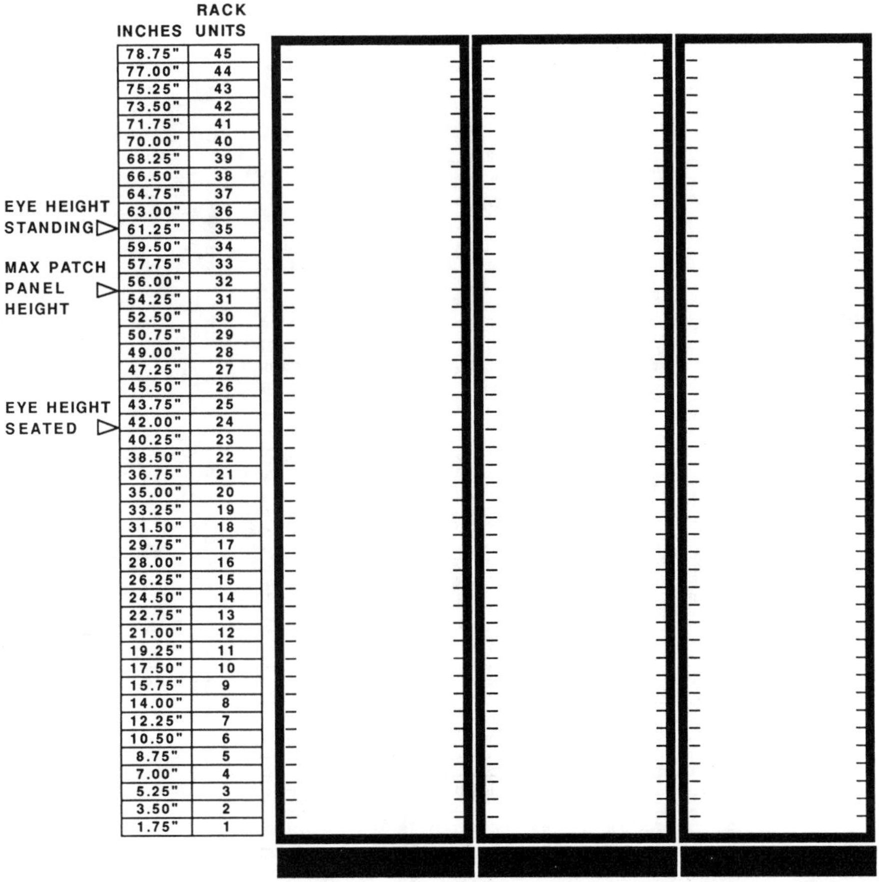

EYE HEIGHT STANDING
MAX PATCH PANEL HEIGHT
EYE HEIGHT SEATED

INCHES	RACK UNITS
78.75"	45
77.00"	44
75.25"	43
73.50"	42
71.75"	41
70.00"	40
68.25"	39
66.50"	38
64.75"	37
63.00"	36
61.25"	35
59.50"	34
57.75"	33
56.00"	32
54.25"	31
52.50"	30
50.75"	29
49.00"	28
47.25"	27
45.50"	26
43.75"	25
42.00"	24
40.25"	23
38.50"	22
36.75"	21
35.00"	20
33.25"	19
31.50"	18
29.75"	17
28.00"	16
26.25"	15
24.50"	14
22.75"	13
21.00"	12
19.25"	11
17.50"	10
15.75"	9
14.00"	8
12.25"	7
10.50"	6
8.75"	5
7.00"	4
5.25"	3
3.50"	2
1.75"	1

Figure 2.2 Hardware location template for a series of rack enclosures being installed at a facility. Deviations from the standard template are shown as drawing details.

- UL-listed type 12 enclosures
- NEMA type 12 enclosures
- NEMAJ type 4 enclosures
- IEC 297-2 specifications
- IEC 297-3 specifications
- IP 55/NEMA type 12/13 enclosures
- DINJ 41494 Part 1

The chassis of most of the electronic equipment used for industrial electronics and professional audio/video have front panel dimensions that conform to the EIA specifications for mounting in standard modular equipment enclosures. Figure 2.4 shows the standard RS-310-C rack-mounting hole dimensions.

Blank panels, drawers, shelves, guides, and other accessories are designed and built to conform to the EIA standards. Rack-mounted hardware for interconnecting and supporting the wiring is also available. Figure 2.5 shows some of the hardware available for use with standard equipment enclosures.

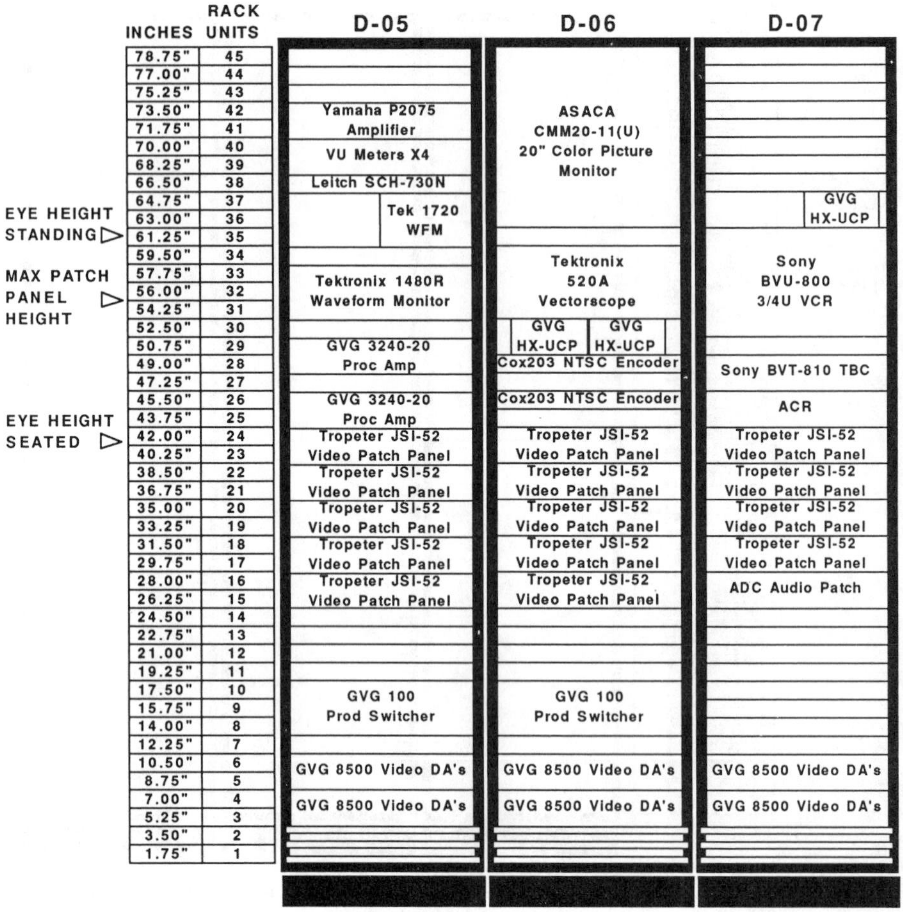

Figure 2.3 Equipment rack drawing for a group of enclosures, showing the overall assembly.

2.2.1 Types of Rack Enclosures

There are two main types of racks. The first is the floor-mounted open frame, shown in Figure 2.6. This EIA standard equipment enclosure consists of two vertical channels (with mounting holes), separated at the top and bottom by support channels. The frame is supported in the free-standing mode by a large base, which provides front-to-back stability. The rack may also be permanently secured to the floor by bolts, eliminating the need for a base. Equipment is mounted directly to the vertical members and is accessible from the front, side, and rear.

The second type of rack is a box frame that is free-standing, with front and, optionally, rear equipment-mounting hardware. This is illustrated in Figure 2.7. This frame can be completely enclosed by installing optional side, rear, top, and bottom panels or access doors. Horizontal brackets mounted on the left and right sides of the

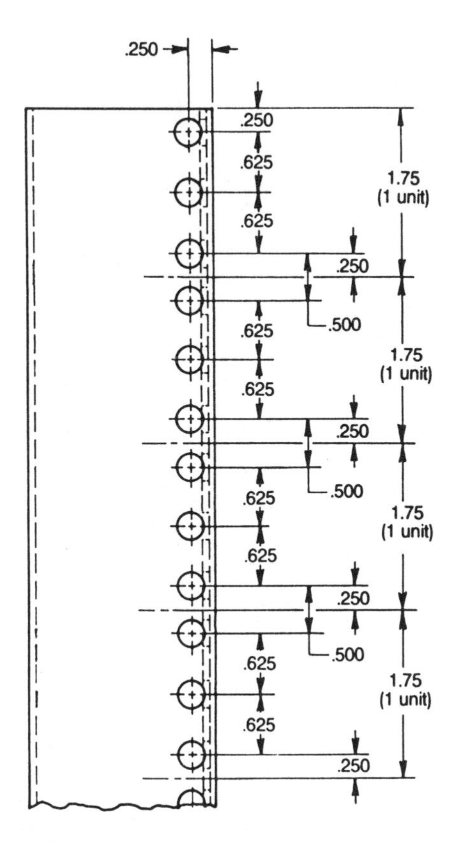

Figure 2.4 Standard equipment-mounting dimensions for RS-310-C rack enclosures.

---

frame increase rigidity and provide support for vertical mounting angles and other accessories.

Standard racks are available in widths of 19-, 24-, and 30-in. The preferred, and most widely used, width is 19-in (482.6 mm). The racks are designed to hold equipment and panels that have vertical heights of 1.75-in (44.45 mm) or more, in increments of 1.75-in. One *rack unit* (RU) is defined as 1.75-in (44.45 mm). The rack height is usually specified in rack units. Holes or slots along the left and right edges of the equipment front panel or support panel are provided for screws to fasten the unit to specific mounting holes in the rack. The mounting hole locations are defined by EIA specification so that equipment mounts vertically only at specific heights—in 1.75-in increments—within the enclosure.

The simplest vertical mounting angles are "L"-shaped and have holes uniformly distributed along their entire length through both surfaces (see Figure 2.8). The holes on one surface are used to attach the angle vertically to the horizontal side members of the rack frame. The horizontal members usually allow the vertical mounting angle to be positioned at different depths from front to rear in the rack. The holes on the

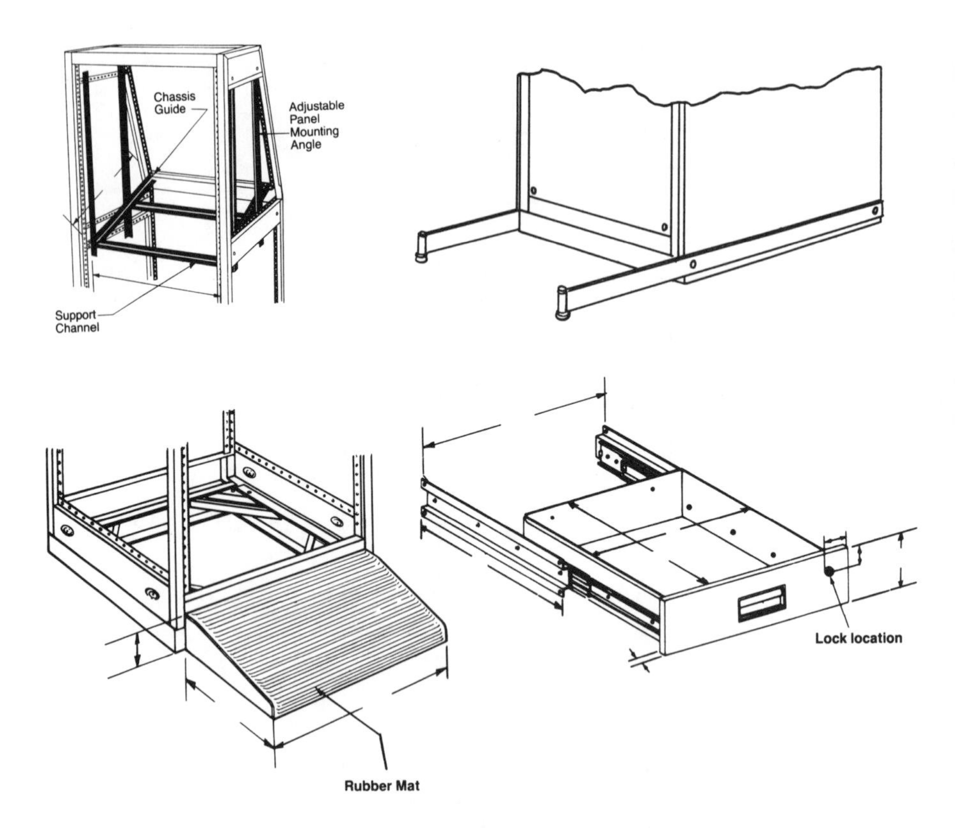

Figure 2.5 Common accessories available for use with rack enclosures. (*Courtesy of Emcor Products.*)

---

other surface of the mounting angle are used for securing the equipment. Mounting holes are arranged in groups of three, centered on each 1.75-in rack-unit interval.

Rack angles of a more complex design may be used when necessary to mount accessories, such as drawers, shelves, and guides that do not use the front-mounting angles and, therefore, must be secured by an alternate means. Figure 2.9 shows one common rack angle of this type.

To secure equipment to vertical supports or other mounting hardware, 10-32 UNF-2B threaded clip nut fasteners are placed at the appropriate clearance holes in the mounting angles for each piece of equipment being mounted. Some rack enclosures provide 10-32 threaded mounting holes on the vertical support channels.

The manufacturers of equipment enclosures offer a wide variety of accessory hardware to adapt their products to varied equipment-mounting requirements. A variety of paint colors and laminate finishes are available. Vertical rack-mount support channels are available unpainted with zinc plating to provide a common ground return for the equipment chassis.

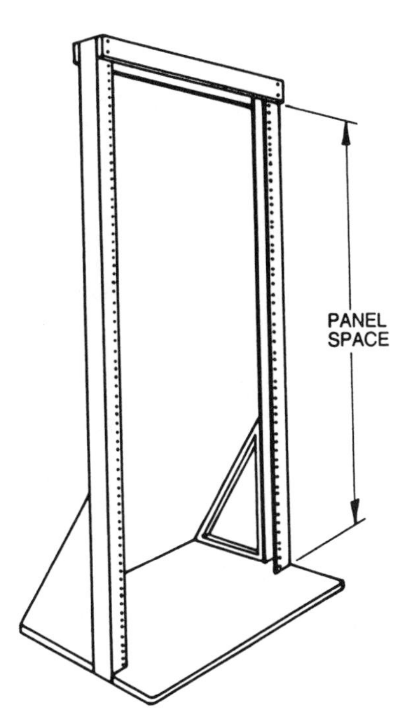

Figure 2.6 Standard open-frame equipment rack.

2.2.2 Rack Configuration Options

Groups of rack frames can be placed side by side in a number of custom configurations. The most common has racks arranged in a row, bolted together without panels between adjacent frames. Side panels can be mounted on each end of the row, resulting in one long enclosure. This is illustrated in Figure 2.10. Side panels can be installed between frames in a row, if necessary, to provide electromagnetic shielding, a heat barrier, or a physical barrier. A rear door, with or without ventilation perforations or louver slots, can be installed to protect the rear wiring and provide a finished appearance. A front door can be used if access to the equipment front panel controls is not necessary. Clear or darkened Plexiglas doors can be used to allow viewing of meters or other display devices, or to showcase some aspect of the technology used in the rack. The use of a clear plexiglass door is shown in Figure 2.11.

A top panel is recommended to protect the equipment inside from falling debris and dust. Bottom panels are usually not installed unless bottom shielding is required.

Standard equipment racks provide flexibility because equipment can be mounted at any height in the rack. Many different-shaped frames are also available. These shapes conform to the same equipment-mounting and -mating dimensions, which permit assembling different frames together. Shapes are available with sloping fronts. Various wedge shapes are common. Racks can be angled with respect to each other by inserting wedge-shaped frames as intermediaries between adjacent frames. Complex consoles for housing control panels and monitoring equipment can be assembled by bolting together the differently shaped frames. With these options, a complex

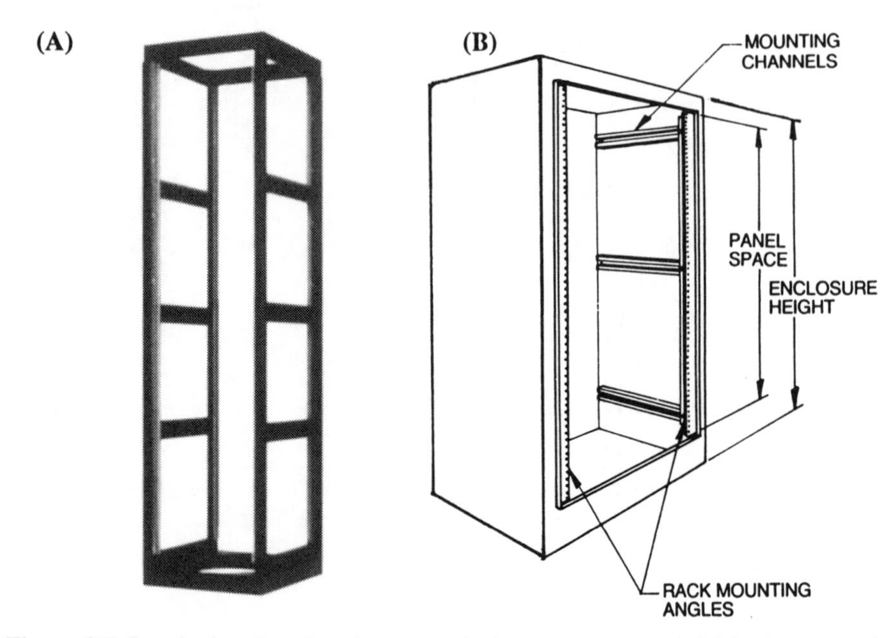

Figure 2.7 Standard enclosed equipment rack: (a) covers removed; (b) basic enclosed rack assembly.

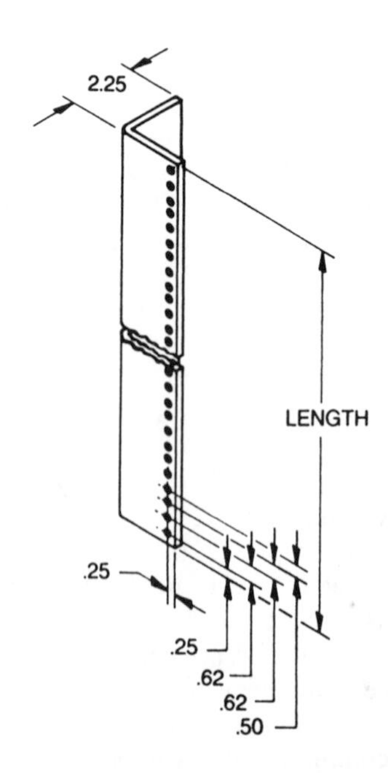

Figure 2.8 "L" equipment mounting bracket, used to support heavy instruments in a rack.

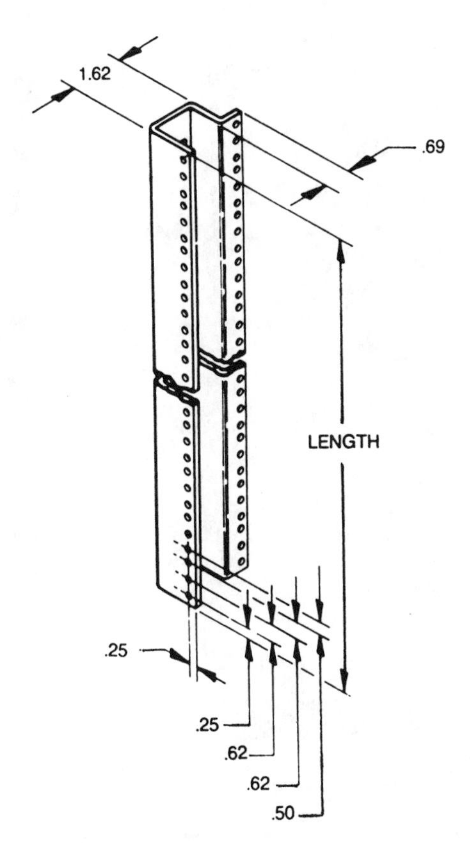

Figure 2.9 Rack angle brace used for mounting shelves, drawers, and other accessory hardware in an enclosure.

---

console shape can be assembled to meet functional and human factor requirements. Figure 2.12 illustrates several of the stock configurations.

Although control consoles can be assembled from standard components that conform to standard enclosure dimensions, in many instances, custom-made consoles are desirable. These are helpful to achieve a more efficient layout for controls, or to develop a more sophisticated appearance within the control room environment. Figure 2.13 shows a custom operating console for a television broadcast center.

Some equipment enclosure manufacturers offer an intermediate step between a stock rack and a custom-made console. By using off-the-shelf rack elements, the customer can specify the exact size and configuration required. After the dimensions have been provided to the manufacturer, the individual supporting rails and frames are cut to specification, and the unit is assembled. Figure 2.14 shows several types of semi-custom rack enclosures.

Figure 2.10 Group of racks arranged in a row to form one large equipment enclosure. (*Courtesy of A. F. Associates.*)

2.2.3 Selecting an Equipment Rack

When selecting the model of rack that will be used in a facility, the physical dimensions and weight of equipment to be mounted will be needed. Specify racks with enough depth to accommodate the deepest piece of equipment that will be installed. At the same time, allow ventilating air to flow freely past and through the equipment. Allow additional clearance at the rear of the equipment chassis for connectors, and allow enough space for the minimum bend radius of the largest cable. Additional depth may also be required for cable bundles that must pass behind a deep piece of equipment.

Select a rack model that has sufficient strength to support the full array of anticipated equipment. Also, allow a margin of error for future expansion.

Select paint and laminate colors and textures for the rack assemblies and hardware. This information should be included in the specifications for the racks, which is included in the project manual.

Figure 2.11 Use of a protective Plexiglas door on an equipment enclosure. (*Courtesy of Emcor Products.*)

2.3 Equipment Rack Layout

When specifying the location of equipment within racks and consoles, give careful consideration to several factors:

- Physical equipment size and weight
- Power consumption
- Ventilation needs
- Mechanical noise

Human factors also must be considered. Equipment placement should be governed by the operational use of the equipment. Human factors that need consideration include:

- Accessibility to controls
- Height with respect to the operating position

Figure 2.12 A selection of stock equipment enclosures designed for specialized installations.

Figure 2.13 Custom console built to meet the specialized needs of the client. (*Courtesy of A. F. Associates.*)

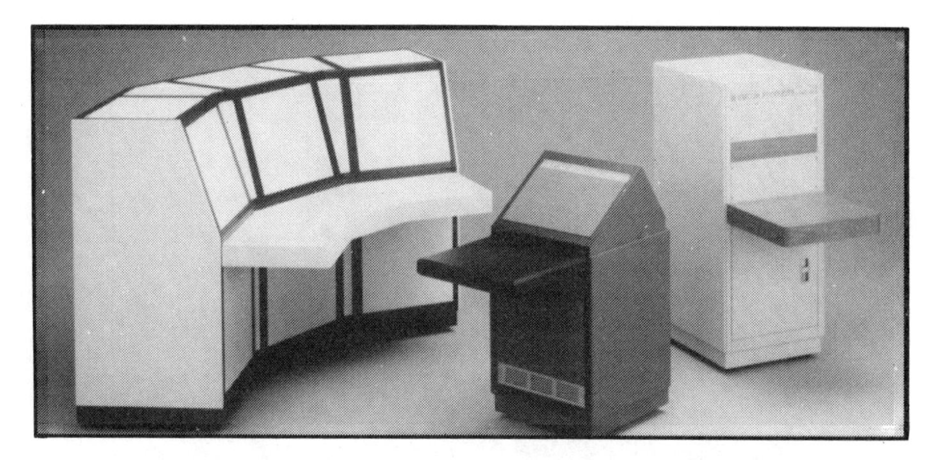

(A) above; (B) below; (C) next page

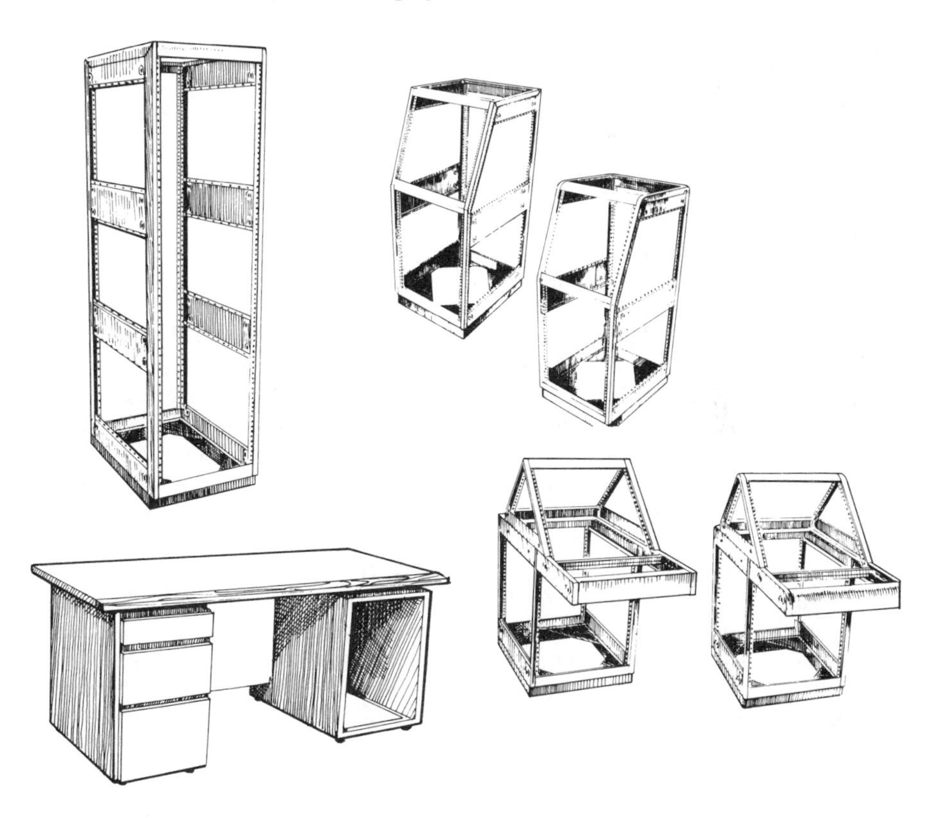

Figure 2.14 Use of stock rack elements to produce a semi-custom equipment enclosure: (a) stock rack-based consoles; (b) artist drawing of various stock designs; (c) video production station built from stock elements. (*Illustrations* a *and* b *courtesy of Emcor Products. Photograph* c *courtesy of A. F. Associates.*)

Figure 2.14(c)

- Line of sight to controls, meters, and display devices, from the operator's point of view
- Reflections on display devices from room lighting or windows
- Noise generated by the equipment

Do not completely fill a given rack with equipment. From a practical point of view, leaving blank spaces will allow for future equipment expansion and replacement.

Provide storage spaces in the racks, if required. For example, if a tape or disk pack is loaded on a machine, a location should be available for holding its container. Rack-mountable shelves and drawers are available in different sizes for this purpose.

Avoid cable clutter by providing easy access to wiring and connections. This will make installation, maintenance, and modifications easier throughout the life of the system.

Place the tops of jack fields at or below eye level. Jack field labels must be readable. The average eye level of males is 65.4-in (1660 mm) and females 61.5-in (1560 mm). Lining up the tops of jack fields that are mounted horizontally across several racks will create a neat appearance. If room is available, place blank panels between patch panels to space them vertically and to allow room for access from the front and rear.

Keep the field as confined as possible to allow the use of the shortest possible patch cords. This is especially critical when using phase-matched video patch panels. These require that patch cords be a fixed, short length.

Provide a pair of rear vertical mounting angles for supporting heavy or deep equipment. Eliminate them if they are not required. Mount heavy equipment in the lower part of the rack to facilitate easier installation and replacement. One exception might be a piece of equipment that generates excessive heat. Mounting it at the top of the rack will allow the heat to escape by convection, without heating other equipment (power supplies are a good example).

2.3.1 Cooling Considerations

It is a normal practice to cool the room in which technical equipment is installed. At the same time, comfort of the personnel in the room must be ensured and usually takes precedence over the comfort of the equipment. Additional steps should be taken to control heat build-up and hot spots within equipment racks and consoles. Use all possible heat-removing techniques within the racks before installing fans for that purpose. Fans cost money, consume power, take up space, are noisy, and will eventually fail. Dust drawn through the fan will collect on something. If that something is a filter, it must be cleaned or replaced periodically. If the dust collects on equipment, overheating may occur. Some steps that can be taken by the system engineer in the design phase include the following:

- Limit the density of heat-producing equipment installed in each rack.
- Leave adequate space for the free movement of air around the equipment. This will help the normal convection flow of air upward as it is heated by the equipment.
- Specify perforated or louvered blank panels above or below heat-producing equipment. A perforated or louvered rear door may also be installed to improve air flow into and out of the rack.
- When alternative equivalent products are available, select equipment that generates the least amount of heat. This will usually result from lower power consumption—a desirable feature.
- If a choice exists among equivalent units, select the one that does not require a built-in fan. Units without fans may be of a low power consumption design, which implies (but does not guarantee) good engineering practice.
- Balance heat loads by placing high heat-producing equipment in another rack to eliminate hot spots.
- Place equipment in a separate air-conditioned equipment room to reduce the heat load in occupied control rooms.
- Remove the outer cabinets of equipment or modify mounting shelves and chassis to improve air flow through the equipment. Consult the original equipment manufacturer, however, before operating a piece of hardware with the cover removed. The cover is often used to channel cooling air throughout the instrument, or to provide necessary electrical shielding.
- When specifying new equipment designs, describe the environment in which the equipment will be required to operate. Stipulate the maximum temperature that can be tolerated.
- Pressurize each rack with filtered cooling air, which is brought in at the bottom of the rack and allowed to flow out only at the top of the rack.
- If the rack is designed to be directly cooled by cold air forced into the rack, consider removing the equipment fans and the housing so that cold air can flow through the equipment directly. This will reduce fan noise, weight, and power consumption. The purpose of a fan is to move air past heat-producing elements. With or without the fan, the air conditioner must still remove the heat from that air. As many as 12 fans are used in some digital processing equipment. Remember that the manufacturer's warranty might be voided if the fans and housing are removed. Check with the manufacturer before making any modifications.
- When forced-air cooling is used, provide a means of adjusting the air flow into each rack to balance the volume of air moving through the enclosures. This

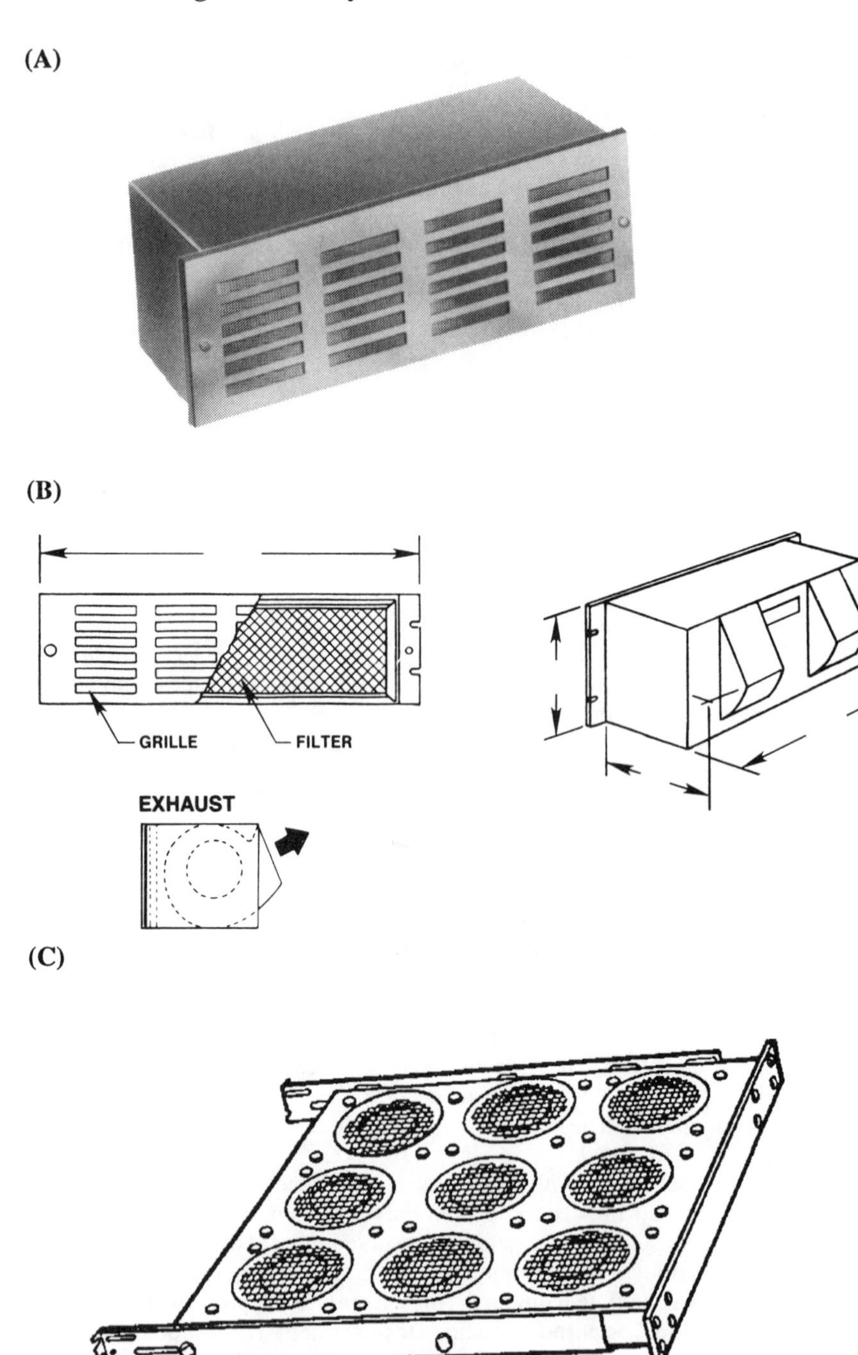

Figure 2.15 Rack accessories used for cooling the equipment enclosure: (a) photo of a rack-mounted blower; (b) detail drawing of a rack-mounted blower; (c) a common type of horizontal cooling shelf constructed of individual fans. (*Photograph* a *courtesy of Emcor Products.*)

will control the amount of cooling and concentrate it in the racks where it is needed most. Adjust the air flow to the minimum required to properly cool the equipment. This will minimize the wind noise produced by air being forced through openings in the equipment.

- Install air directors, baffles, or vanes to direct the air flow within the rack. This strategy works for controlling convection and forced-air flow.

Provide a minimum of 3 ft (1 m) clearance at the rear of equipment racks. Besides enabling the enclosure door to swing fully open, this will facilitate efficient cooling and easy equipment installation and maintenance.

If required in a given installation, cooling fans and devices are available for equipment racks. Common types are shown in Figure 2.15.

2.4 Single-Point Ground

Equipment racks and peripheral hardware must be properly grounded for reliable operation. Single-point grounding is the basis of any properly designed technical system ground network. Fault currents and noise should have only one path to the facility ground. Single-point grounds can be described as *star* systems, whereby radial elements circle out from a central hub. A star system is illustrated in Figure 2.16. Note that all equipment grounds are connected to a *main ground point*, which is then tied to the facility ground system. Multiple ground systems of this type can be cascaded as needed to form a *star-of-stars*. The object is to ensure that each piece of equipment has one ground reference. Fault energy and noise then are efficiently drained to the outside earth ground system.

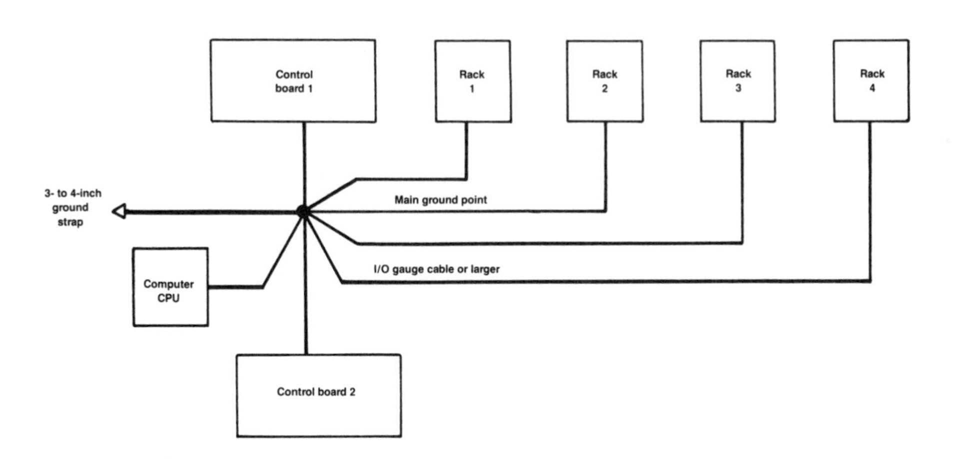

Figure 2.16 Typical facility grounding system. The *main facility ground point* is the reference from which all grounding is done at the plant. If a bulkhead entrance panel is used, it will function as the main ground point.

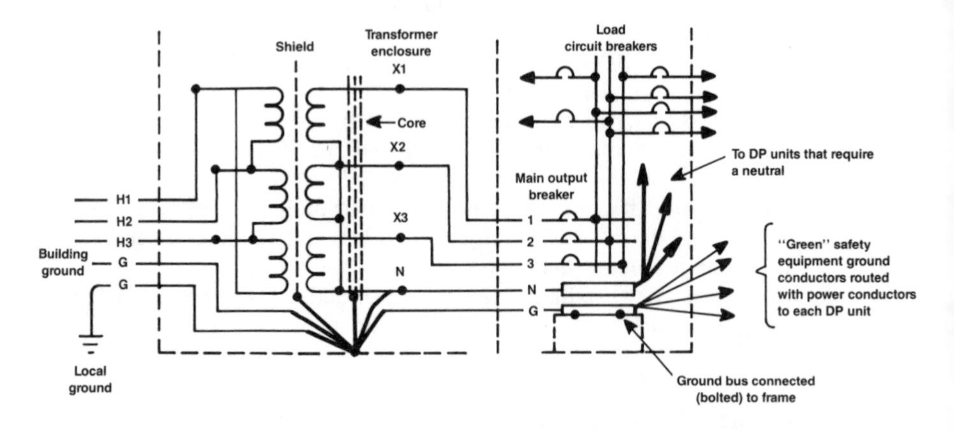

Figure 2.17 Single-point grounding applied to a power-distribution system. (Adapted from: Federal Information Processing Standards Publication No. 94, *Guideline on Electrical Power for ADP Installations*, U.S. Department of Commerce, National Bureau of Standards, Washington, DC, 1983.)

2.4.1 Technical Ground System

Figure 2.17 illustrates a star grounding system as applied to an ac power-distribution transformer and circuit-breaker panel. Note that a central ground point is established for each section of the system: one in the transformer vault and one in the circuit-breaker box. The breaker ground ties to the transformer vault ground, which is connected to the building ground system. Figure 2.18 shows single-point grounding applied to a data processing center. Note how individual equipment groups are formed into a star grounding system, and how different groups are formed into a star-of-stars configuration. A similar approach can be taken for a data processing center using multiple modular power center (MPC) units. This is shown in Figure 2.19. The terminal mounting wall is the reference ground point for the facility.

Figure 2.20 shows the recommended grounding arrangement for a typical broadcast or audio/video production facility. The building ground system is constructed using heavy-gauge copper wire (no. 4 gauge or larger) if the studio is not located in an RF field, or a wide copper strap (3-in minimum) if the facility is located near an RF energy source. The copper strap is required because of the *skin effect* (see Chapter 3). A common method of determining the required size of the ground strap (in inches) inside the building is to specify the minimum width of the strap as 1.5 percent of its length. For example, if the total grounding run from the perimeter ground system to the farthest piece of equipment is 350 ft, use a 5.25 in ground strap. For short runs in an RF field, do not use a ground strap that is less than 3-in wide.

Run the strap or cable from the perimeter ground to the main facility ground point. Branch out from the main ground point to each major piece of equipment, and to the various equipment rooms. Establish a *local ground point* in each room or group of racks. Use a separate ground cable for each piece of equipment (no. 12 gauge or larger). Figure 2.21 shows the grounding plan for a communications facility. Equipment grounding is handled by separate conductors tied to the *bulkhead panel* or entry plate.

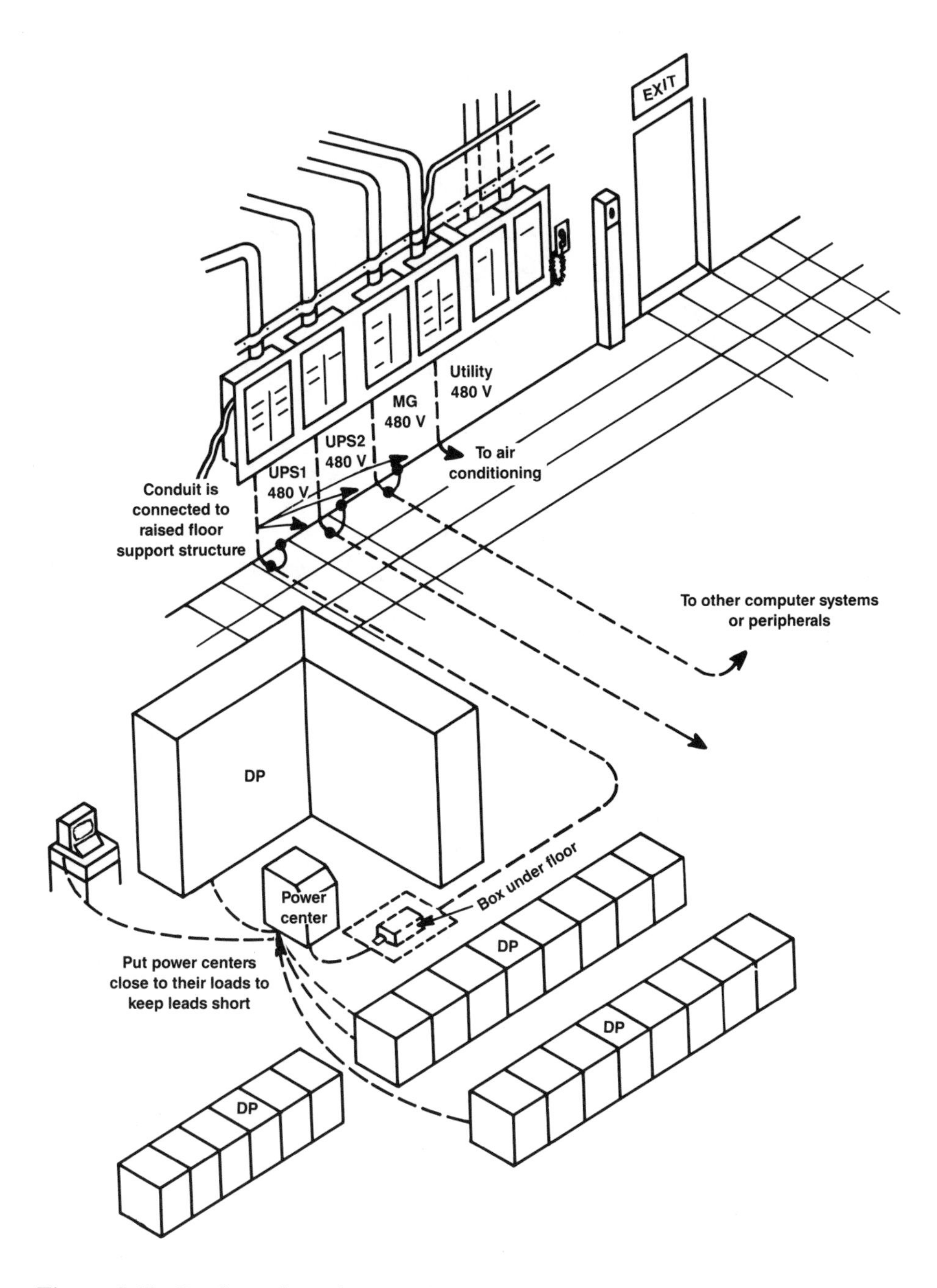

Figure 2.18 Configuration of a star-of-stars grounding system at a data processing facility. (Adapted from: Federal Information Processing Standards Publication No. 94, *Guideline on Electrical Power for ADP Installations*, U.S. Department of Commerce, National Bureau of Standards, Washington, DC, 1983.)

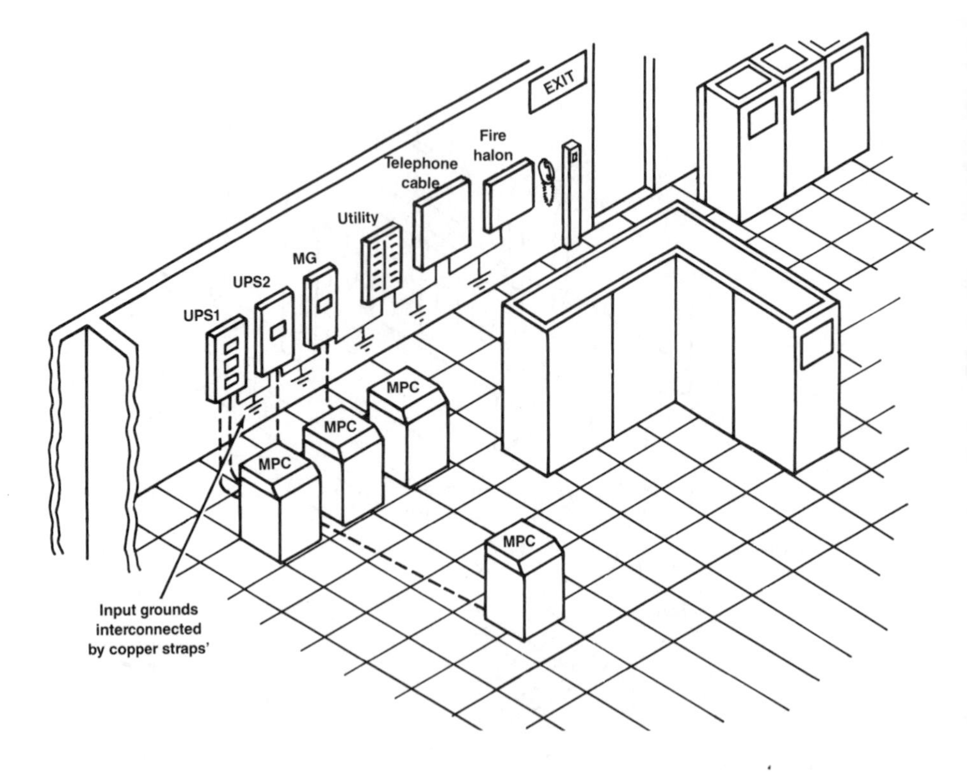

Figure 2.19 Establishing a star-based single-point ground system using multiple modular power centers. (Adapted from: Federal Information Processing Standards Publication No. 94, *Guideline on Electrical Power for ADP Installations*, U.S. Department of Commerce, National Bureau of Standards, Washington, DC, 1983.)

(Bulkhead panels are discussed in Chapter 9.) A *halo* ground is constructed around the perimeter of the room. Cable trays are tied into the halo. All electronic equipment is grounded to the bulkhead to prevent ground-loop paths. Figure 2.22 shows a top-down view of a bulkhead system ground.

The ac line ground connection for individual pieces of equipment often presents a built-in problem for the system designer. If the equipment is grounded through the chassis to the equipment room ground point, a ground loop may be created through the green-wire ground connection when the equipment is plugged in. The solution involves careful design and installation of the ac power distribution system to minimize ground-loop currents, while providing the required protection against ground faults. Some equipment manufacturers provide a convenient solution to the ground-loop problem by isolating the signal ground from the ac and chassis ground. This feature offers the user the best of both worlds: the ability to create a signal ground system and ac ground system free of interaction and ground-loops.

It should be emphasized that the design of a ground system must be considered as an integrated package. Proper procedures must be used at all points in the system. It takes only one improperly connected piece of equipment to upset an otherwise perfect ground system. The problems generated by a single grounding error can vary from

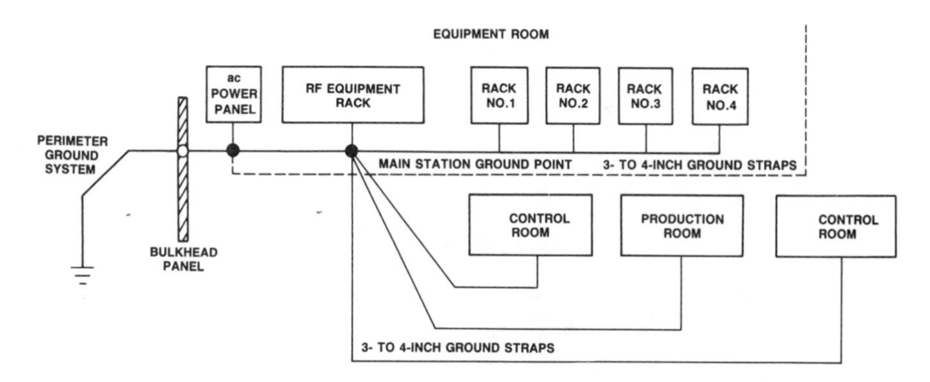

Figure 2.20 Typical grounding arrangement for individual equipment rooms at a communications facility. The ground strap from the main ground point establishes a *local ground point* in each room, to which all electronic equipment is bonded.

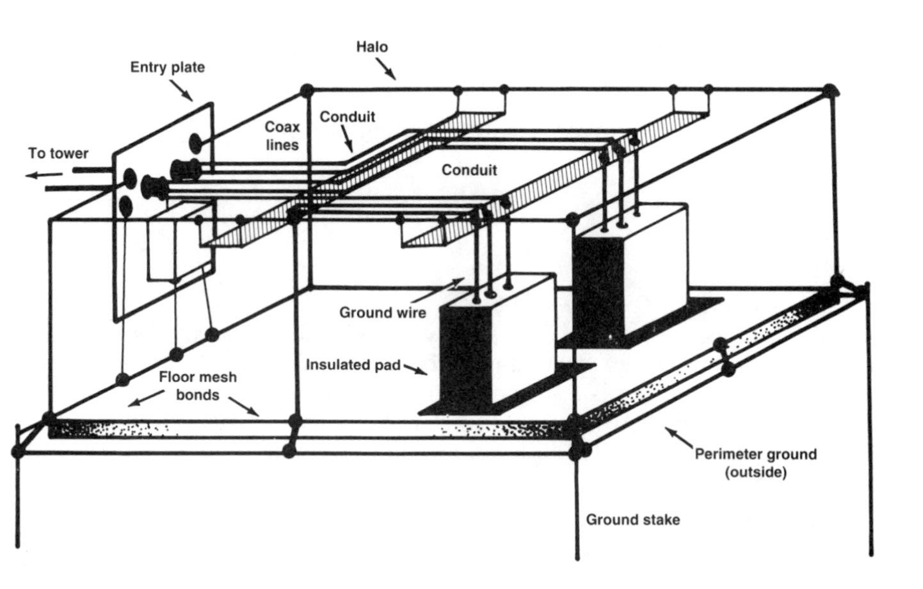

Figure 2.21 Bulkhead-based ground system, including a grounding halo.

trivial to significant, depending on where in the system the error exists. This consideration naturally leads to the concept of ground-system maintenance for a facility. Check the ground network from time to time to ensure that no faults or errors have occurred. Any time new equipment is installed or old equipment is removed from service, give careful attention to the possible effects that such work will have on the ground system.

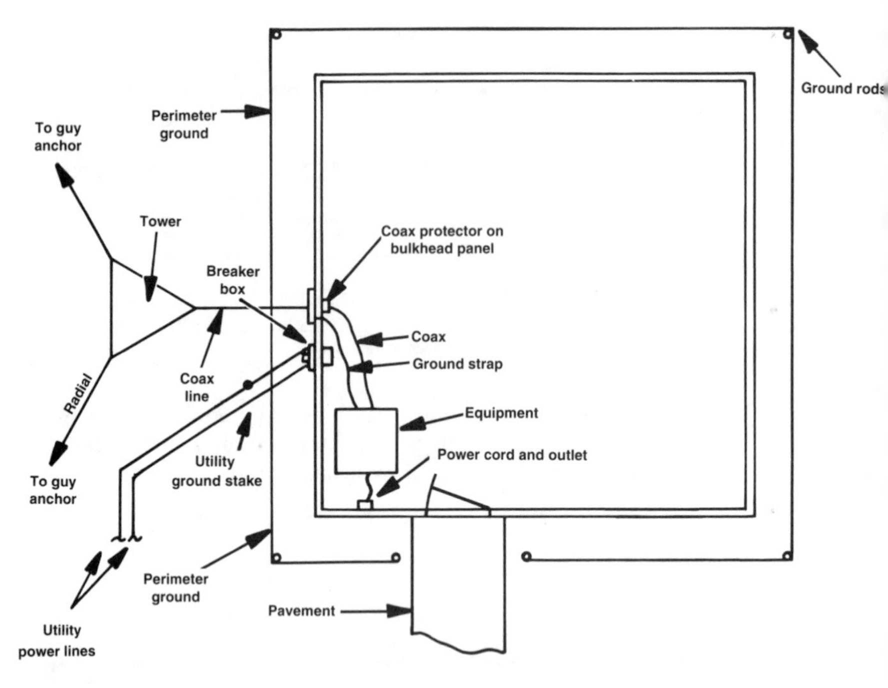

Figure 2.22 A bulkhead ground system integrating all elements of a communications facility.

2.4.1.A *Grounding Conductor Size*

The NEC and local electrical codes specify the minimum wire size for grounding conductors. The size varies, depending on the rating of the current-carrying conductors. Code typically permits a smaller ground conductor than hot conductors. It is recommended, however, that the same size wire be used for ground lines and hot lines. The additional cost involved in the larger ground wire is often offset by the use of one size of cable. Further, better control over noise and fault currents is achieved with a larger ground wire.

Separate insulated ground wires should be used throughout the ac distribution system. Do not rely on conduit or raceways to carry the ground connection. A raceway interface that appears to be mechanically sound may not provide the necessary current-carrying capability in the event of a phase-to-ground fault. Significant damage may result if a fault occurs in the system. When the electrical integrity of a breaker panel, conduit run, or raceway junction is in doubt, fix it. Back up the mechanical connection with a separate ground conductor of the same size as the current-carrying conductors. Loose joints have been known to shower sparks during phase-to-ground faults, creating a fire hazard. Secure the ground cable using appropriate hardware. Clean paint and dirt from attachment points. Properly label all cables.

Structural steel, compared with copper, is a poor conductor at any frequency. At dc, steel has a resistivity 10 times that of copper. As frequency rises, the *skin effect* is more pronounced because of the magnetic effects involved. (The skin effect is discussed in Section 3.3.3.) Further, because of their bolted, piecemeal construction,

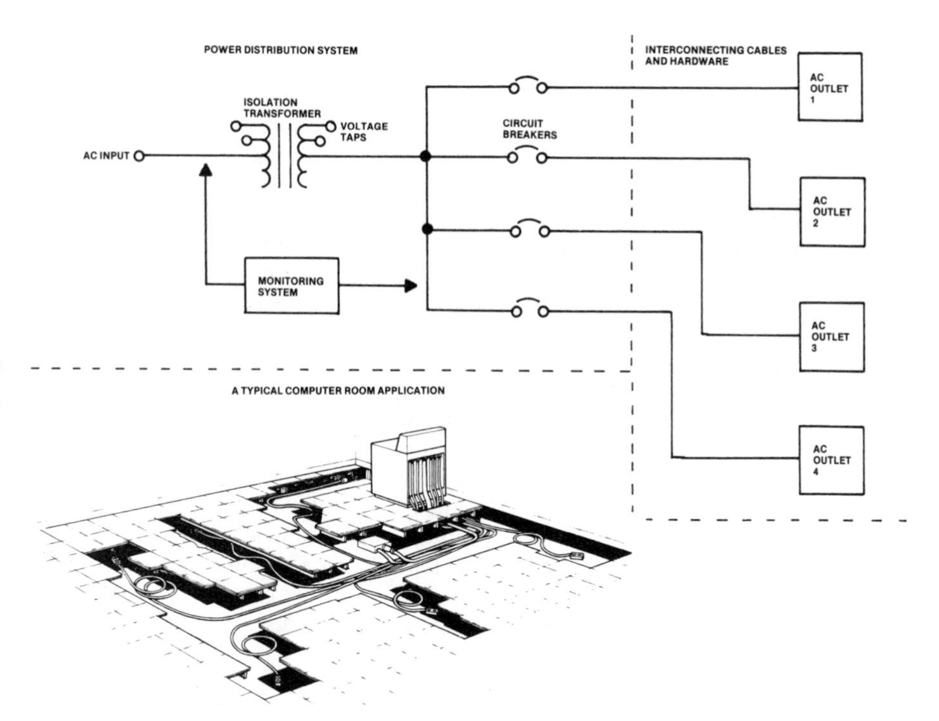

Figure 2.23 The basic concept of a computer-room modular power center. Both single- and multi-phase configurations are available. When ordering an MPC, the customer can specify cable lengths and terminations, making installation quick and easy.

---

steel racks and building members should not be depended upon alone for circuit returns.

2.4.2 Power-Center Grounding

A modular power center (MPC), commonly found in computer-room installations, provides a comprehensive solution to ac power distribution and ground-noise considerations. Such equipment is available from several manufacturers, with various options and features. A computer power distribution center generally includes an isolation transformer designed for noise suppression, distribution circuit breakers, power supply cables, and a status monitoring unit. The system concept is shown in Figure 2.23. Input power is fed to an isolation transformer with primary taps to match the ac voltage required at the facility. A bank of circuit breakers is included in the chassis, and individual pre-assembled and terminated cables supply ac power to the various loads. A status monitoring circuit signals the operator of any condition that is detected outside normal parameters.

The ground system is an important component of the MPC. A unified approach, designed to prevent noise or circulating currents, is taken to grounding for the entire facility. This results in a clean ground connection for all on-line equipment.

The use of a modular power center can eliminate the inconvenience associated with rigid conduit installations. Distribution systems also are expandable to meet future facility growth. If the plant is ever relocated, the power center can move with it. MPC units usually are expensive. However, considering the installation costs by a licensed electrician of circuit-breaker boxes, conduit, outlets, and other hardware on-site, the power center approach may be economically attractive. The use of a power center also will make it easier to design a standby power system for the facility. Many computer-based operations do not have a standby generator on site. Depending on the location of the facility, it may be difficult or even impossible to install a generator to provide standby power in the event of a utility company outage. However, by using the power center approach to ac distribution for computer and other critical-load equipment, an uninterruptible power system may be installed easily to power only the loads that are required to keep the facility operating. With a conventional power distribution system—where all ac power to the building or a floor of the building is provided by a single large circuit breaker panel—separating the critical loads from other nonessential loads (such as office equipment, lights, and air conditioning/heating equipment) can be an expensive detail.

2.4.3 Isolation Transformers

One important aspect of an MPC is the isolation transformer. The transformer serves to:

- Attenuate transient disturbances on the ac supply lines.
- Provide voltage correction through primary-side taps.
- Permit the establishment of an isolated ground system for the facility served.

Whether or not an MPC is installed at a facility, consideration should be given to the appropriate use of an isolation transformer near a sensitive load.

The ac power supply for many buildings often originates from a transformer located in a basement utility room. In large buildings, the ac power for each floor may be supplied by transformers closer to the loads they serve. Most transformers are 208 Y/120 V three-phase. Many fluorescent lighting circuits operate at 277 V, supplied by a 480 Y/277 V transformer. Long feeder lines to data processing (DP) systems and other sensitive loads raise the possibility of voltage fluctuations based on load demand and ground-loop-induced noise.

Figure 2.24 illustrates the preferred method of power distribution in a building. A separate dedicated isolation transformer is located near the DP equipment. This provides good voltage regulation and permits the establishment of an effective single-point star ground in the DP center. Note that the power distribution system voltage shown in the figure (480 V) is maintained at 480 V until it reaches the DP step-down isolation transformer. Use of this higher voltage provides more efficient transfer of electricity throughout the plant. At 480 V, the line current is about 43 percent of the current in a 208 V system for the same conducted power.

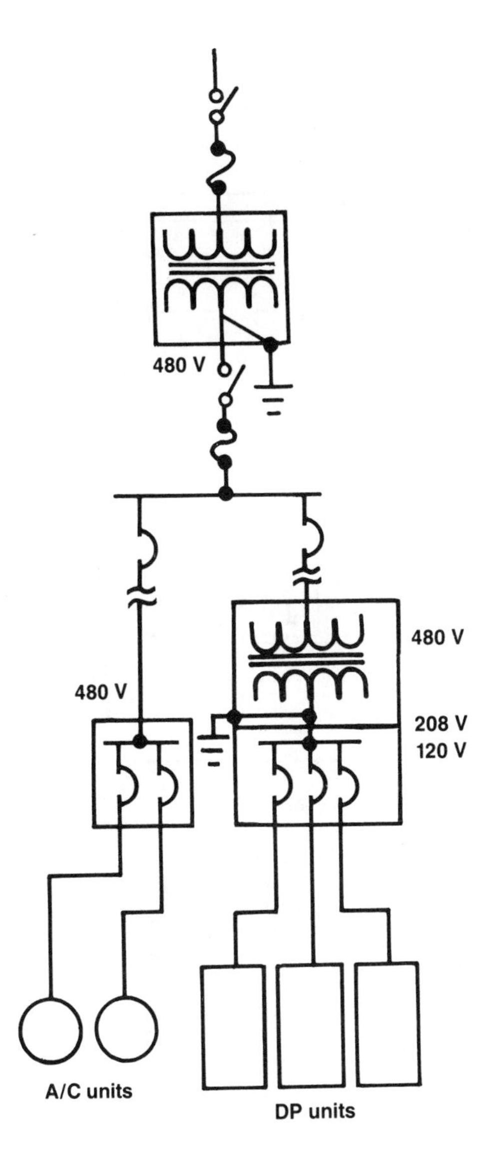

Figure 2.24 Preferred power distribution configuration for a DP site. (Adapted from: Federal Information Processing Standards Publication No. 94, *Guideline on Electrical Power for ADP Installations*, U.S. Department of Commerce, National Bureau of Standards, Washington, DC, 1983.)

2.4.4 Grounding Equipment Racks

The installation and wiring of equipment racks must be planned carefully to avoid problems during day-to-day operations. Figure 2.25 shows the recommended approach. Bond adjacent racks together with 3/8 to 1/2in-diameter bolts. Clean the

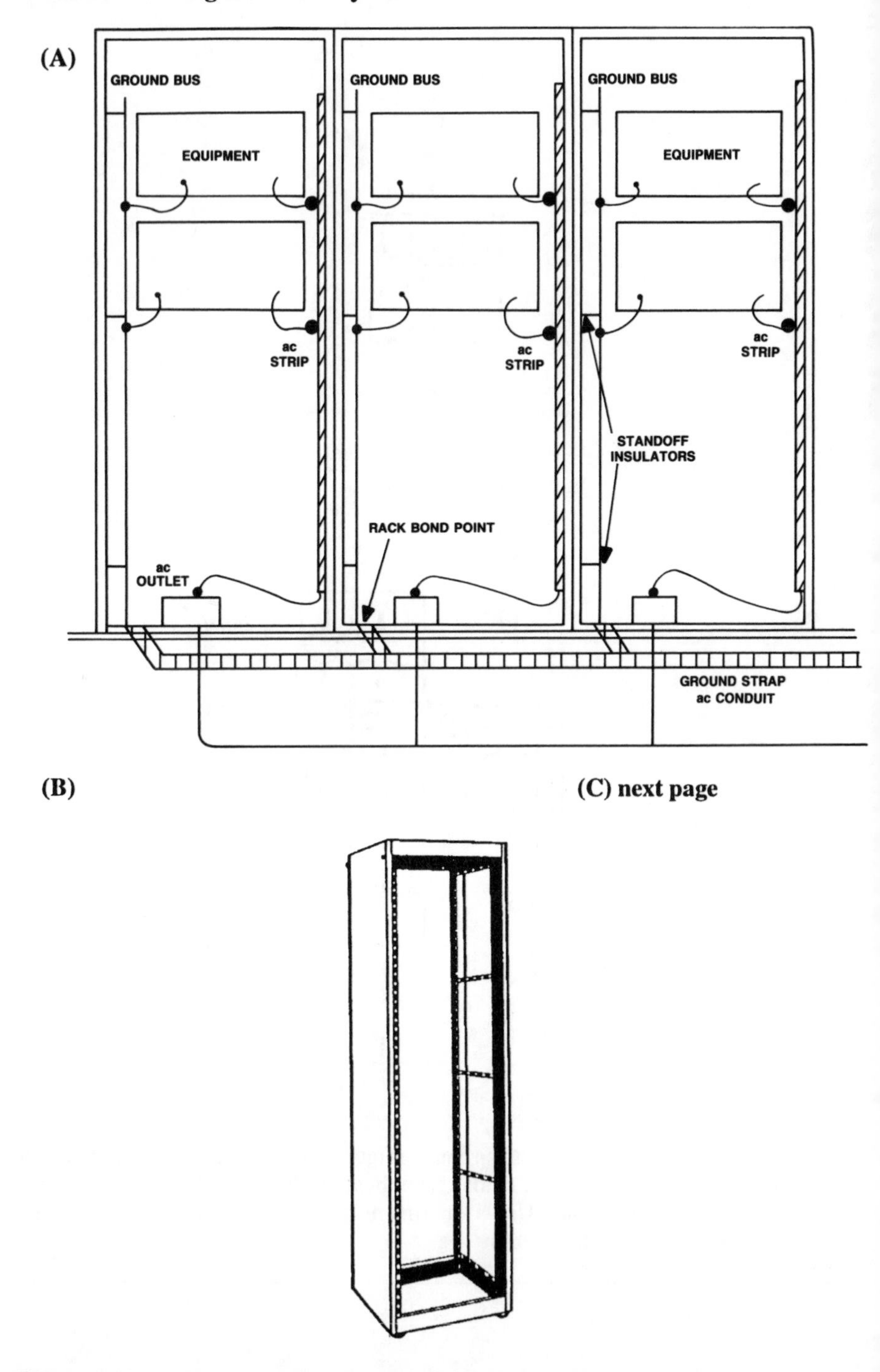

Figure 2.25 Recommended grounding method for equipment racks: (a) overall grounding scheme; (b) detail of rack support rails; (c) detail of ac power strip mounting. To make assembly of multiple racks easier, position the ground connections and ac receptacles at the same location in all racks.

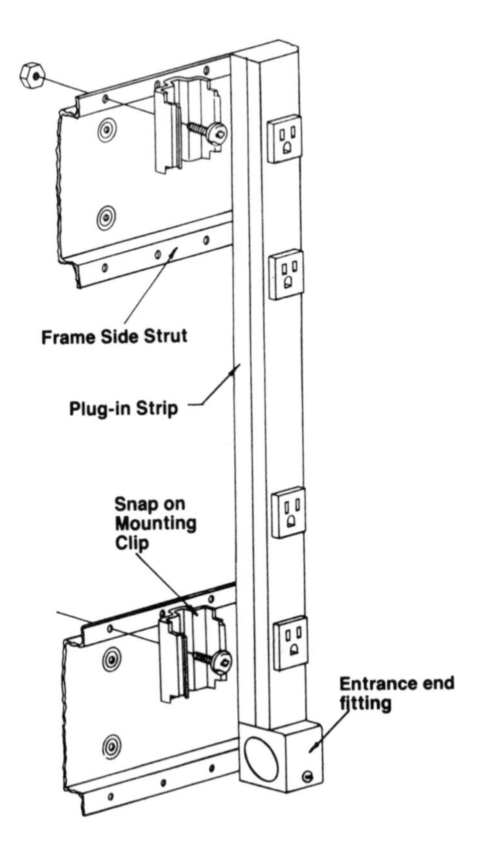

Figure 2.25(c)

---

contacting surfaces by sanding down to bare metal. Use lock washers on both ends of the bolts. Bond racks together using at least six bolts per side (three bolts for each vertical rail).

Run a ground strap from the *main facility ground point*, and bond the strap to the base of each rack. Spot-weld the strap to a convenient spot on one side of the rear portion of each rack. Secure the strap at the same location for each rack used. A mechanical connection between the rack and the ground strap may be made using bolts and lock washers, if necessary. Be certain, however, to sand down to bare metal before making the ground connection. Because of the importance of the ground connection, it is recommended that each attachment be made with a combination of crimping and silver-solder.

Install a vertical ground bus in each rack (as illustrated in Figure 2.25). Use about 1 1/2-in wide, 1/4-in thick copper busbar. Size the busbar to reach from the bottom of the rack to about 1 ft short of the top. The exact size of the busbar is not critical, but it must be sufficiently wide and rigid to permit the drilling of 1/8-in holes without deforming.

Mount the ground busbar to the rack using insulated standoffs. Porcelain standoffs commonly found in high-voltage equipment are useful for this purpose. Porcelain standoffs are readily available and reasonably priced. Attach the ground busbar to the

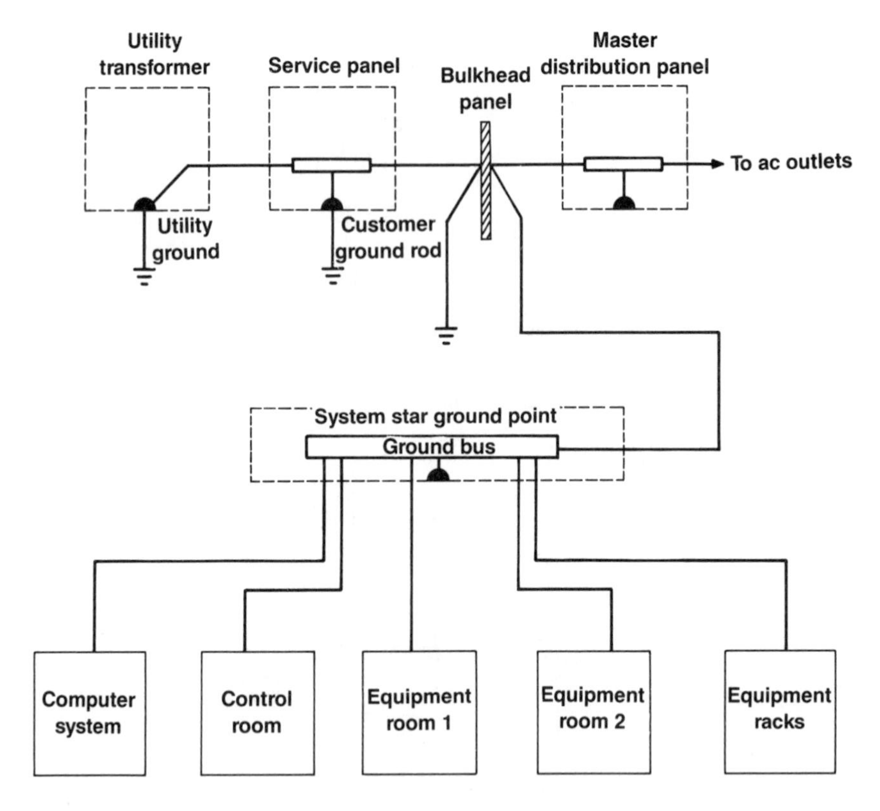

Figure 2.26 Equivalent ground circuit diagram for a medium-size commercial/industrial facility.

---

rack at the point that the facility ground strap attaches to the rack. Silver-solder the busbar to the rack and strap at the same location in each rack used.

Install an ac receptacle box at the bottom of each rack. Isolate the conduit from the rack. The easiest approach is to use an insulated bushing between the conduit and the receptacle box. With this arrangement, the ac outlet box can be mounted directly to the bottom of the rack near the point that the ground strap and ground busbar are bonded to the rack. An alternative approach is to use an orange-type receptacle. This type of outlet isolates the green-wire power ground from the receptacle box. Use insulated standoffs to mount the ac outlet box to the rack. Bring out the green-wire ground, and bond it to the rack near the point that the ground strap and ground busbar are silver-soldered to the rack. The goal of this configuration is to keep the green-wire ac and facility system grounds separate from the ac distribution conduit and metal portions of the building structure. Carefully check the local electrical code to ensure that such configurations are legal.

Although the foregoing procedure is optimum from a signal-grounding standpoint, note that under a ground fault condition, performance of the system may be unpredictable if high currents are being drawn in the current-carrying conductors supplying the load. Vibration of ac circuit elements resulting from the magnetic field effects of high-current-carrying conductors is insignificant as long as all conductors are within

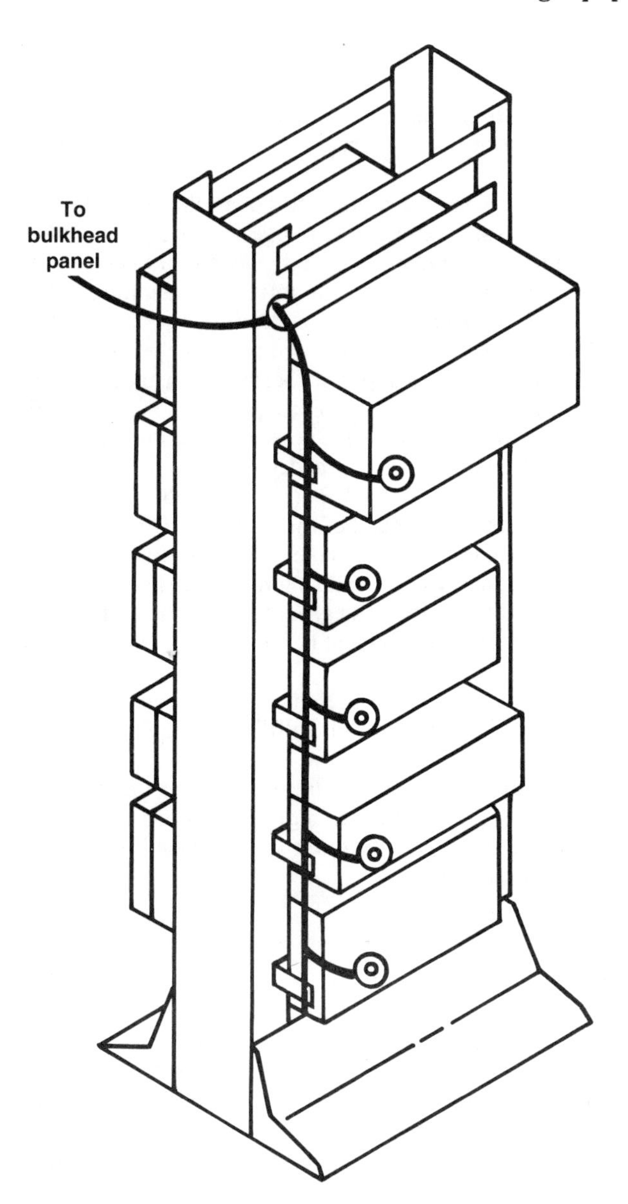

Figure 2.27 Ground bus for an open-frame equipment rack.

the confines of a given raceway or conduit. A ground fault will place return current outside of the normal path. If sufficiently high currents are being conducted, the consequences can be devastating. "Sneak" currents from ground faults have been known to destroy wiring systems that were installed exactly to code.

The fail-safe wiring method for equipment-rack ac power is to use orange-type outlets, with the receptacle green-wire ground routed back to the breaker-panel star ground system. Insulate the receptacle box from the rack to prevent conduit-based noise currents from contaminating the rack ground system. Try to route the power conduit and facility ground cable or strap via the same path, if such a compromise configuration is necessary. Remember to keep metallic conduit and building structures insulated from the facility ground line, except at the bulkhead panel (main grounding point).

Mount a vertical ac strip inside each rack to power the equipment. Insulate the power strip from the rack using porcelain standoffs. Power equipment from the strip using standard three-prong grounding ac plugs. Do not defeat the safety ground connection. Equipment manufacturers use this ground to drain transient energy.

Mount equipment in the rack using normal metal mounting screws. If the location is in a high-RF field, clean the rack rails and equipment panel connection points to ensure a good electrical bond. This is important because, in a high-RF field, detection of RF energy can occur at the junctions between equipment chassis and the rack.

Connect a separate ground wire from each piece of equipment in the rack to the vertical ground busbar. Use no. 12 gauge stranded copper wire (insulated) or larger. Connect the ground wire to the busbar by drilling a hole in the busbar at a convenient elevation near the equipment. Fit one end of the ground wire with an enclosed-hole solderless terminal connector (no. 10-sized hole or larger). Attach the ground wire to the busbar using appropriate hardware. Use an internal-tooth lock washer between the busbar and the nut. Fit the other end of the ground wire with a terminal that will be accepted by the equipment. If the equipment has an isolated signal ground terminal, tie it to the ground busbar.

Figure 2.26 shows each of the grounding elements that are discussed in this section integrated into one diagram. This approach fulfills the requirements of personnel safety and equipment performance.

Follow similar grounding rules for simple one-rack equipment installations. Figure 2.27 illustrates the grounding method for a single open-frame equipment rack. The vertical ground bus is supported by insulators, and individual jumpers are connected from the ground rail to each chassis.

2.4.5 Computer Floors

Many large technical centers, particularly DP facilities, are built on raised "computer floors." A computer cellular floor is a form of ground plane. Grid patterns on 2 ft centers are common. The basic open grid electrically functions as a continuous ground plane for frequencies below approximately 20 MHz. To be effective, the floor junctions must be bonded together. The mating pieces should be plated to prevent corrosion, oxidation, and electrolytic (galvanic) action.

Floor tiles are typically backed with metal to meet fire safety requirements. Some tiles are of all-metal construction. The tiles, combined with the grounded grid structure, provide for effective electrostatic discharge (ESD) protection in equipment rooms.

Figure 2.28 illustrates the interconnection guidelines for a grounded raised computer floor. Note that the grid structure is connected along each side of the room to a ground ring. The ground ring, in turn, is bonded to the room's main ground conductor. Note also that all cabling enters and leaves the facility in one area, along one wall, forming a bulkhead panel for the room.

In a small facility, where one computer-floor-based center feeds various peripheral equipment, connection of the grounding bulkhead to structural steel, conduit, and raceways would not be made. Grounding would be handled by the main facility ground

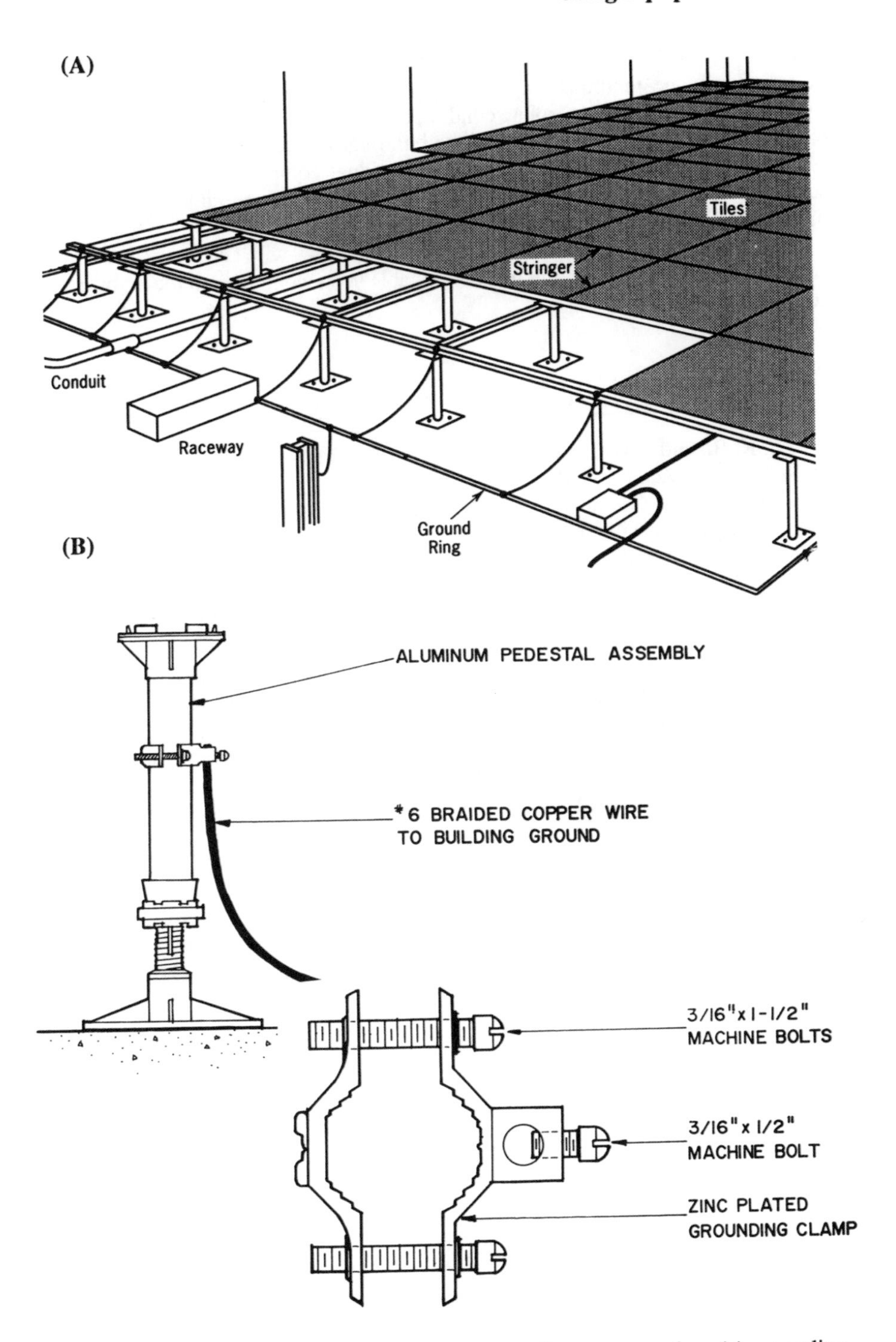

Figure 2.28 Grounding system for raised computer floor construction: (a) grounding of metal supports for raised floor; (b) ground wire clamp detail.

conductor. If the layout is sufficiently simple, and cable trays and conduit do not overlap each other, textbook single-point grounding is practical. In large facilities, however, it is impractical to isolate conduit, cable trays, and the main facility ground conductor from the structural steel of the building. In such cases, it is necessary to bond these elements together outside of individual equipment rooms. Within the rooms, however, maintain the single-point ground scheme. Bond any cable tray or conduit entering the room to the bulkhead panel.

In a large facility, when possible, use common paths—with adequate separation to prevent noise on signal-carrying lines—for cable trays, conduit, and the main facility ground conductor.

2.5 Bibliography

Benson, K. B., and J. Whitaker: *Television and Audio Handbook for Engineers and Technicians*, McGraw-Hill, New York, 1989.

Block, Roger: "The Grounds for Lightning and EMP Protection," PolyPhaser Corporation, Gardnerville, NV., 1987.

Davis, Gary, and Ralph Jones: *Sound Reinforcement Handbook*, Yamaha Music Corporation, Hal Leonard Publishing, Milwaukee, 1987.

Fardo, S., and D. Patrick: *Electrical Power Systems Technology*, Prentice-Hall, Englewood Cliffs, NJ, 1985.

Federal Information Processing Standards Publication No. 94, *Guideline on Electrical Power for ADP Installations*, U.S. Department of Commerce, National Bureau of Standards, Washington, D.C., 1983.

Lanphere, John: "Establishing a Clean Ground," *Sound & Video Contractor* magazine, Intertec Publishing, Overland Park, KS, August 1987.

Lawrie, Robert: *Electrical Systems for Computer Installations*, McGraw-Hill, New York, 1988.

Morrison, Ralph, and Warren Lewis: *Grounding and Shielding in Facilities*, John Wiley & Sons, New York, 1990.

Mullinack, Howard G.: "Grounding for Safety and Performance," *Broadcast Engineering* magazine, Intertec Publishing, Overland Park, KS, October 1986.

Whitaker, Jerry: *AC Power Systems*, CRC Press, Boca Raton, FL, 1991.

CHAPTER 3

WIRING PRACTICES

3.1 Introduction

All signal-transmission media impair—to some extent—an input electrical signal as it is transmitted, whether analog or digital. Foremost among the impairments is attenuation. The distance over which transmission is possible is determined technically by the threshold sensitivity of the signal receiver. Subjectively, the maximum distance is determined by user-established specifications for tolerable signal bandwidth reduction and S/N (signal-to-noise ratio) increase. Noise in this analysis is a generic term that includes Gaussian noise present in all active components in the transmission system, unwanted signals (crosstalk) coupled from parallel signal-transmission circuits, EMI (electromagnetic interference), and RFI (radio frequency interference) from the total environment through which the signal passes.

All other transmission impairments can be grouped within a generic term of *non-linearities*. These include passband frequency-response flatness deviations, harmonic distortion, and aberrations detected as frequency-specific differences in signal gain and phase. The methods used to interconnect various pieces of equipment, and the hardware used to make the interconnection, determine largely how the overall system will operate. Proper cable installation and termination requires skill and experience. To ensure that the installation will be of high quality and have a neat, organized appearance, the system engineer should specify the practices to be followed by installers.

Installation specifications should be included in the project manual to guide the installers and to ensure good workmanship and adherence to industry standards. Specify how the wiring is to be bundled, supported, and routed within the racks. Group cables into bundles that are held together by cable ties or another method of harnessing. Crosstalk between cables carrying different types and levels of signals can be minimized by isolating the cables into separate groups for video, pulse, audio, control, data, and power. Audio cable should be further subdivided into the following categories:

- Low level (below -20 dBm)

- Medium level (-20 to +20 dBm)
- High level (above +20 dBm)

Control cables and cables carrying dc can be bundled together.

Specify wire and cable types and colors, and identify each on drawings and cable schedules with a unique identifying number or code.

3.1.1 Electrical Properties of Conductors

At the heart of any facility is the cable used to tie distant parts of the system together. Conductors are rated by the American Wire Gauge (AWG) scale. The smallest is no. 36; the largest is no. 0000. There are 40 sizes in between. Sizes larger than no. 0000 AWG are specified in *thousand circular mil* units, referred to as "MCM" units (M is the roman numeral expression for 1,000). The cross-sectional area of a conductor doubles with each increase of three AWG sizes. The diameter doubles with every six AWG sizes.

Most conductors used for signal and power distribution are made of copper. Stranded conductors are used where flexibility is required. Stranded cables usually are more durable than solid conductor cables of the same AWG size.

Resistance and inductance are the basic electrical parameters of concern in the selection of wire for electronic systems. Resistivity is commonly measured in ohm-centimeters (Ω-cm). Table 3.1 lists the resistivity of several common materials.

Ampacity is the measure of the ability of a conductor to carry electrical current. Although all metals will conduct current to some extent, certain metals are more efficient than others. The three most common high-conductivity conductors are:

- Silver, with a resistivity of 9.8 Ω/circular mil-foot
- Copper, with a resistivity of 10.4 Ω/cmil-ft
- Aluminum, with a resistivity of 17.0 Ω/cmil-ft

The ampacity of a conductor is determined by the type of material used, the cross-sectional area, and the heat-dissipation effects of the operating environment. Conductors operating in free air will dissipate heat more readily than conductors placed in a larger cable or in a raceway with other conductors. Table 3.2 lists key specifications for larger-size cables typically used in ac power distribution.

3.1.1.A *Effects of Inductance*

Current through a wire results in a magnetic field. All magnetic fields store energy, and this energy cannot be changed in zero time. Any change in the field takes a finite length of time to occur. Inductance (H) is the property of opposition to changes in energy level. The inductance of equipment interconnection cables is usually a distributed parameter.

Voltage drop in a conductor is a function of resistance and inductance. The skin effect (see Section 3.3.3) and circuit geometry affect both parameters. For example, the inductance of a #10 conductor is approximately 3.5 μH/100-in. At 1 MHz, this translates to a resistance of 22 Ω/100-in.

Table 3.1 Resistivity of common materials.

Material	Resistivity (μΩ-cm)
Silver	1.468
Copper	1.724
Aluminum	2.828
Steel	5.88
Brass	7.5

Table 3.2 Primary specifications for large-size wire.

Wire (AWG)	Area (cmil)	Number of cond.	Diameter each conductor (in)	dc resistance Ω/1,000 ft	
				Copper	Aluminum
12	6,530	1	0.0808	1.62	2.66
10	10,380	1	0.1019	1.018	1.67
8	16,510	1	0.1285	0.6404	1.05
6	26,240	7	0.0612	0.410	0.674
4	41,740	7	0.0772	0.259	0.424
3	52,620	7	0.0867	0.205	0.336
2	66,360	7	0.0974	0.162	0.266
1	83,690	19	0.0664	0.129	0.211
0	105,600	19	0.0745	0.102	0.168
00	133,100	19	0.0837	0.0811	0.133
000	167,800	19	0.0940	0.0642	0.105
0000	211,600	19	0.1055	0.0509	0.0836
250	250,000	37	0.0822	0.0431	0.0708
300	300,000	37	0.0900	0.0360	0.0590
350	350,000	37	0.0973	0.0308	0.0505
400	400,000	37	0.1040	0.0270	0.0442
500	500,000	37	0.1162	0.0216	0.0354
600	600,000	61	0.0992	0.0180	0.0295
700	700,000	61	0.1071	0.0154	0.0253
750	750,000	61	0.1109	0.0144	0.0236
800	800,000	61	0.1145	0.0135	0.0221
900	900,000	61	0.1215	0.0120	0.0197
1,000	1,000,000	61	0.1280	0.0108	0.0177

3.2 Coaxial Cable

Of all of the cable types used to interconnect a given facility, coaxial cable usually represents the greatest challenge. (Note that fiber-optic lines are covered separately in Chapter 8.) The unique properties of coax permit use over a broad range of frequencies, offering distinct advantages for the system engineer.

3.2.1 Transporting Energy at High Frequencies

The motion of electrical energy requires the presence of an electric field and a magnetic field. Any two conductors can direct the flow of energy. The basic geometry for energy transport is two parallel conductors, as illustrated in Figure 3.1. The transmission line exhibits distributed capacitance C and distributed inductance L along its length. When the switch in the diagram is closed, current begins to flow, charging the capacitance. This current also establishes a magnetic field around both conductors. The energy in these two fields is supplied at a fixed rate. The voltage wave propagates down the line at a fixed velocity, given by the following equation:

$$V = \sqrt{L \times C}$$

The velocity in the conductors is typically about one-half the speed of light.

Energy is stored on the line, and, as energy is added, it must be transported past any existing storage. This requires an electrical field and a magnetic field behind the wavefront. The current I that flows in the line is given by the following equation:

$$I = \frac{V}{\sqrt{\frac{L}{C}}}$$

$$I = \frac{V}{Z}$$

where:
Z = the characteristic impedance of the line in Ω

If the transmission line were cut at some point and terminated in an impedance Z, energy would continue to flow on the line as if it had infinite length. When the wavefront reaches the termination, energy is dissipated per unit time rather than being stored per unit time.

The transmission line principles presented here represent an ideal circuit. In a practical transmission line, many factors contribute to losses and some radiation, including the following:

- Skin effect
- Dielectric and conductive losses
- Irregularities in geometry

These factors change with the frequency of the transported wave.

When a transmission line is not terminated in its characteristic impedance, reflections of the transported wave will occur. When a signal reaches an open circuit on the line, the total current flow at the open point must be zero. A reflected wave, therefore, is generated that cancels this current. If, on the other hand, a signal reaches a short circuit on the line, a reflected wave is generated that cancels the voltage. Reflections

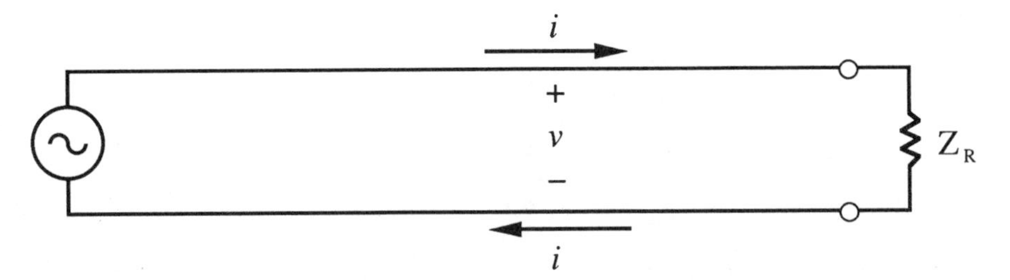

Figure 3.1 Basic transmission line circuit.

of these types return energy to the source. The signal at any point along the line is a composite of the initial signal and any reflections.

Sinusoidal signals are assumed when the input impedance of a transmission line is discussed. The input impedance is determined by the following:

- Characteristic impedance of the line
- Terminating impedance
- Applied frequency
- Length of the line

Reflected energy reaching the source modifies the voltage-current relationship. On short unterminated lines, the input impedance can vary significantly. If the reflected wave returns in phase with the input signal, no current will flow; the input impedance is infinite. If the input signal returns 90° out of phase, the line will appear as a pure reactive load to the source.

3.2.2 Operating Principles

A coaxial transmission line consists of concentric center and outer conductors that are separated by a dielectric material. When current flows along the center conductor, it establishes an electric field. The electric flux density and the electric field intensity are determined by the dielectric constant of the dielectric material. The dielectric material becomes polarized with positive charges on one side and negative charges on the opposite side. The dielectric, therefore, acts as a capacitor with a given capacitance per unit length of line. Properties of the field also establish a given inductance per unit length, and a given series resistance per unit length. If the transmission line resistance is negligible and the line is terminated properly, the following formula describes the characteristic impedance (Z_0) of the cable:

$$Z_0 = \sqrt{\frac{L}{C}}$$

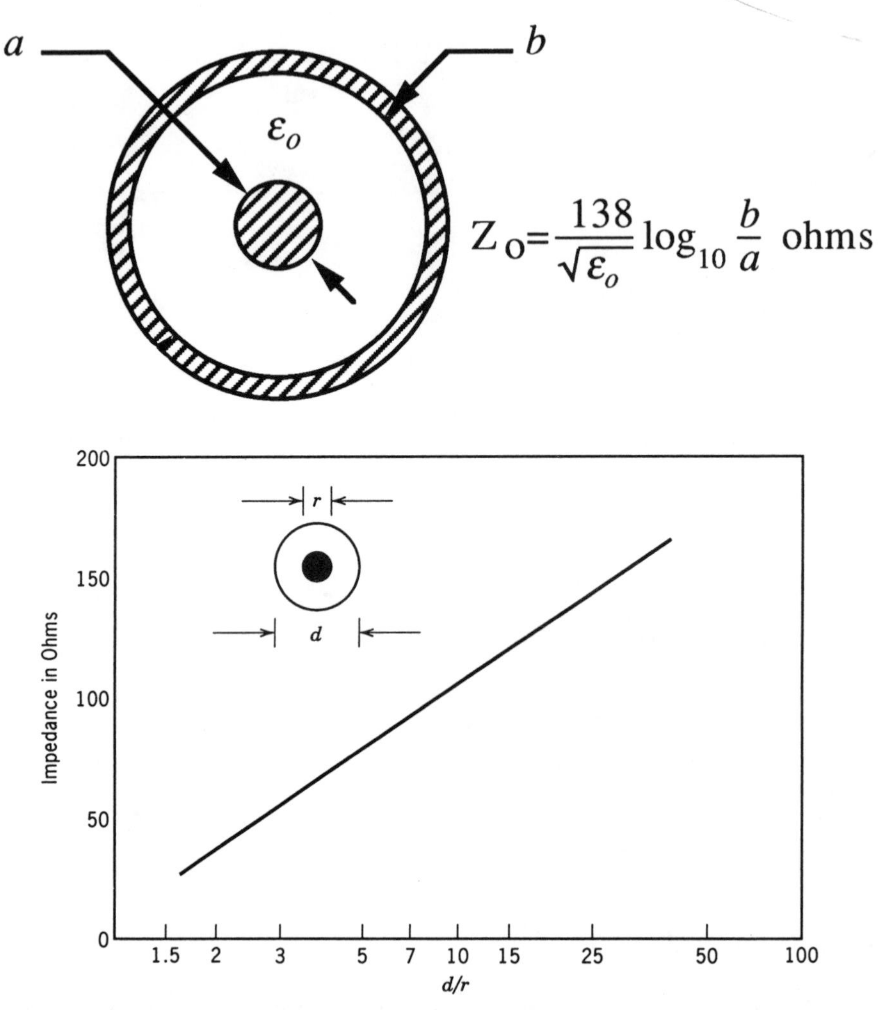

Figure 3.2 The interdependence of coaxial cable physical dimensions and characteristic impedance.

---

where:
L = inductance in H/ft
C = capacitance in F/ft

Coaxial cables typically are manufactured with 50 Ω or 75 Ω characteristic impedances. Other characteristic impedances are possible by changing the diameter of the center and outer conductors. Figure 3.2 illustrates the relationship between characteristic impedance and the physical dimensions of the cable.

Trade num.	Std. avail. lengths (ft)	AWG, strand.	Insulation type	Nom. OD (in)	Num. of shields	Shield type	Nom. imp. (ohms)	Nom. vel. of prop.	Nom. cap. (pF/ft)	Attn. at 10 MHz (dB) 100 ft
8279	100, 500, 1,000	23 (7x32)	Polyethylene	0.220	Single	Tinned braid, copper	75	66%	21	1.3
8281	500, 1,000	20 (solid)	Polyethylene	0.304	Double	Tinned braid, copper	75	66%	21	0.78
9231	500, 1,000	20 (solid)	Polyethylene	0.304	Double	Tinned braid, copper	75	66%	21	0.78

Figure 3.3 Data sheet for common types of coaxial video cable.

3.2.3 Installing Coaxial Cable

In order to meet its performance specifications, coaxial cable must be installed properly to avoid mechanical stress and damage, which can alter its characteristic impedance and, therefore, the signal it carries. At high frequencies, the change in characteristic impedance resulting from damage or compression of the cable will cause high-frequency components to be reflected to the source. The reflected signal will be added to the instantaneous amplitude of the transmitted video or data signal and cause it to be distorted. A typical data sheet for common types of coaxial cable is shown in Figure 3.3.

To avoid possible damage to a cable, its minimum bend radius must not be exceeded. Design conduit and cable trays with the minimum bend requirements in mind. The recommended bend radius of coaxial cable for a single permanent bend is 10 times the cable diameter. In installations where the cable will be repeatedly flexed, the minimum bend radius is 15 times the cable diameter. Provide a loop of slack cable to prevent sharp transitions at the point of bending. Various methods of strain relief are available, which can be used to limit the bend radius of the cable at the point where it flexes.

When pulling cables through conduit, the mechanical stress must be distributed evenly over each cable. Do not exceed the maximum allowable pulling tension for the weakest cable in the conduit, or the conductors may be stretched or broken. For cables with copper conductors, the allowable tension is 40 percent of the breaking strength. This point is the maximum pulling tension that may be applied without stretching the copper center conductor. The maximum pulling tension specification for a given cable is available from the cable manufacturer.

If necessary, use a lubricant to reduce friction in conduit. Dry compounds, such as talc and powdered soapstone, are available. Liquids and pastes may also be used. Any lubricating compound must be compatible with the cable jacket material.

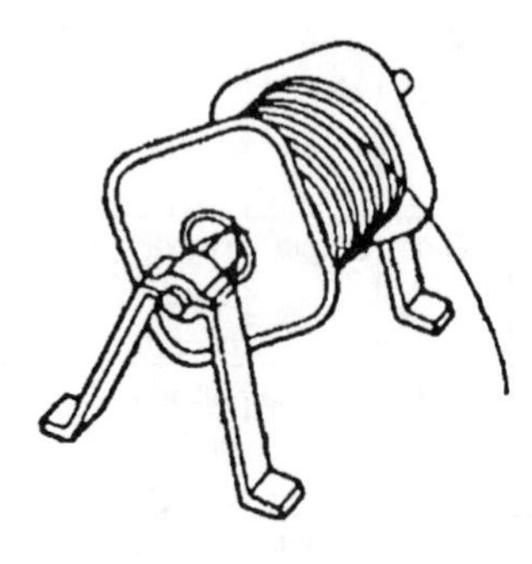

Figure 3.4 The proper method to pull cable from a reel.

Pull coaxial cable by the braid. Pull the cables with a steady tension to avoid jerking the conductors. Grips or clamps, such as *Kellum grips*[1], should be used to pull cables. These grips use the Chinese finger puzzle principle to grip the cable, distributing the pulling tension evenly throughout the cable. They are reusable and easy to install. Other pulling devices are also available. Spring scales and similar tension-measuring devices can be used to ensure that the tension limit of the cable is not exceeded.

When cable is pulled over the flange of a stationary reel, the cable will be twisted 360° for every revolution around the spool, causing kinks in the cable. This twisting can damage the conductors and make the cable difficult to pull. The reel must be mounted on an arbor so that the cable can be pulled from a revolving reel (see Figure 3.4).

Twisting also can be avoided by specifying that the cable be supplied in carton put-ups. The cable is laid into these cartons, not wound on reels. That way, the cable, when drawn from an opening in the carton, will not be twisted. The cartons can be stacked on each other and need less space and set-up time than arbor let-offs. No other let-off equipment is required. Inertia spills, where the reel spins, dumping cable onto the floor, are eliminated when using this type of cable packaging. Figure 3.5 shows one such carton put-up.

In a permanent system installation, use continuous unbroken lengths of cable between devices. If a coaxial cable must be spliced, use coaxial cable connectors that are designed for that specific cable. Male and female cable end connectors or two male connectors with a dual female adapter are available. The proper connectors will maintain the coaxial configuration, impedance, and shielding of the cable with minimum discontinuity. If the splice will be exposed to high humidity or immersed in water, encapsulate the splice in a sealant/encapsulant, such as Scotchcast or RTV (room temperature vulcanized) silicone rubber, to prevent infiltration of moisture. (Scotchcast is a registered trademark of 3M.) If the cable has a polyethylene jacket, use fine sandpaper to roughen its surface before applying the sealant to provide good adhesion.

1 **Manufactured by the Kellum Division of Hubbell, Stonington, CT.**

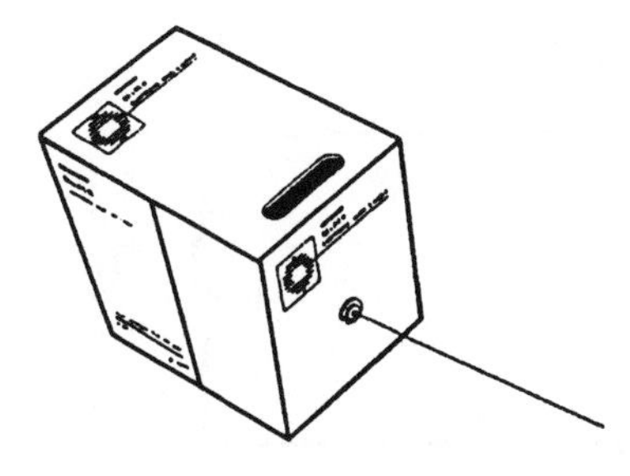

Figure 3.5 A boxed put-up, used to prevent kinking of cable during unreeling.

3.2.4 Installation Considerations

Do not run video cable in the same wire tray with power cables. Electromagnetic coupling of the 60 Hz current in power lines can induce hum in the video signal. The mixing of signal and power cables in the same cable tray may also violate local and national electrical codes.

Lightning protection is required where cables enter a building. Use coaxial-type lightning arrestors for this purpose. They provide a method for safely connecting the shield of a cable to ground during a lightning discharge.

In locations where cables must be strung between two poles or buildings, determine whether the cable can support its own weight across the span. The sag-vs.-span specification for a cable is usually available from the manufacturer. Use a steel *messenger cable* to support the line if the span is longer than the cable can support. Special hardware is available to secure signal cables to the messenger line, which will support the load.

When cable is to be stored outside, seal the ends of the cable to prevent moisture from entering and damaging it. Take care when installing cable in areas where water is present or can accumulate. If the line is going to be pulled into a conduit or tray that may be filled with water, seal the cable end first. Water can enter the cable through a tear in the jacket. The jacket must be protected during the pulling process.

Cold temperatures will cause the materials used to make most coaxial cables stiffen. At very cold temperatures, the jacket may become brittle and crack when the cable is flexed. If the cable has just been brought in from a cold area or is being installed in an unheated building, store it in a heated area before it is installed. The heat will make the cable more flexible and easier to pull. A portable heater can be used to warm the cable at the pull site. Keep the heat from being applied directly to the coax by using baffles or diffusers.

The National Electrical Code requires that signal-carrying cables conform to certain rules enacted to prevent electrical shock to humans and fire hazards. Research the applicable rules, and follow them.

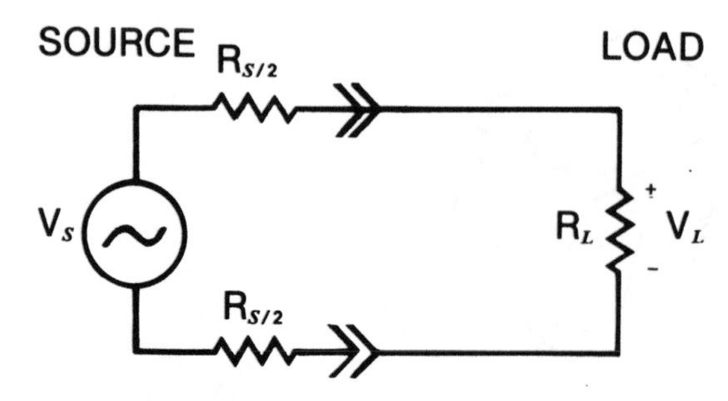

Figure 3.6 A basic source and load connection. No grounds are indicated, and both the source and the load float.

3.3 Grounding Signal-Carrying Cables

Proper ground system installation is the key to minimizing noise currents on signal-carrying cables. Audio, video, and data lines are often subject to ac power noise currents and RFI. The longer the cable run, the more susceptible it is to disturbances. Unless care is taken in the layout and installation of such cables, unacceptable performance of the overall system may result.

3.3.1 Analyzing Noise Currents

Figure 3.6 shows a basic source and load connection. No grounds are present, and both the source and the load float. This is the optimum condition for equipment interconnection. Either the source or the load may be tied to ground with no problems, provided only one ground connection exists. *Unbalanced* systems are created when each piece of equipment has one of its connections tied to ground, as shown in Figure 3.7. This condition occurs if the source and load equipment have unbalanced (single-ended) inputs and outputs. This type of equipment uses chassis, or common, ground for one of the conductors. Problems are compounded when the equipment is separated by a significant distance.

As shown in the figure, a difference in ground potential causes current flow in the ground wire. This current develops a voltage across the wire resistance. The ground-noise voltage adds directly to the signal. Because the ground current is usually the result of leakage in power transformers and line filters, the 60 Hz signal gives rise to hum of one form or another. Reducing the wire resistance through a heavier ground conductor helps the situation, but it cannot eliminate the problem.

By amplifying the high side and the ground side of the source and subtracting the two to obtain a *difference signal*, it is possible to cancel the ground-loop noise. This is the basis of the *differential input* circuit, illustrated in Figure 3.8. Unfortunately, problems still may exist with the unbalanced-source-to-balanced-load system. The reason centers on the impedance of the unbalanced source. One side of the line will have a slightly lower amplitude because of impedance differences in the output lines. By creating an output signal that is out of phase with the original, a balanced source

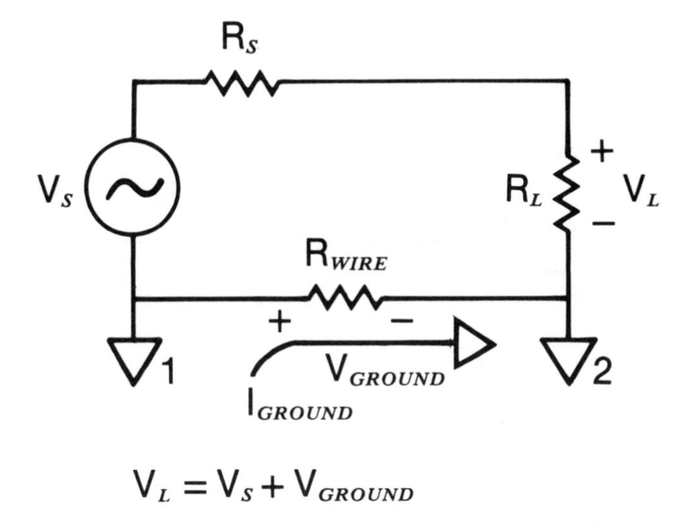

Figure 3.7 An unbalanced system in which each piece of equipment has one of its connections tied to ground.

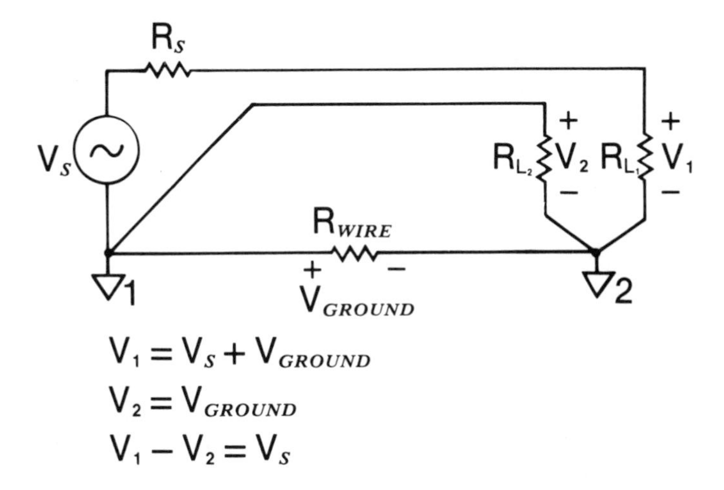

Figure 3.8 Cancellation of ground-loop noise by amplifying both the high and ground side of the source and subtracting the two signals.

can be created to eliminate this error (see Figure 3.9). As an added benefit, for a given maximum output voltage from the source, the signal voltage is doubled over the unbalanced case.

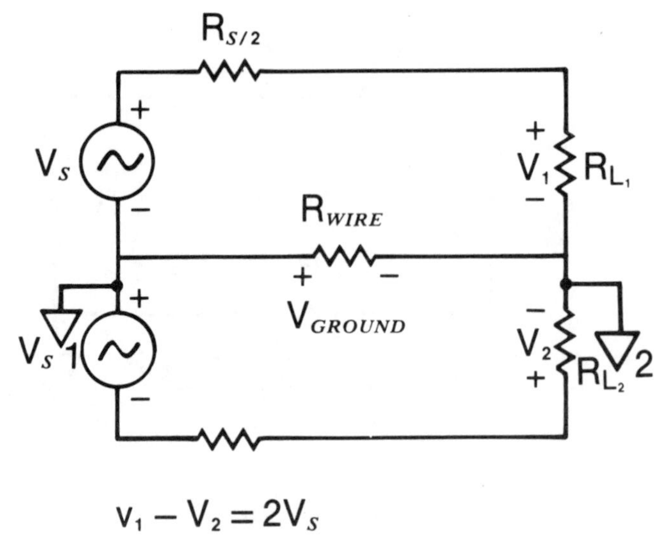

Figure 3.9 A balanced source configuration, where the inherent amplitude error of the system shown in Figure 3.8 is eliminated.

3.3.2 Types of Noise

Open (non-coaxial) wiring can couple energy from external fields. These fields result from power lines, signal processes, and RF sources. The extent of coupling is determined by the following:

- Loop area between conductors
- Cable length
- Cable proximity
- Frequency
- Field strength

Two basic types of noise can appear on ac power, audio, video, and computer data lines within a facility: *normal mode* and *common mode*. Each type has a particular effect on sensitive load equipment. The normal-mode voltage is the potential difference that exists between pairs of power (or signal) conductors. This voltage also is referred to as the *transverse-mode* voltage. The common-mode voltage is a potential difference (usually noise) that appears between power or signal conductors and the local ground reference. The differences between normal-mode and common-mode noise are illustrated in Figure 3.10.

The common-mode noise voltage will change depending on what is used as the ground reference point. Often, it is possible to select a ground reference that has a minimum common-mode voltage with respect to the circuit of interest, particularly if the reference point and the load equipment are connected by a short conductor. Common-mode noise can be caused by electrostatic or electromagnetic induction.

A single common-mode or normal-mode noise voltage is rarely found. More often than not, load equipment will see both types of noise signals. In fact, unless the facility

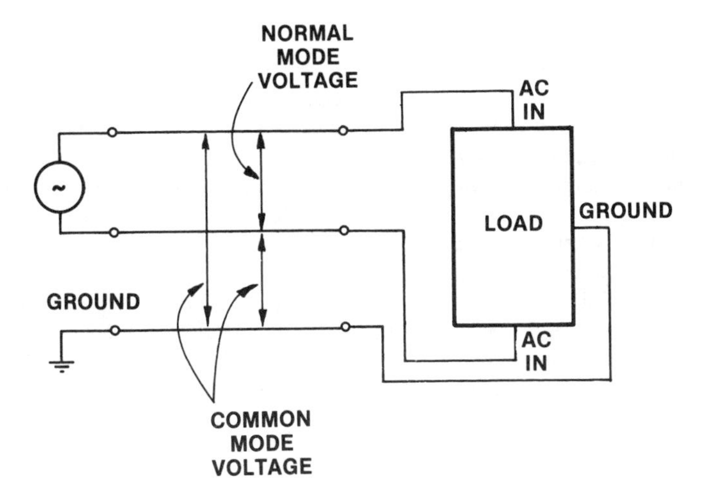

Figure 3.10 The principles of normal-mode and common-mode noise voltages as they apply to ac power circuits.

wiring system is unusually well balanced, the noise signal of one mode will convert some of its energy to the other.

Common-mode and normal-mode noise disturbances typically are caused by momentary impulse voltage differences among parts of a distribution system that have differing ground potential references. If the sections of a system are interconnected by a signal path in which one or more of the conductors is grounded at each end, the ground offset voltage can create a current in the grounded signal conductor. If noise voltages of sufficient potential occur on signal-carrying lines, normal equipment operation can be disrupted (see Figure 3.11).

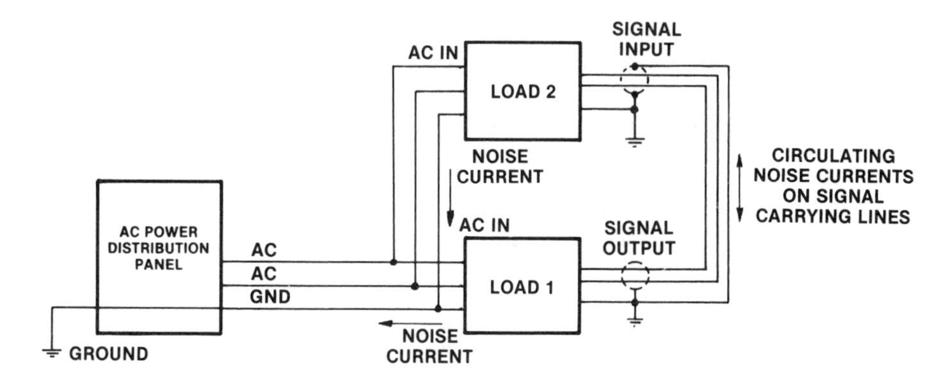

Figure 3.11 An illustration of how noise currents can circulate within a system because of the interconnection of various pieces of hardware.

3.3.3 Skin Effect

Low-level signal cables are particularly susceptible to high-frequency noise energy because of the *skin effect* of current-carrying conductors. When a conductor carries an alternating current, a magnetic field is produced, which surrounds the wire. This field is expanding and contracting continually as the ac current wave increases from zero to its maximum positive value and back to zero, then through its negative half-cycle. The changing magnetic lines of force cutting the conductor induce a voltage in the conductor in a direction that tends to retard the normal flow of current in the wire. This effect is more pronounced at the center of the conductor. Thus, current within the conductor tends to flow more easily toward the surface of the wire. The higher the frequency, the greater the tendency for current to flow at the surface. The depth of current flow is a function of frequency and is determined from the following equation:

$$d = \frac{2.6}{\sqrt{\mu \times f}}$$

where:
d = depth of current in mils
μ = permeability (copper = 1, steel = 300)
f = frequency of signal in MHz

It can be calculated that at a frequency of 100 kHz, current flow penetrates a conductor by 8 mils. At 1 MHz, the skin effect causes current to travel in only the top 2.6 mils in copper, and even less in almost all other conductors. Therefore, the series impedance of conductors at high frequencies is significantly higher than at ac power line frequencies. This makes low-level signal-carrying cables particularly susceptible to disturbances resulting from RFI.

Both skin effect and self-inductance combine to reduce current flow in a conductor as the frequency is increased. If the *loop area* of the circuit is large, the self-inductance will also be large. In facilities exhibiting uncontrolled geometries, where the return path for current is undefined, the effects of self-inductance will dominate over the skin effect.

Because current penetration is a function of permeability, steel exhibits a greater skin effect than copper. This difference, however, disappears at high frequencies, because permeability rapidly falls off as frequency is increased. At frequencies greater than 250 kHz, the impedance of steel and copper will be about the same, and the inductance effects will dominate.

It follows that in a facility constructed with steel beams, the steel provides a better conductive path than a copper ground strap because of the large surface areas afforded by the structural steel. This large surface area reduces inductance, which is an important factor in controlling high-frequency noise. A grid formed by structural steel usually provides a better ground system than can be achieved by installing copper conductors, provided the steel elements are welded together. Bolted construction can lead to unpredictable performance, which is likely to deteriorate with time.

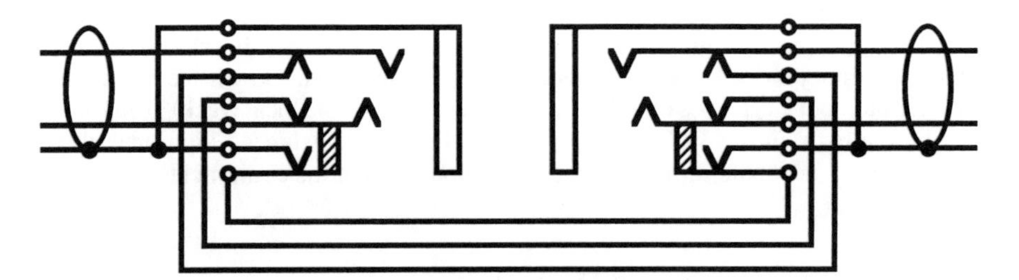

Figure 3.12 Patch-panel wiring for seven-terminal normaling jack fields. Use patch cords that connect ground (sleeve) at both ends.

3.3.4 Patch-bay Grounding

Patch panels for audio, video, and data circuits require careful attention to planning to avoid built-in grounding problems. Because patch panels are designed to tie together separate pieces of equipment, often from remote areas of a facility, the opportunity exists for ground loops. The first rule of patch-bay design is to never use a patchbay to switch low-level (microphone) signals. If mic sources must be patched from one location to another, install a bank of mic-to-line amplifiers to raise the signal levels to 0 dBm before connecting to the patchbay. Most video output levels are 1 V P-P, giving them a measure of noise immunity. Data levels are typically 5 V. Although these line-level signals are significantly above the noise floor, capacitive loading and series resistance in long cables can reduce voltage levels to a point that noise becomes a problem.

Newer-design patch panels permit switching of ground connections along with signal lines. Figure 3.12 illustrates the preferred method of connecting an audio patch panel into a system. Note that the source and destination jacks are *normalled* to establish ground signal continuity. When another signal is plugged into the destination jack, the ground from the new source is carried to the line input of the destination jack. With such an approach, jack cords that provide continuity between sleeve (ground) points are required.

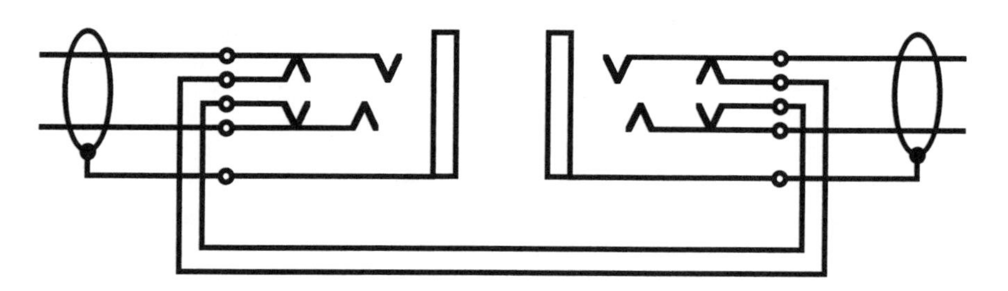

Figure 3.13 Patch-panel wiring for conventional normaling jack fields. Use patch cords that connect ground (sleeve) at both ends.

If only older-style conventional jacks are available, use the approach shown in Figure 3.13. This configuration will prevent ground loops, but, because destination shields are not carried back to the source when normaling, noise will be higher. Bus all destination jack sleeves together, and connect them to the local (rack) ground. The wiring methods shown in Figures 3.12 and 3.13 assume balanced input and output lines with all shields terminated at the load (input) end of the equipment.

3.4 Cable Harness Hardware

Permanent installation of interconnected equipment requires that some orderly means be employed for routing and supporting the cables. To provide strain relief on the connectors and a neat installation, wires and cables should be bound or harnessed into bundles running between system components and secured to some type of supporting structure or frame. Cables carrying like or related signals should be grouped together to prevent crosstalk.

3.4.1 Cable Ties

Individual cable ties provide the easiest and most flexible means of binding cable bundles. There are many factors to consider when selecting the proper cable tie for each application. Generally, the environmental concerns are limited to the effects of extreme heat and, in the case of mobile installations, extreme cold, moisture, and ultraviolet light (from sunlight). The physical properties of plastic materials normally degrade during exposure to high temperature because of oxidation. The maximum temperature for successful service depends on the material used and environmental conditions. Initially, plastics become more flexible and weaker when exposed to high temperatures. After a period of time, oxidation may occur, which will cause the material to become brittle, making plastic cable ties more susceptible to failure from impact and vibration. Low-temperature exposure will also make most plastics more brittle during exposure, but little permanent degradation of the properties remains when the material is returned to room temperature.

Mechanical stress also affects the life of a cable tie. As the bundle diameter is reduced, the tie experiences more bending stress. A thick strap on a small diameter represents a high-stress condition. If the tie is under high load, this will add more stress to the tie body. A thinner tie will have shortened life because surface cracks will penetrate the thickness of the tie faster. Applications subject to high vibration will result in impact stresses, which can cause surface cracks to propagate.

Several other external factors affect the life of a cable tie. Chemical exposure can degrade the tie material. This is the most detrimental factor to the life of a tie. Direct sunshine, high altitudes, and high temperature also decrease the life of the cable tie. Dry environments cause nylon 6/6 ties to become more brittle. High humidity and temperatures can result in degradation because of hydrolysis in nylon.

Hand-tensioning of a cable tie can result in too little or too much tension. Tie-wrap tensioning tools are available from tie manufacturers. These can be adjusted to apply the proper tension for each type and size of tie. One operation applies tension to the tie and cuts the tail of the tie when the proper tension has been reached.

Figure 3.14 shows some of the tie-wrap products commonly available from several suppliers. Use of a tensioning tool is illustrated in Figure 3.15.

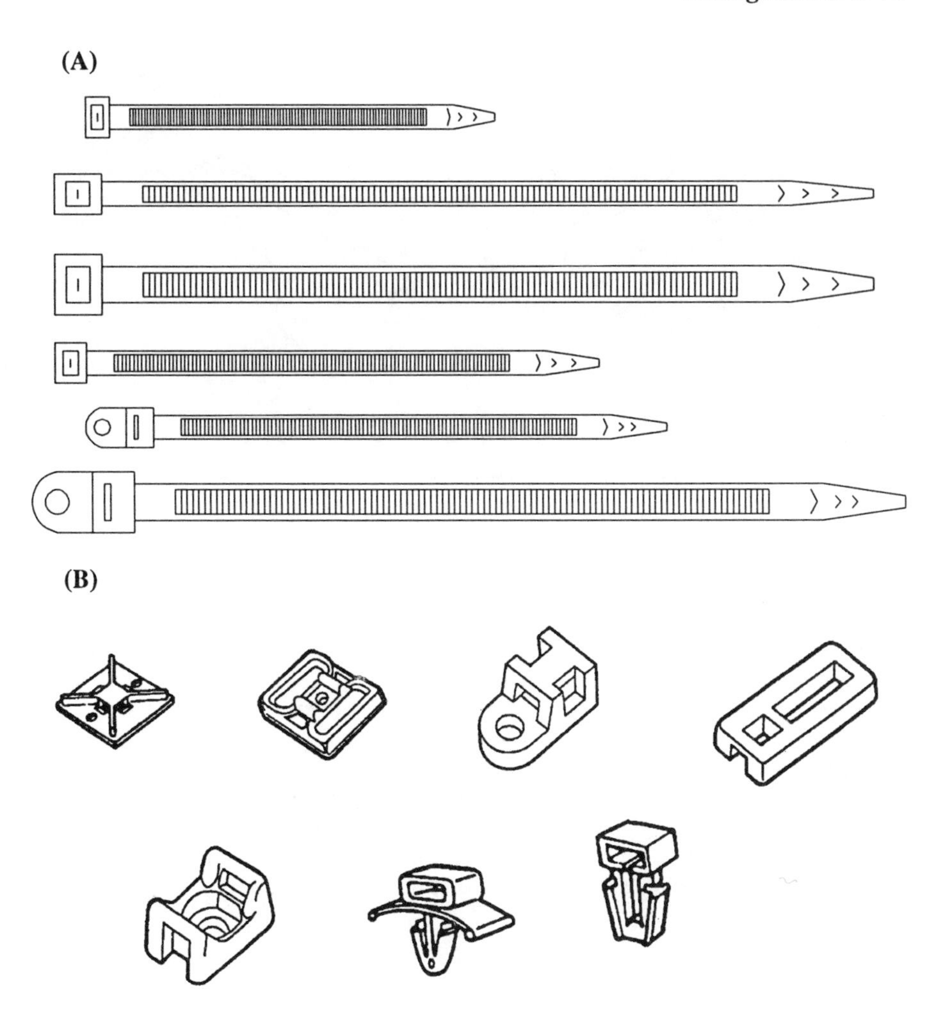

Figure 3.14 A selection of tie-wrap products available from several manufacturers: (a) tie-wraps of various sizes; (b) mounting clamps.

A wide variety of cabling methods may be applied in a given facility. The type used will be dictated by the following:

- Number and size of cables to be secured
- Type of cable support used (cable tray, conduit, or clamps)
- Exposure to vibration or harsh environmental conditions
- Voltages and currents being transported

Figure 3.16 shows some of the more common cabling products available for small- to medium-sized runs.

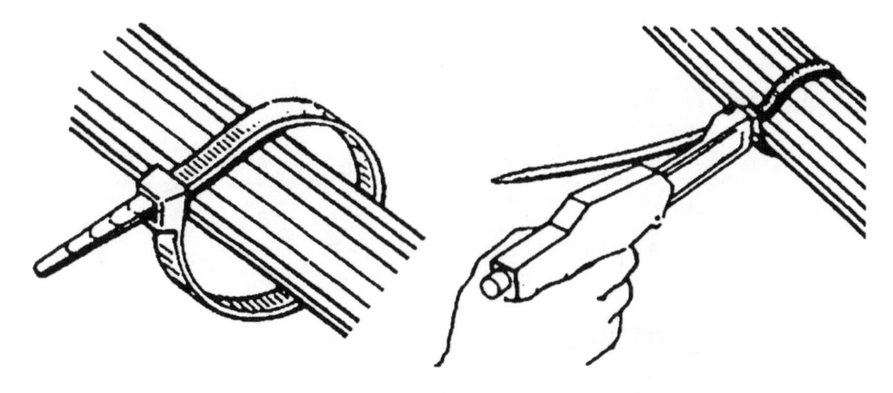

Figure 3.15 Use of a tie-wrap tensioning tool on a cable bundle.

3.4.2 Twist-On Cable Wrap

Twist-on cable wrap protects cables against rough surfaces. It absorbs vibration and reduces impact damage, insulates cables, and resists abrasion. The product twists on like tape, making it easy to apply. This also facilitates breakouts of single or multiple wires and rerouting of replacement wires, as shown in Figure 3.16 (lower left).

Twist-on cable wrap products are available in polyethylene, nylon, and flame-retardant polyethylene for bundle sizes from 1/16-in (1.6 mm) to 7-in (178 mm) diameters. The polyethylene and nylon material is available in black and natural colors. Flame-retardant polyethylene is typically available in its natural color only. Products are also available that conform to military specifications.

3.4.3 Braided Sleeving

Braided sleeving is designed to protect wire bundles, harnesses, and cable assemblies from mechanical and environmental damage. The high-tensile-strength, damage-resistant filament braid protects against rough surfaces, sharp edges or corners. The product offers the following benefits:

- Resists scuffing and abrasion
- Cushions against vibration and damage from impact
- Prevents condensation while allowing complete drainage
- Dissipates heat and moisture
- Does not degrade in most fluid environments

Sleeved cable bundles are more organized and more attractive than other cabling techniques. The expandable, open-weave construction works like a Chinese finger puzzle. When pushed over a cable bundle, the weave expands. When pulled taught, it tightens around the bundle. Braided sleeving, woven from polyester and flame retardant polyester yarn, slips on quickly and is self-fitting over irregular shapes and contours. It fits a wide range of diameters, simplifying inventory requirements. The open weave allows for easy inspection of sleeved components and easy break out of

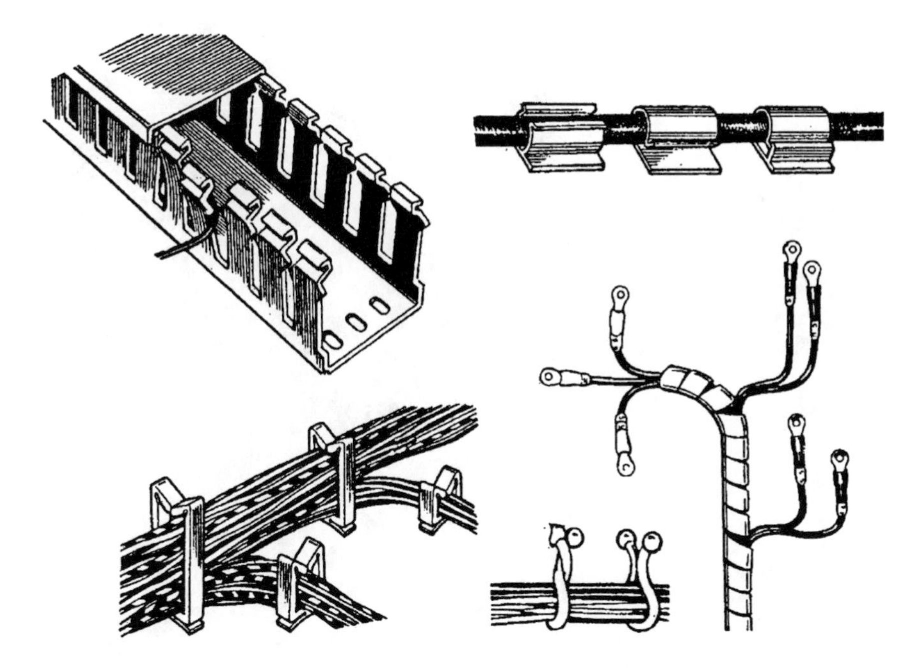

Figure 3.16 Some of the common methods of cabling small- to medium-sized runs in a facility.

---

individual cables. In spite of its high tensile strength, braided sleeving maintains full cable flexibility, even at low temperatures. It also does not kink like tubing when bent. It can be flexed repeatedly without damage. Because of its low surface friction, wire-pulling compounds are not required for installation. Applying slight tension after application makes for an even tighter fit. Figure 3.17 illustrates the use of braided sleeving in an equipment housing.

3.5 Cable Identification and Marking

Cable-jacketing materials are available in a variety of colors, which can be used to identify different signal types. In large quantities, and for a fee, suppliers will apply colored striping onto the cable. Using this approach, many more color codes are possible for cable identification. Because of the limited number of stock colors, they can only be used to identify the type of signal certain cables are carrying, for example, video, sync, or data. Red, green, and blue cables might be used to identify the three cables in an RGB bundle used to carry component video signals. Colored tape, available in many widths, can also be used to distinguish particular cables.

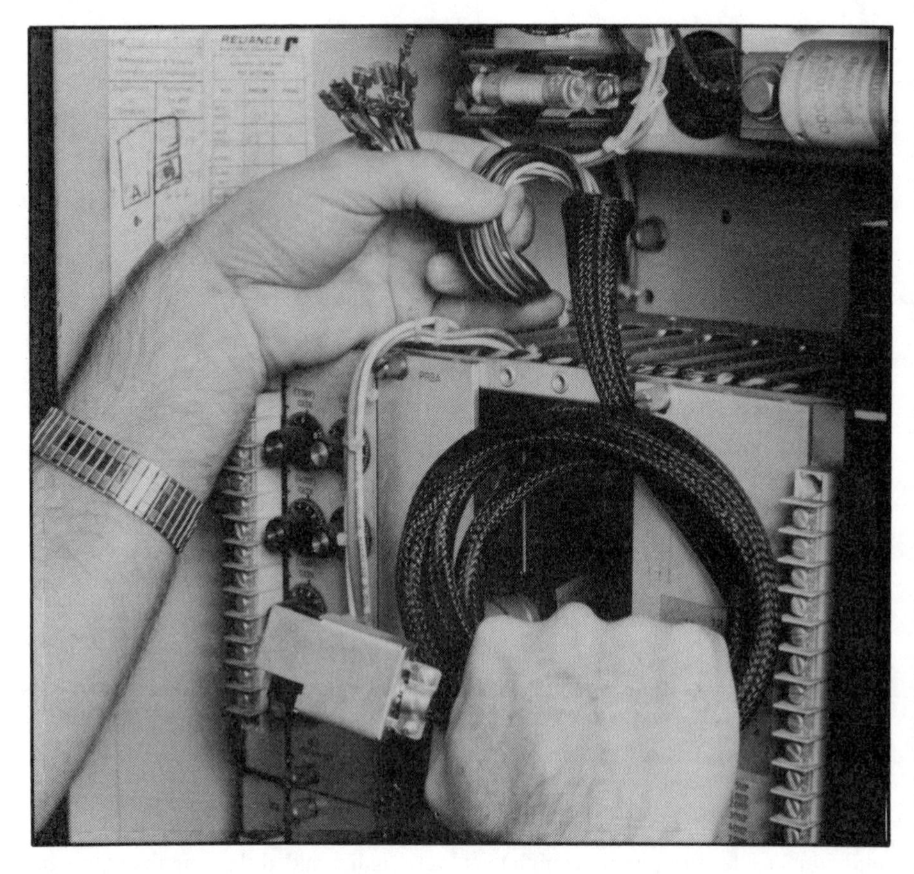

Figure 3.17 Braided cable sleeving used in an equipment housing.

3.5.1 Wire Markers

There are several types of wire markers available for placing identification information onto cables and wires. They include heat-shrinkable tubing or sleeving, wraparound adhesive tape labels, and write-on cable ties. Each type has its own advantages and disadvantages.

Heat-shrinkable sleeving is available in several materials (PVC and polyolefin), diameters, and colors. When heated with a heating tool, the sleeving can be reduced (recovered) to one-third its supplied (expanded) size. Identification information is written onto the sleeving by hand using a pen and permanent ink. Automatic and hand-operated tools are available for printing white or black number and letter codes on dark- or light-colored sleeving, respectively. Print wheels must be set manually for each code. The manual tool is acceptable for small quantities, but it is impractical when large quantities are required. The ink used must be smear-resistant and permanent in all cases. Use high-carbon, noncorrectable fabric ribbon with impact printers. Depending on the sleeve material, the markings may be permanent upon printing before shrinking. The markings on some sleeving becomes permanent only after shrinking.

Heat-shrinkable markers are available cut to length, in several diameters, flattened, and mounted with adhesive tape in a ladder configuration on perforated (pin feed or tractor feed) guide strips for feeding directly from the box through a dot-matrix or daisy-wheel printer or typewriter.

A computer can to be used to print sequential numbers or more elaborate labels from a cable schedule database. Software for this purpose is usually available from the manufacturers of the guide-mounted sleeving. A flat-file or relational-type database program can be used to maintain a cable schedule, and can easily be formatted to print wire markers. With the appropriate software, large typefaces can be used to make the labels more readable.

The thin strips of adhesive tape holding the tubing to the tractor guide strips on each side should allow easy removal of each sleeve. Remove the printed sleeve from the guide strips by grasping the sleeve close to the tape and pulling forward and downward in a tearing motion. Another way is to peel off the top adhesive tape on both sides and lift the sleeve off the carrier. Slip the sleeve over the wire or cable. Use a standard heating tool to shrink the printed sleeve when a tight fit is needed. With a shrink ratio of 3:1, the sleeving can be slipped over a connector and shrunken to fit the cable after the connector has been installed. Heat-shrinkable wire markers are the most durable and the best looking of the available markers.

Sheets of heat-shrinkable markers (*Bradysleeve*[2]), which are applied by crimping and joining two layers of material to form a matrix of marker boundaries on the sheet, are available. The left and right edges of the sheets are perforated for the tractor feed of an impact printer. After printing, the markers are torn apart for use. The manufacturer offers a printer with software and a high-volume applicator machine that automatically separates the markers into strips. These strips are positioned and prepared to accept the wire to be marked. A thermostatically controlled heat-shrink oven is used to shrink the markers onto cables as they are fed into the belt drive of the device. Because the applicator feeds the markers in order of printing, it may not be practical where cable marking is not necessarily done sequentially.

A disadvantage of this type of marker is that if it is heated and shrinks too much, the stress can rip the crimp joint apart, rendering the marker useless.

At the very least, the marker should include a number that identifies the cable on the cable-pull schedule. Descriptive information, such as signal type, source, and destination, can also be included on the label. Cables to be used on temporary remote field applications should also be labeled with the company name.

3.5.2 Wrap-Around Adhesive Tape Wire Markers

Wrap-around adhesive tape wire markers are available in pre-printed and write-on strips, mounted on cards, from several manufacturers. Pre-printed numbered markers are also available on small spools for easy dispensing from a marker dispenser. Write-on markers are available in many sizes mounted on tractor-feed backing for printing on a dot-matrix or daisy-wheel printer or typewriter. One end of each marker is transparent and, when wrapped, is intended to cover the white printed portion to protect the ink from smudging.

---

2 Manufactured by Brady, Inc.

Adhesive wire markers are easy to install, but they are not as durable as heat-shrinkable sleeve markers. They require meticulous care during installation for proper alignment and to avoid contaminating the adhesive. If oil from the skin gets on the adhesive, the label will eventually unwrap from the cable. Though advertised as permanent, these markers will eventually begin to peel at the end. If handled frequently or flexed, they may come off. If bent, the marker will kink, which can cause it to begin to peel off.

3.5.3 Write-On Cable Ties

Cable ties are available that have an enlarged flat write-on area near the female end. They are used for marking cables and cable bundles. The cable identification information is written using permanent ink. Such marker ties are sturdy and easy to install. However, the information has to be individually written or stamped onto each marker. Ties cannot be printed in bulk using a computer.

3.5.4 Equipment Identification Labels

Several options are available for identifying equipment. Letraset adhesive lettering (available from most office, drafting, or art supply stores) is available in many typefaces, sizes, and colors. They can also be peeled off easily when changes must be made.

Plastic molded lettering is available in many styles, sizes, and colors. Larger letters can be screwed on, but gluing is easier and works for all sizes.

Custom-engraved, rectangular laminated plastic tags can also be used for identifying equipment. They are available in several colors, and different-size lettering can be used. They can be glued or screwed in place.

The surface of the equipment must be clean and free of oil before applying any of these labels.

3.6 Bibliography

Benson, K. B., and J. Whitaker: *Television and Audio Handbook for Engineers and Technicians*, McGraw-Hill, New York, 1989.

Block, Roger: "The Grounds for Lightning and EMP Protection," PolyPhaser Corporation, Gardnerville, NV, 1987.

Davis, Gary, and Ralph Jones: *Sound Reinforcement Handbook*, Yamaha Music Corporation, Hal Leonard Publishing, Milwaukee, 1987.

Fardo, S., and D. Patrick: *Electrical Power Systems Technology*, Prentice-Hall, Englewood Cliffs, NJ, 1985.

Federal Information Processing Standards Publication No. 94, *Guideline on Electrical Power for ADP Installations*, U.S. Department of Commerce, National Bureau of Standards, Washington, DC, 1983.

Lanphere, John: "Establishing a Clean Ground," *Sound & Video Contractor* magazine, Intertec Publishing, Overland Park, KS., August 1987.

Lawrie, Robert: *Electrical Systems for Computer Installations*, McGraw-Hill, New York, 1988.

Morrison, Ralph, and Warren Lewis: *Grounding and Shielding in Facilities*, John Wiley & Sons, New York, 1990.
Mullinack, Howard G.: "Grounding for Safety and Performance," *Broadcast Engineering* magazine, Intertec Publishing, Overland Park, KS., October 1986.
Whitaker, Jerry: *AC Power Systems*, CRC Press, Boca Raton, FL, 1991.
———: *Maintaining Electronic Systems*, CRC Press, Boca Raton, FL, 1991.

CHAPTER 4

INTERCONNECTING VIDEO SYSTEMS

4.1 Introduction

The requirements of a video facility vary greatly, depending on the application. The simplest system might consist of one camera, two tape machines, and two monitors. At the other extreme is a major market broadcast facility that may include 10 cameras or more, 20 to 30 different types of tape machines, production switchers, special effects generators, radio relay links, and a high-powered transmitter. The larger the installation, the greater the interdependence of each element.

Planning and building a video facility is complicated by the number of signals involved. In addition to video and (usually) two channels of audio, timing, control, time code, tally, communications, and other ancillary signals are also typically required for proper operation of the plant. To further complicate system design, the relationship between various systems may change on a regular basis. Video projects range from a simple, but fast-paced, newscast to complicated multi-layering commercial production. The changing needs of the facility have led to plants being built around central *machine rooms*, where videotape recorders (VTRs) and graphics equipment may be assigned as required. Routing of the necessary signals is accomplished by a *routing switcher*. Figure 4.1 illustrates a simplified television production center. Note the interdependence of the various elements.

4.1.1 Video Synchronization

When video sources are combined, either through electronic videotape editing or mixing, all inputs must be timed to scan the images in exact synchronism. Mixing requires that each video source be synchronized vertically and horizontally, and that the color subcarrier phase of the reference burst be synchronized among all the sources.

Horizontal sync and the subcarrier must both be precisely timed, and must maintain a constant subcarrier-to-horizontal (SC/H) phase relationship.

4.1.1.A *Master Sync Generator*

The master sync generator is the video clock for the entire facility. The generator determines the absolute frequency stability of the plant. Often, two separate sync generators feeding an automatic changeover switch are employed for redundancy. Many types of source equipment include an internal sync generator. This feature provides regeneration of the required pulses at the proper time for internal use. This feature also permits the output video time to be set arbitrarily with respect to the reference, simplifying system timing adjustment. The signal typically used to synchronize individual systems with the rest of the plant is composite *color-black*.

4.1.2 Zero-Time Concept

Proper timing of a video facility is crucial to proper operation of the system. The most common concept in timing is the *zero-time* point. This is a location within the system where all composite video sources are precisely synchronized. The location can be either at the input of the production switcher, at the input to the video routing system, or another appropriate point. Zero-time means that there is no time difference between sync and subcarrier of any two video inputs. To achieve a zero-time point from a single sync generator for a given facility, the sum of the delays of the reference signals from the sync generator through the source equipment to the zero-time point must be equal for all paths. The following elements affect the delay:

- Propagation delay through the cables
- Delay of timing pulses from the generator to the video source
- Delay in any pulse or video distribution amplifier (DA) through which the signal flows
- Electronic delay within the video source equipment

Delay compensation can be accomplished by making the delay of each path from the generator to the zero-time point equal to the delay of the longest path. This compensation can be realized by one or more of the following:

- Delaying the pulses to the source equipment and rotating the subcarrier phase
- Delaying the composite reference signal (typically color-black) to the source equipment
- Delaying the composite video signal from the source equipment to the zero-time point

Delay within a given path can be increased by using one of the following techniques:

- *Regenerative subcarrier distribution amplifier (DA).* A regenerative subcarrier DA receives continuous-wave subcarrier signals and provides a constant level of subcarrier at the output, regardless of the input level. It further allows a complete 360° rotation of phase.

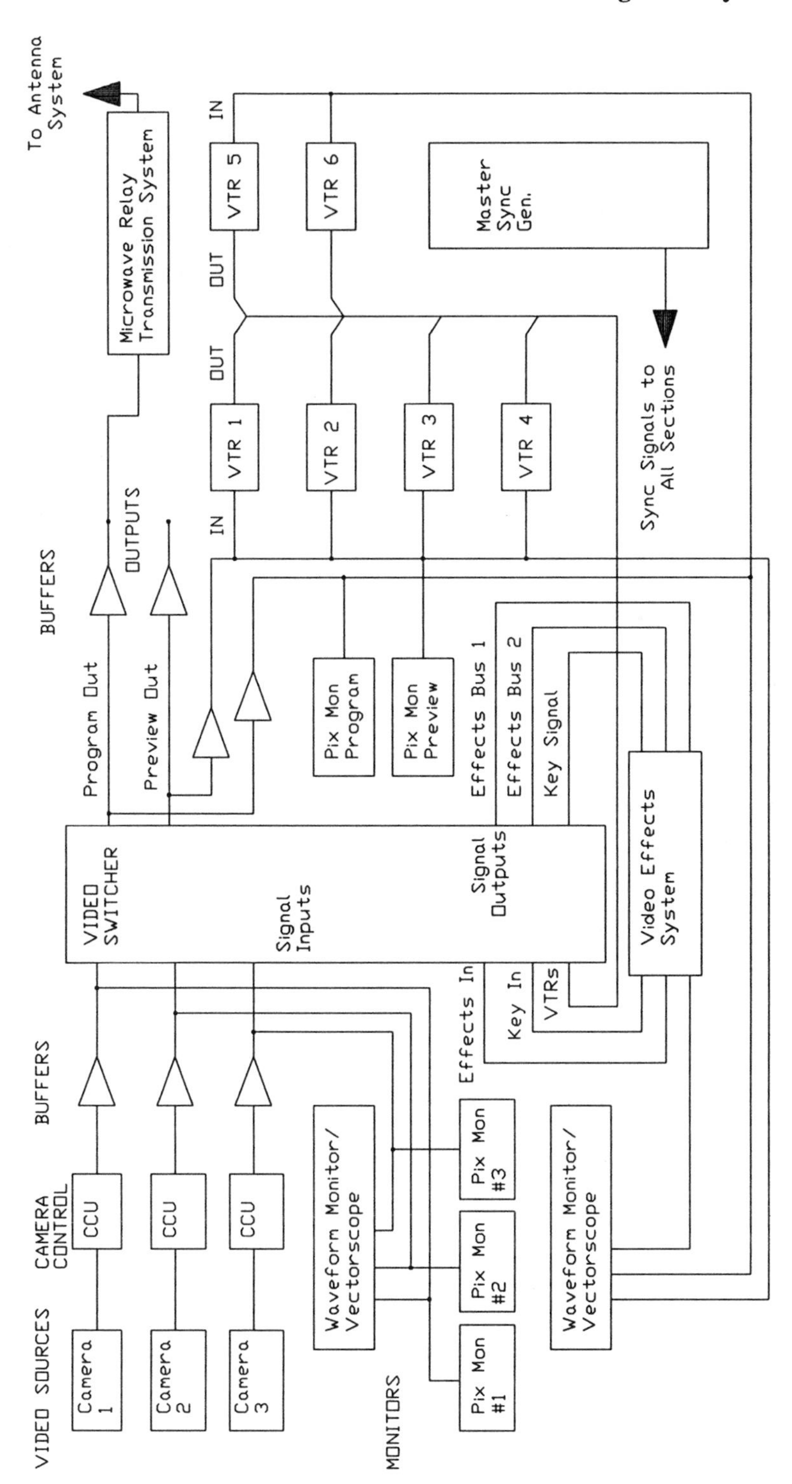

Figure 4.1 Flow diagram of a television production center.

- *Regenerative pulse DA.* A regenerative pulse DA receives pulses, strips the pulse at the 50 percent amplitude point, delays each edge by an equal amount, and then regenerates the proper amplitude negative-going pulse.
- *Video delay DA.* A video delay DA couples a normal video DA with a broadband, low group delay-generating delay line.
- *Slave sync generator* (discussed previously in Section 4.1.1).
- *Installation of excess coaxial cable.* This approach matches timing between elements by assuring that the physical equivalent cable distance between each component of the system is identical.

4.1.2.A *Area-Timing*

Area-timing is a modification of the zero-timing concept. A slave sync generator is used to time a smaller subgroup of source equipment, such as a VTR room or production suite. The area can be timed in a traditional manner, and the timing of the entire facility can be adjusted to match the timing of the master sync generator. Using a slave generator provides the added benefit that it can run independent of the master generator.

4.2 Selecting Coaxial Video Cable

The function of a video cable is to carry signals from the source to the destination with a minimum of degradation. System requirements dictate the choice of cable type to be used for each application. The materials and construction of a cable determine its effect on a video signal. Proper handling and installation practices can prevent damage and guarantee good performance and long life.

4.2.1 Cable Characteristics

Video signals are typically transported by coaxial cable with a characteristic impedance of 75 Ω. Matching the 75 Ω input and output impedance of the equipment ensures a good-quality signal transmission with no reflections. For best results, precision video cable should be used. Precision video cable is sweep-tested by the manufacturer to assure that the cable meets published specifications.

The center conductor of coaxial cable used for video signals should be solid copper in order to assure a low dc resistance. Copper- or silver-covered steel wire, which is designed for RF applications, should not be used. If either the conductor or the shield has a high dc resistance, the frequency response will roll off at low frequencies and cause distortion of the video signal. Direct current flows through the entire cross section of a conductor, but at higher frequencies the skin effect phenomenon causes current to concentrate at or near the surface of the conductor. Because of the higher resistance of the steel core compared with copper, the copper-covered steel cable will exhibit a higher resistance at low frequencies down to dc. As a result, the video waveform will be distorted. Low-frequency components will decrease in amplitude, and high-frequency transitions will remain high, resulting in overshoot or undershoot at the edges of sync pulses and at any other transitions in the video waveform. In the extreme, the resulting effect is a peaking of the waveform, which shows up as spikes at the edges of the sync pulses and at fast transitions in the picture. Figure 4.2 shows the attenuation characteristics of two types of RG-59 cable.

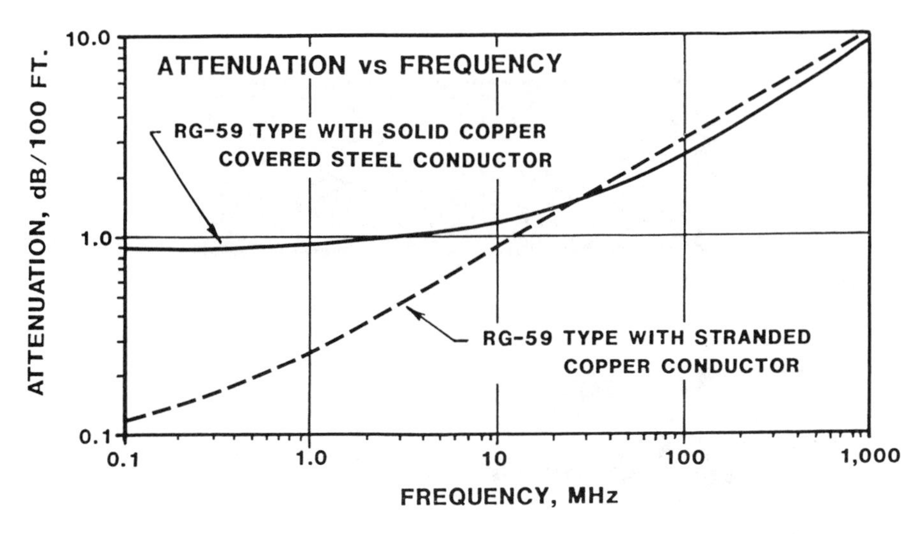

Figure 4.2 Attenuation characteristics of two types of RG-59 cable.

Distortion of the vertical and/or horizontal sync pulses can cause synchronization problems. Any spikes in the picture portion of the waveform, which extend down below blanking, can falsely trigger sync separator circuits in some hardware. Video DAs are not designed to equalize a signal exhibiting low-frequency roll-off. Therefore, do not employ a DA to compensate for use of the wrong type of cable.

When a cable is to be used in an application where repeated flexing is anticipated, specify a stranded-center conductor, which is more flexible. This will allow the cable to bend more easily, resulting in longer life.

4.2.1.A *Shield*

The shield braid should provide physical coverage of 90 percent or more to protect against electromagnetic interference. Various types of braided coaxial cable shield are illustrated in Figure 4.3. Some cables designed for CATV applications use aluminum-foil tape shielding, which provides very good shield coverage. However, because of its severe low-frequency roll-off below 1 MHz, this should not be used for baseband video signal applications. Only use cable with solid-copper braid shielding.

4.2.1.B *Signal Loss*

Coaxial cable attenuation varies with length and frequency. The signal loss must be considered in the facility design process. Coaxial cable attenuation is specified by the manufacturer in dB/100 ft or dB/100 m at one or more frequencies. Figure 4.2 shows a typical attenuation chart. The attenuation for a given length of cable can be calculated using this published data. If the attenuation is too high for the application, a larger (lower-loss) coaxial cable should be selected. Cable loss may also be compensated for through the addition of an equalizing DA.

Frequency response is affected by the distributed capacitance of the cable, which causes signal energy to roll off at high frequencies. Low-capacitance cable will exhibit less roll-off at high frequencies. An equalizing DA can be used to compensate for this high-frequency roll-off.

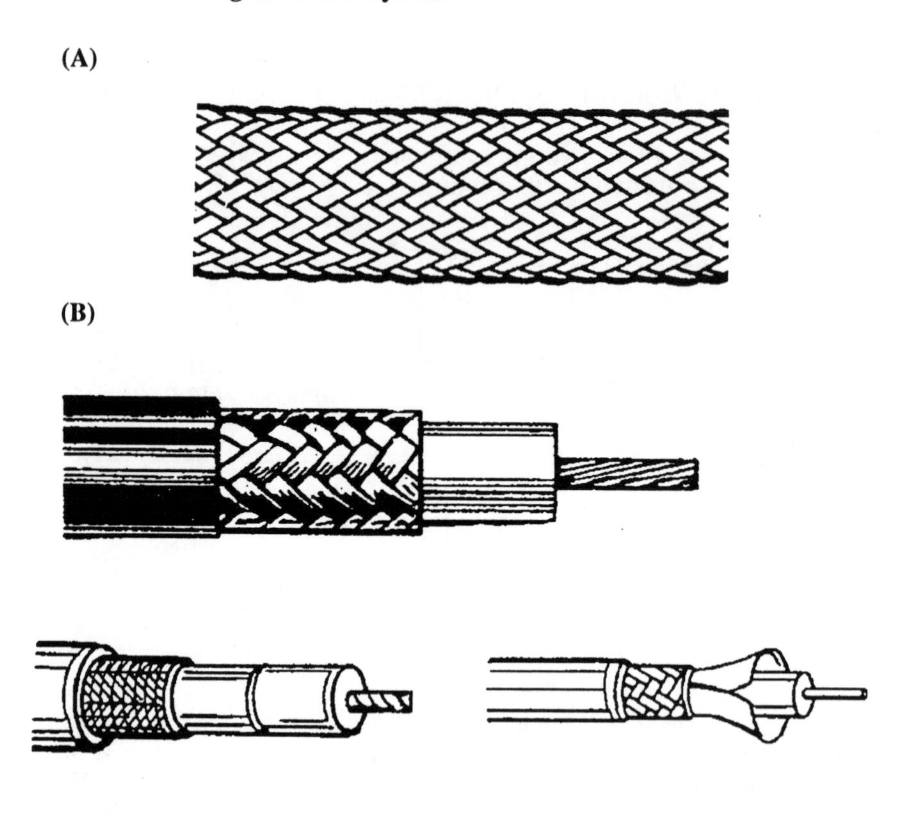

Figure 4.3 Coaxial cable braided shield: (a) basic braid weave; (b) three common applications of a braided shield.

In some applications, as in the case of closed-circuit television, power may be supplied to a camera or other system through the coaxial cable. A low-resistance conductor and shield are required to keep the voltage drop within acceptable limits. It may be necessary to increase the power supply voltage to compensate for this voltage drop in long cable runs.

4.2.1.C *Cable Jacket*

Video cables are available with several types of protective jacket materials. Each has particular properties that make it appropriate for specific environments. PVC-jacketed cable costs less than other materials and is used for indoor wiring. Because of its lower replacement cost, however, PVC-jacketed cables may be specified for outdoor use for temporary installations where the cable is expected to wear out from repeated mechanical abuse before sunlight and water can cause noticeable performance deterioration.

Polyethylene or high-density polyethylene has good water-resistance properties. It is the best choice for instances where the cable may be exposed to moisture or immersed in water, such as in underground conduits, or where it may be buried directly or used outdoors. Polyethylene-jacketed cable is available in many colors. This makes it easy to identify the signal type.

Most polyethylene and PVC coaxial cables have a maximum temperature rating of 60° - 80°C. In some cases, this rating may not be high enough. Teflon cable may be required in areas where a high-ambient temperature is anticipated. (Teflon is a registered trademark of Dupont.) Teflon cable is available with temperature ratings up to 260°C and can be used in or around steam piping or heating ducts. NEC-approved Teflon plenum cables are required for installation in plenums, air ducts, or air returns, where a fire may cause the cable to burn and give off hazardous toxic fumes that could circulate through the air conditioning ducts into other rooms. Teflon cable can be laid directly in false-ceiling air-plenum areas without the need for expensive metal conduit. This cable is also resistant to chemicals but is not suitable where it may be exposed to radiation. The cost of Teflon is approximately 8 to 10 times greater per pound than PVC insulations.

4.2.2 Cable-Rating Standards

The National Electrical Code (NEC) requires that signal-carrying cables conform to certain rules enacted to prevent fire hazards and electric shock to humans. The code has adopted strict requirements regarding smoke emission and flame propagation, resulting from a number of tragic high-rise fires and deaths from toxic smoke. These revisions have affected the design and construction of electronic cable. Some of the tests required for video cables include:

- Conductor dc resistance
- Insulation resistance
- Heat aging properties
- Cold bend properties
- Smoke emission and flame propagation

The predominant NEC code articles applicable to video cables include the following:

- Article 725—remote control, signaling, and power-limited circuits
- Article 760—fire-protective signaling systems
- Article 770—optical-fiber cables
- Article 800—communications circuits
- Article 820—CATV systems

The four major UL classifications are:

- General purpose (no suffix)
- Plenum (P)—for use in return air plenums, ducts, and environmental air areas
- Riser (R)—for use in vertical shafts
- Residential (X)—limited use in dwellings and certain raceways

4.2.2.A *Acceptable Products*

Belden 8281 precision video cable has been a standard by which other video cables are judged by many end users. Specifications for 8281 are as follows:

- Inner conductor 20 AWG, solid
- Dielectric outside diameter 0.198 in (5.03 mm)
- Shield double braid, coverage 96 percent
- Jacket outside diameter 0.305 in (7.75 mm)
- Propagation velocity 66 percent
- Capacitance 21.0 pF/ft (69 pF/m)
- Attenuation at 10 MHz 0.78 dB

4.2.3 Coaxial Cable for Component Video Signals

Several 75 Ω coaxial cables can be combined into a single outer jacket for carrying separate component video signals, such as RGB. This is a good way to keep cable runs neat and uncluttered. Because timing of the signal components is critical, the length of each of the component cables must match to within a few nanoseconds. As with any cable, 75 Ω connectors are recommended to minimize the mismatch. Several manufacturers offer pre-assembled multiple-component cables in various lengths. A typical coaxial multi-conductor cable is illustrated in Figure 4.4.

4.2.4 Video Cable for Temporary Installations

Video *remote* cables are designed for applications where it is necessary to combine video, audio, intercom, and power lines in a single flexible cable to interconnect camera equipment and a recorder or base station. These cables are usually supplied with the camera equipment. Custom cable assemblies are also available.

Other examples of multiple wire cables include *snakes*, which are umbilicals used to interconnect separate equipment modules and control cables.

The construction of a remote cable has to address the technical and physical requirements of the application. Although small-diameter cables are used, individual cables must have low loss. All video, audio, and power cables require high-density braided shields to minimize crosstalk while maintaining the necessary flexibility. Winding pitch and insulation characteristics also affect flexibility. The outer jacket must resist wear and stay flexible in cold outdoor temperatures. On heavy cables with large connectors, reliable and effective strain relief is required where the cable enters the connector body.

4.2.5 RF Cable

It may be desirable to distribute one or more channels of video with audio around a facility or to offices within a building for monitoring or communications purposes. A master antenna television (MATV) system may be the most practical means. To keep the cost of cable low, 75 Ω coaxial cable, with a copper-covered steel center conductor, may be used. Because of the skin effect, high-frequency currents tend to flow at or

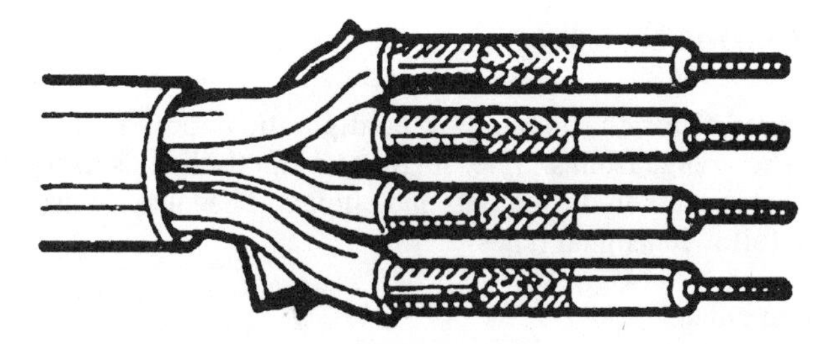

Figure 4.4 Coaxial multi-conductor cable.

near the surface of the conductor. Because the RF signal current does not flow through the steel core, losses caused by the higher resistance do not have a significant effect. RG-59 type 75 Ω cable is a good choice.

Radio communications equipment uses RF coaxial cable with a characteristic impedance of 50 Ω. Again, because of the high frequencies used, low-cost cable with a copper-covered steel center conductor may be specified. RG-58 type 50 Ω cable is a good choice.

4.3 Video Interconnection Components

Efficient signal switching, routing, and distribution form the basis of a productive video center. Switching and distribution schemes differ considerably in configuration to suit a variety of applications. Control of the switching functions may be centralized or distributed, performed locally or remotely, and may follow a number of logic protocols.

4.3.1 Distribution Amplifier

The function of a DA is straightforward: to amplify an input signal for distribution to one or more loads, and to provide isolation between the loads. DA performance specifications should be at least better than those demanded of the overall system. This is especially important where the signal will be routed through many DAs.

Looping a signal cable from the *looping input* of one device to the input of one or more other devices is common practice, but do this with caution. Each device adds its load to the others in the path. Know the limitations of looping and—in particular—the limitations and specifications of the devices involved. This can help to avoid signal degradation.

The cables going into and out of the DAs in a large system are among the most densely packed group of cables in any facility. For a complex network, the number of cables can be massive. For this reason, leave space between every one or two DA chassis to provide clearance for cable installation and removal.

4.3.2 Routing Switcher

Routing switchers differ considerably in configuration and size to suit a variety of applications. They range from simple, single-bus input selectors to complex, master-grid systems that use hundreds of inputs and outputs. They may be used to switch one or more of the following signal types:

- Program video
- Program audio
- Intercommunication signals
- Control signals
- Data
- Time code signals

Control of switching functions may be centralized or distributed, depending on the facility's requirements.

Multiple-bus routing switchers are used primarily to make a large number of sources available to multiple destinations. A *master grid* concept may be employed, where all sources appear as inputs to a large matrix with any number of switching buses. Alternatively, smaller matrices may be used, each performing a specific function. The master-grid approach offers the greatest flexibility because all sources are available at each destination. System timing is simplified as well, because the lengths of program paths are predictable.

Figure 4.5 illustrates the video signal path through a typical routing switcher. Inputs are distributed to elements of the switching matrix by a *fan-out amplifier*. This provides the required number of isolated outputs to feed the crosspoints. Matrix expansion is accomplished by adding fan-out amplifiers, bus drivers, and additional crosspoints. The output circuit provides a program output and monitor output. The routing switcher illustrated represents an input-oriented design, in which the number of inputs is significantly greater than the number of outputs.

A multiple-bus routing switcher is generally controlled in one of two ways:

- On a dedicated bus-by-bus basis
- Full matrix *xy* control

In some applications, both methods may be used. Control is further divided into the following categories:

- Centralized, where all matrix assignments are made at a central control panel or by a central computer.
- Distributed, where matrix assignments are made at individual workstations (studios or control rooms).

Routing switcher manufacturers offer a variety of control configuration options, designed to meet specific needs.

4.3.3 Video Patch Panel

The jacks commonly used in patch panels in the United States conform to Western Electric standard dimensions. The number of insertion cycles a jack can endure should

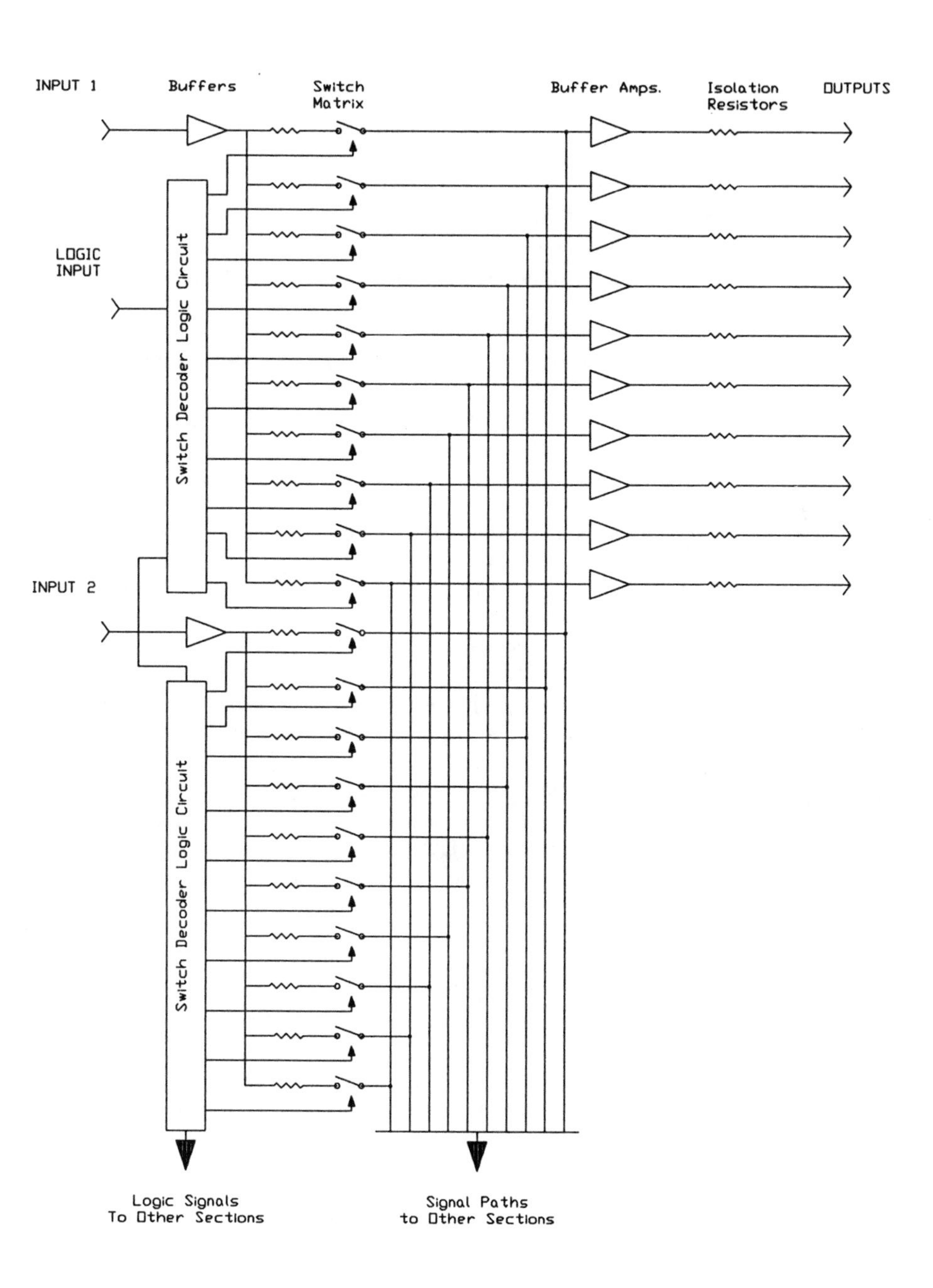

Figure 4.5 Multiple-bus video routing switcher.

be rated in the tens of thousands. The factors affecting the life and reliability of a jack include contact wear and failure of the termination switch. Desirable features include the following:

- Contacts fully isolated from the panel.
- Sealed metal housing to keep out contaminants and provide EMI protection.
- Easy replacement from the front of the panel.
- Low VSWR (below 600 MHz).
- High signal isolation (40 dB).
- 75 Ω characteristic impedance.
- Wide designation strips, making it easier to label the field and to allow more flexibility in selecting names that will fit on the labels.

If a patch cable is inserted in the signal path of a timed video system, it will delay the signal by an amount determined by its length and physical properties. The patch thereby alters the timing of the signal path. This can be avoided by using *phase-matched normal-through* patch panels. The design of these patch panels anticipates the delay caused by a fixed length of patch cable by including that length in the loop-through circuit.

With phase-matched panels, the normaling connection in each connector module includes a length of cable that provides a fixed delay through the panel, usually 3 ft (0.914 m). If a patch cord of the same length as the internal cable is used to make connections between patch points, the delay will be the same as that of the normal-through delay; there will be no change in the timing of the signals passing through the patch panel. When a patch cord is plugged in, it is substituted for the loop cable through the switching mechanism normally used in normalled patch connectors. Thus, critical timing relationships can be maintained. Figure 4.6 shows a phase-matched patch panel.

In a normal uncompensated patch panel, when a cable is used to patch between two points on the panel, the length of the patch cord is added to that of the cables connected to the patch. The additional cable length delays the signal by approximately 1.52 ns/ft (5 ns/m). To avoid the delay problems associated with conventional patch panels, phase-matched normal-through video patch panels should be used.

If phase-matched patch panels are used, all of the patch cords must be the same length as the delay built into the patch panel. Obviously, if all of the patch cords must be the same short length for the phase-matched panel, it would not be possible to patch between panels that are separated by a longer distance than the cord can reach. This limitation should be considered when laying out patch panels in a rack.

Color-coded cables may be specified. When different-length patch cords are specified, different colors can be used to distinguish one length from another.

4.3.4 BNC Connector

BNC connectors are used on professional video equipment. Specify 75 Ω, not 50 Ω, connectors where the best video performance is required and where high-bandwidth video signals are anticipated. Fifty-ohm connectors designed for 50 Ω coaxial cable are commonly used on 75 Ω cable for baseband video signal transmission. Above 10 MHz, however, the physical dimensions of signal-carrying components, such as cable, connectors, and other hardware, begin to noticeably affect the characteristic impedance of the line, and the reflection coefficient of the line and its elements. The abrupt

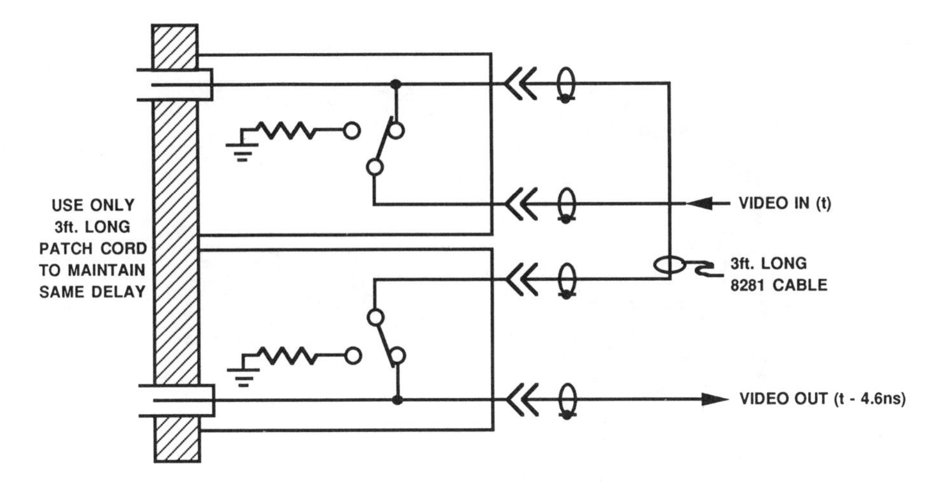

Figure 4.6 A phase-matched video patch panel.

---

transition at a 50 Ω connector attached to a piece of 75 Ω cable will cause high-frequency components to be reflected back down the cable. This may not noticeably affect the waveform of a video signal with a 5 MHz bandwidth, but above 10 MHz, reflections caused by the mismatch will begin to distort the waveshape and affect picture quality. The bandwidth of a high-end, 800-line studio television camera will extend beyond 10 MHz. High-definition television (HDTV), with a bandwidth that extends to 30 MHz, and computer graphics, which may extend to 300 MHz, are especially critical. BNC connectors are available in a number of styles, as shown in Figure 4.7.

Select the right connector to fit the dimensions of the cable being terminated. When using crimp-type connectors, avoid using the wrong-size crimp ring. It may distort and not provide the correct pressure on the cable when crimped. The center pin should be gold plated. Gold is less susceptible to corrosion and is a good conductor. The outer contact should be made of beryllium copper, instead of phosphor bronze or spring brass. Because of its higher elasticity, beryllium copper is more reliable and can withstand more flexing cycles before mechanical failure occurs.

Strip coaxial cables using a good coaxial cable stripper. Such tools cut all of the cable elements to the specified length for the connector being used. Follow the connector manufacturer's instructions for the proper cut dimensions. Make sure the shield braid is cut clean and is not dragged along by the cutting blade. All of the strands of the braid should be the same length. A common type of coaxial cable preparation tool is illustrated in Figure 4.8.

4.3.4.A *Dual Crimp-Type Connectors*

Assemble a dual crimp-type connector using the following steps:

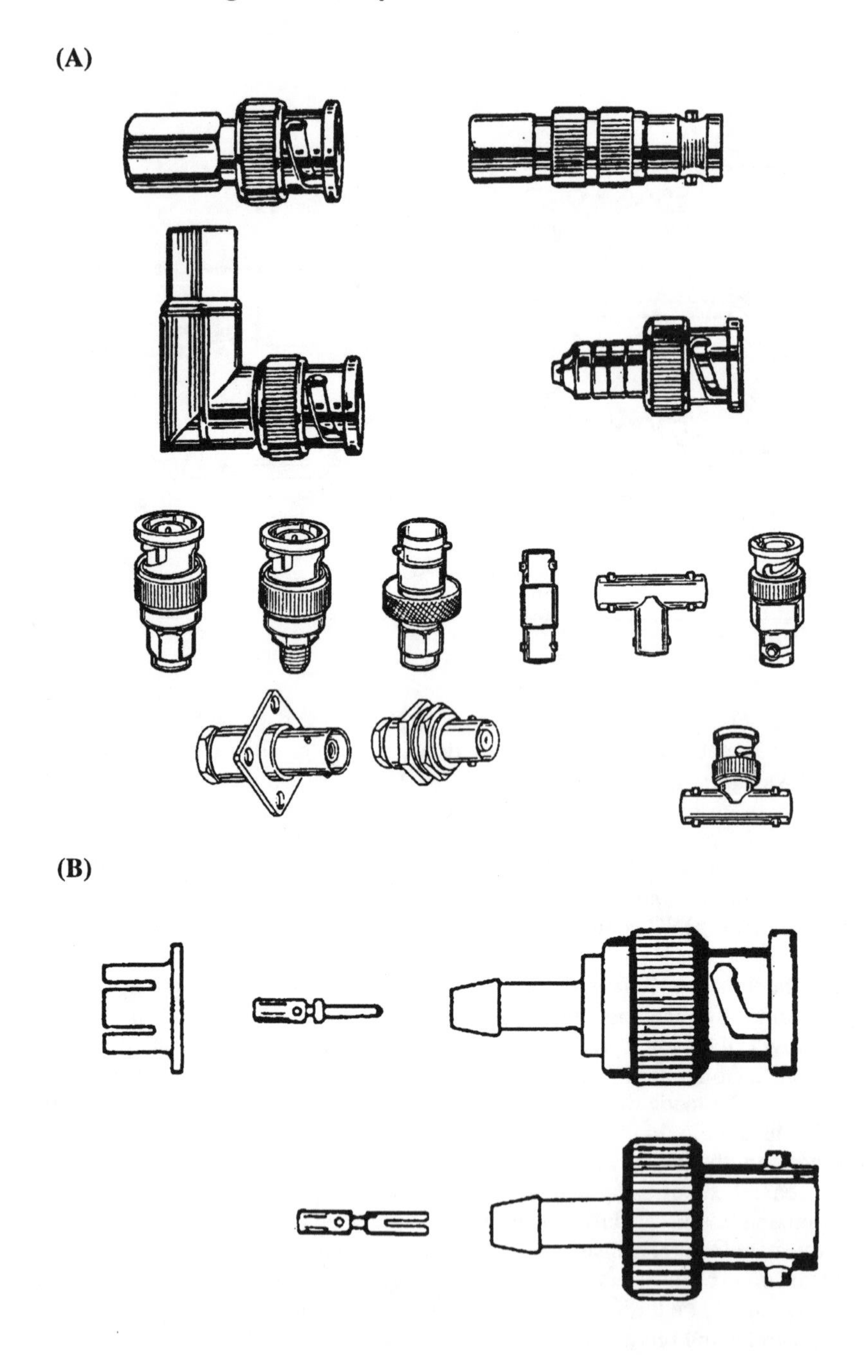

Figure 4.7 Various types of BNC connectors available for video applications: (a) general connector types; (b) typical connector construction.

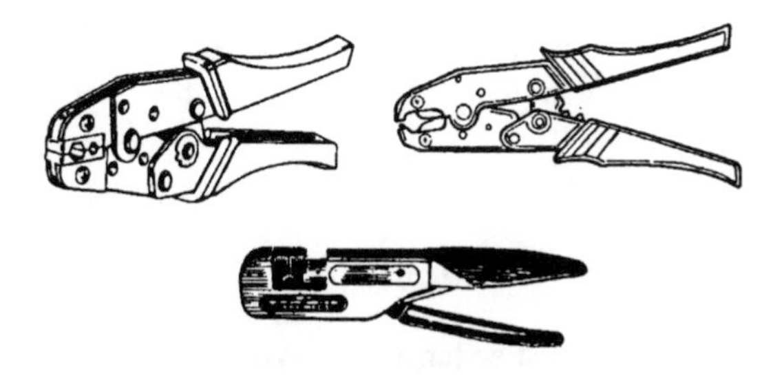

Figure 4.8 Coaxial cable preparation and crimp tool.

- Push the center pin onto the center conductor of the cable.
- Make sure the end of the conductor can be seen through the inspection hole in the pin.
- Crimp the pin onto the conductor using the recommended crimping tool.
- Slip the crimp sleeve onto the cable for later use.
- Insert the pin and cable into the connector until it snaps into place.
- Flair the braid to allow it to pass over the connector body easily.
- Push the connector onto the cable, guiding the shield braid over the knurled portion of the connector. This will keep the braid as neat as possible. The shielding should just touch the connector body or have a slight clearance.
- Slide the crimp sleeve forward over the shield braid and up against the connector body. No shield braid should be visible or protrude around the sleeve.
- Crimp the sleeve using the recommended crimping tool.

4.3.4.B *Screw-Type Connector*

Assemble a screw-type connector using the following steps:

- Push the center pin onto the center conductor of the cable.
- Solder the pin to the center conductor so there is no gap between the pin and the insulation.
- Make sure that no solder protrudes above the surface of the pin to hinder insertion into the connector.
- Slide the compression fitting onto the cable.
- Slip the ferrule over the braid, up against the cable outer jacket.
- Fold the braid back over the ferrule and smooth it out.
- Cut off any excess braid.
- Insert the pin and cable into the connector until it snaps into place.
- Screw on the compression fitting, using a wrench to tighten.

After installing the connector onto the cable, inspect it. Make sure the center pin is secure and straight and protrudes by the correct amount. It should snap into place when the connector is pushed into the cable. If the cable was not pushed in far enough before crimping, the center pin will not be protruding sufficiently to make contact with the center pin of the mating connector. Hold the cable, and pull the connector as hard as you can. It should not slip or come off of the cable. Check to make sure that none of the shield braid is sticking out around either edge of the connector crimp ring.

As an added assurance, the continuity of the finished cable should be tested using a continuity checker or ohm meter.

4.3.5 Video Line Terminator

The end of a transmission line must be terminated with its characteristic impedance to maintain proper signal level and to prevent reflections that distort the signal waveform. At video frequencies, the accuracy and construction of the terminator are important. Improper termination in a complex system can cause signal level differences that can make calibration difficult, and may be interpreted as equipment problems. These level errors may be compounded to the point where it is difficult or impossible to properly adjust signal levels throughout the system.

Video signals from source equipment are distributed throughout a plant using coaxial cable. In a complex system, these signals may be routed through several cascaded DAs, routing switchers, production switchers, and any number of image-processing devices. They may be fed into and out of the same routing switcher two or more times. Proper operation of the facility requires that certain conventions be accepted during plant design and construction.

- The output impedance of each source and the input impedance of each destination is exactly 75 Ω.
- The video level at the input or output of any piece of equipment, and within interconnecting cables, is at the nominal level of 1 V (140 IRE units) from sync tip to peak white. It is 700 mV from blanking to peak white for noncomposite signals.

The latter will not be required unless the output resistance of every piece of equipment and the resistance of every termination is exactly the same.

An unterminated coaxial cable produces the obvious effect of increasing the video level to twice its nominal value. Terminations close to, but not exactly, 75 Ω have a smaller and more elusive effect on video levels that initially may not even be noticed.

Figure 4.9 shows a common element found in most video systems. A video DA receives its input from a reliable calibrated source and is terminated at its looping input with an accurate 75 Ω terminator. It is also assumed that the DA output resistors R1 through R6 are exactly 75 Ω. The six outputs of the amplifier are each terminated with a ±5 percent tolerance terminator. Each of the terminators shown has a normal resistance value within ±5 percent of the 75 Ω value expected.

The DA's gain is set to produce a level of 140 IRE units peak-to-peak at output 1. In this illustration, because of the different resistance values of the terminators at the other outputs, the levels are all different from the level at output 1. They are also different from each other. Notice that because each termination is half of a voltage divider, the total range of the level errors at all of the outputs is only ±2.5 percent, even though each termination has a tolerance of ±5 percent. In this respect, the termination is an integral part of the source device. The value of T1 is just as critical as the value of R1.

At output 6, the 0.1 percent termination produces a level difference of 0.5 mV, or less than 0.1 IRE units. This worst-case error is 10 times better than the possible result of using a 1 percent termination and 50 times better than using a 5 percent termination.

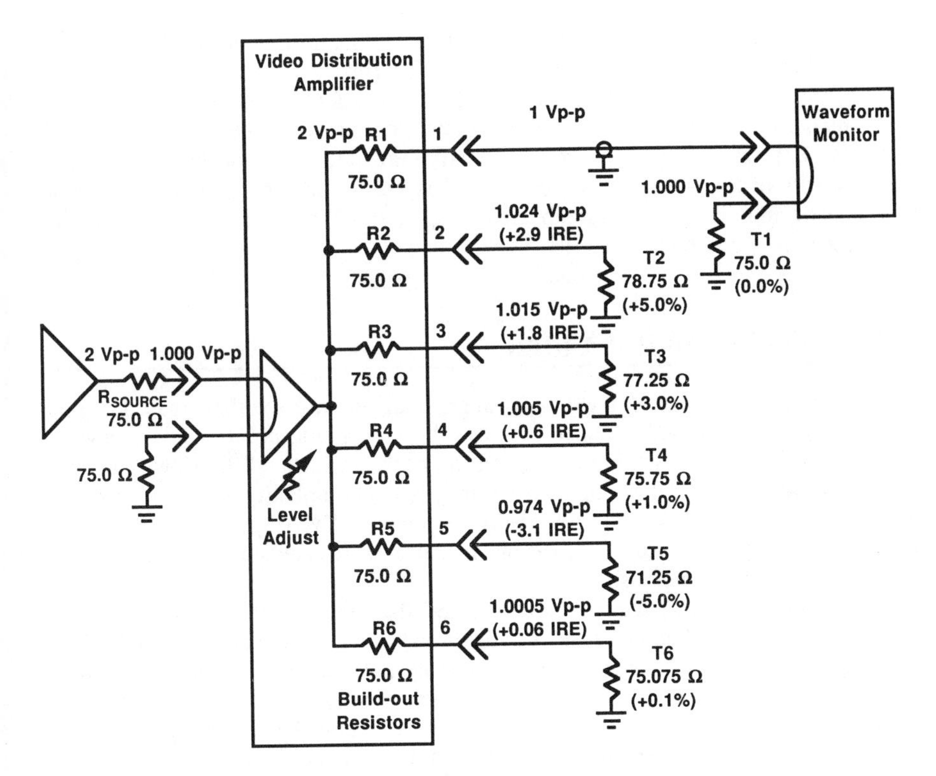

Figure 4.9 Illustration of the effects of termination resistance tolerance.

Using terminations with a 0.1 percent tolerance minimizes level differences caused by incorrect termination values. Assuming that cable lengths have been properly specified, by using 0.1 percent tolerance terminations throughout the system, all levels can be adjusted to match each other within the accuracy and resolution of the test equipment used. Any change in levels can be attributed to either drifting or unstable equipment, or unauthorized adjustments by personnel.

The performance and reliability of a terminator are determined by its construction. The resistive element should be metal film with 0.1 percent tolerance, 1/4 or 1/2 W, with a temperature coefficient of less than 50 PPM. The temperature coefficient of carbon resistor elements can cause a larger change in resistance over the operating temperature range than would result from the original tolerance. Using carbon components can make it difficult to adjust and maintain system levels as the operating temperature changes. The resistor should be soldered to the center pin and the body, not crimped. Crimped contacts can corrode over time, increasing the series resistance of the terminator.

The center pin should be gold plated. Gold is less susceptible to corrosion and has low resistance. A 75 Ω connector body should be used in order to avoid impedance mismatch at high frequencies. Fifty-ohm bodies may be acceptable in simple systems, but in complex facilities—where the signal may pass through multiple cascaded elements—the 75 Ω body is preferable. The body impedance becomes an important consideration at frequencies higher than 10 MHz, where the discontinuity at the connector will begin to cause reflections. When dealing with wide-bandwidth component or high-definition television signals, 75 Ω terminations are mandatory.

The termination is a vital component in any video system; it deserves the same care and consideration given to more complex system components. To avoid video level differences, the system engineer should specify that all terminations be precision 0.1 percent units of high-quality construction. This will eliminate a common cause of problems during initial system setup and testing, and will provide long-term benefits in the form of operational reliability.

4.4 Serial Digital Video[1]

The image quality of an analog video signal gradually degrades with increased system complexity and increased distance between the origination point and the viewing point. After the analog signal has been created, each device that follows will, at best, only minimally degrade the signal. As the length and complexity of the video signal path increases, the quality of the video further decreases. Even digital devices inserted into the analog path will cause distortions because of repeated analog-to-digital and digital-to analog conversions.

With digital systems, the image signal and the digital transmission signal are separate, in the sense that the image is undisturbed as long as the digital signal can be recovered. Fully digital systems will produce no noticeable degradation until they reach the "breaking point," where performance rapidly degrades as disturbances increase. Each time the signal is serialized in such a system, the digital signal is newly generated, minimizing the chance of reaching a failure condition.

4.4.1 Serial Data vs. Parallel Data

Serial digital video equipment offers many advantages. First, it uses coax, not bulky multi-conductor cables. This means that facilities can use existing wiring, and that new cable installations are less expensive than parallel digital video. Further, parallel digital cables have a maximum usable length of approximately 50 m (the result of timing skew between bits). Serial links of more than 300 m are practical. Serial digital, however, also presents some new challenges. System engineers must understand and deal with these challenges to achieve reliable operation.

Serial has high-bandwidth requirements, and noise and distortion affect digital transmission systems differently than more familiar analog systems. These differences can surprise the system engineer who is first venturing into serial digital video. A basic understanding of digital transmission system characteristics is essential when planning an installation and developing a strategy for maintaining it.

---

1 Portions of this section were adapted from the following: Ken Ainsworth, "Monitoring the Serial Digital Signal," *Broadcast Engineering* magazine, Intertec Publishing, Overland Park, KS, November 1991; and Marc S. Walker, "Distributing Serial Digital Video," *Broadcast Engineering* magazine, Intertec Publishing, Overland Park, KS, February 1992.

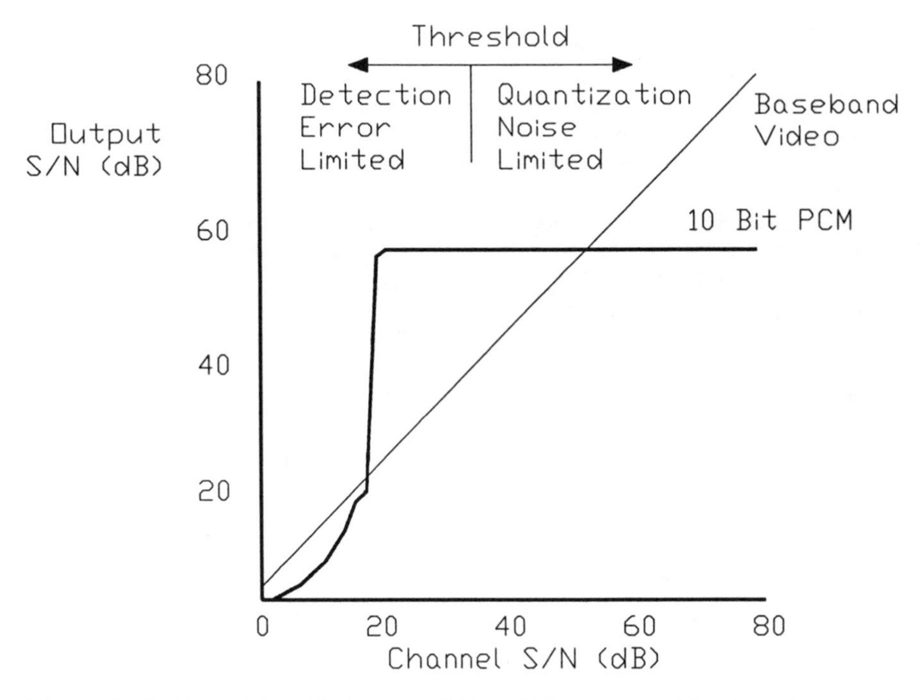

Figure 4.10 The relationship between S/N and bit error rate. The detection error is practically zero when the S/N exceeds 15 - 20 dB. Beyond this point, quantization noise dominates. This is called the *saturation region*.

4.4.2 Understanding PCM

Most serial digital systems currently use pulse code modulation (PCM) to transmit data. In a serial system, each video sample is coded into a sequence of 10 binary symbols. This requires considerably more bandwidth than the original signal. However, as long as the receiver circuitry can differentiate a binary 1 from a 0, additive noise does not degrade the recovered waveform. This allows PCM signals to be repeated and regenerated as needed without degradation, which simplifies distribution.

PCM adds a measure of noise and distortion to the original message in two ways:

- Quantization noise
- Detection error

Quantization noise results from representing the analog signal by a finite number of discrete levels. With uniform quantization, 10-bit representations yield a little more than 60 dB S/N. Eight bits yield approximately 48 dB. Detection errors occur when a receiver incorrectly interprets a transmitted pulse. This can result from low S/N or distortion in the channel. In most PCM systems, the relationship between S/N and the probability of detection error is steep. A change of a few decibels in S/N can cause the error rate to change significantly. This concept is illustrated in Figure 4.10.

In a typical PCM system, the detection error is practically zero when the S/N exceeds 15 to 20 dB. Beyond this point, quantization noise dominates. Designers typically set up PCM systems to operate in this quantization noise-limited region, sometimes referred to as the *saturation region*. The flatness of the output S/N in the saturation region gives PCM considerable tolerance to changes in channel S/N. This situation, however, has at least one configuration drawback: Insensitivity to input S/N provides no warning if the system is close to entering the detection error-limited region. One of the challenges for the system engineer is to determine where the link is operating with respect to this threshold point.

4.4.3 Serial Error Rate

An operating serial digital video system requires an unusually low error rate. Although *bit error rates* (BERs) of 1×10^{-6} are adequate for many digital transmissions, serial systems demand a BER of less than 1×10^{-9}, and sometimes less than 1×10^{-13}. There are several reasons for this. The scrambling system used in serial transmission multiplies channel errors by six at the receiver. A single channel error has a 50 percent probability of causing an error in the most significant bit (MSB), and a 70 percent probability of causing an MSB or MSB-1 error.[2] Therefore, one out of two channel errors will cause a visible picture defect, seen as a short black or white dash. Picture monitors spatially map the video. This makes even occasional errors visible. Also, technical quality standards in most professional video facilities are high. Although no error rate standards for serial currently exist, acceptable performance will probably dictate between one error/minute and one error/day. For 143 Mb/s NTSC, this corresponds to channel bit error rates of 2.3×10^{-10} and 1.6×10^{-13}, respectively.

4.4.4 Operating Principles

Serial digital video is created by converting parallel digital video into a serial form and transmitting it one bit at a time. Figure 4.11 illustrates the process, beginning with an analog video input. Because video is represented with 10-bit words, the serial clock rate is 10 times that of the parallel clock. There are three common rates:

- 143 Mb/s for NTSC composite
- 177 Mb/s for PAL composite

---

2 K. M. Ainsworth and G. D. Andrews. "Evaluating Serial Digital Video Systems," presented at the Fourth International Conference on Television Measurements, June 20-21, 1991.

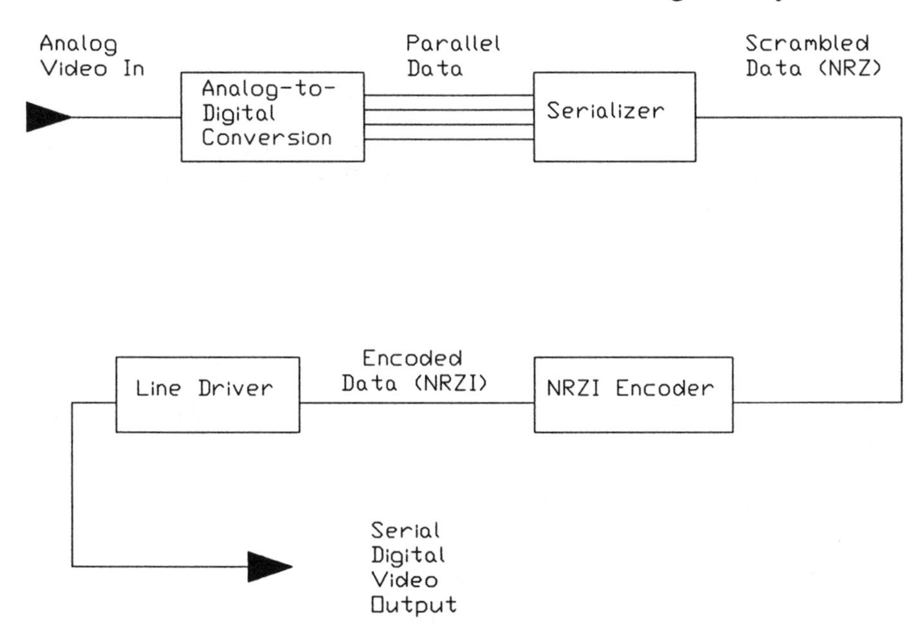

Figure 4.11 Basic block diagram of the serial digital conversion process.

- 270 Mb/s for 4:2:2 component

A typical serial system consists of five major component parts:

- Transmit co-processor
- Serializer
- Transmission channel
- Deserializer
- Receive co-processor

The transmit co-processor adds a line rate synchronizing signal to the incoming parallel video. This *timing reference signal* (TRS) synchronizes the serializer and deserializer. TRS is only added to composite video (signals present in the 4:2:2 component format perform this function). The serial signal has space allocated for ancillary data, such as AES/EBU digital audio. The co-processor can insert this data.

The serializer latches the parallel data from the co-processor and shifts it out one bit at a time, using a phase-locked clock running at 10 times the parallel clock rate. Serial systems always use 10-bit representations. If 8-bit video is input, the serializer sets the two LSBs to zero.

Next, the serial data is scrambled. The scrambler alters the data to enhance clock recovery at the receiver and to reduce the signal dc component. The scrambled signal goes to a coder that represents 1s as data transitions and 0s as no transition. An output stage provides the power to drive the channel, nominally 75 Ω coax.

The loss in metallic cable, expressed in decibels, is proportional to the square root of frequency. For example, if the loss at 50 MHz is 5 dB, then the loss at 200 MHz will be approximately 10 dB. This loss characteristic is specified for the serial channel.

Cable lengths up to about 300 m are the most common. Other channels, such as optical fiber, may be used as long as a compatible receiver equalizer is employed.

At the deserializer, an equalizer compensates for transmission channel losses. Equalization is usually automatic, based on signal amplitude and presumed loss characteristics for the channel. Next, a clock recovery circuit extracts timing information from the signal and uses it to lock up a 10 times parallel rate oscillator. This clocks the deserializer. After a conversion stage, the signal is fed to a shift register, which performs the serial-to-parallel conversion. Figure 4.12 shows the block diagram of a serial digital receiver system.

If the deserialized signal is composite, then nonvideo data, such as TRS and ancillary data, are removed by the receiver co-processor. The co-processor might decode this data or strip it off, depending on the application. Because the 4:2:2 component standard allows ancillary data, stripping is not required for component video.

4.4.4.A *Bandwidth Requirements*

The *non-return to zero inverted* (NRZI) channel coding used in serial video contains significant spectral energy up to the clock frequency. For 143 Mb/s NTSC, this is approximately 30 times the bandwidth of the baseband signal. This wide bandwidth is a result of the PCM representation of the signal, and it places severe demands on the channel (usually coax) and receiver. Reflections in the channel can distort the transmitted pulses and increase detection errors. Also, coax can be quite lossy at higher frequencies and, therefore, requires equalization at the receiver for everything but short runs.

4.4.4.B *Results of Channel Impairments*

Experience with 143 Mb/s systems has shown that pulse amplitude and large channel reflections result in serious performance impairments. Pulse amplitude variations affect some automatic receiver equalizers and directly alter the signal power in the channel. Large reflections raise the detection error threshold.

High-quality double-shielded coax is important, particularly at the receive end of the link, where signal level may be low. Inferior coaxial cable can add unnecessary noise. Less important parameters include rise time, jitter, and small channel reflections (such as those resulting from 50 Ω connectors). Jitter may become an issue when using a number of serial regenerators.

Receiver performance is generally the largest single factor affecting link dynamic range. Optimizing the previous parameters can make a good receiver better.

4.4.5 Coaxial Cable Considerations

Most system engineers are familiar with the high-frequency losses that occur in a coaxial cable carrying analog signals. The loss in decibels per unit of cable length increases with the square root of the increase in frequency. There is also a slight time delay difference between low-frequency and high-frequency signals in coaxial cable. Combined with frequency response losses, these phase shifts cause pulse rounding, smear, and high-frequency rolloff. Similar distortions occur with a serial digital signal waveform, but they are more severe because of the higher frequencies involved. When the effects of the distortion become longer in time than one bit-interval (3.7 ns for 270 Mb/s), a pulse becomes affected by the preceding pulse or pulses. This effect is referred to as *intersymbol interference* (ISI). Figure 4.13 illustrates how intersymbol

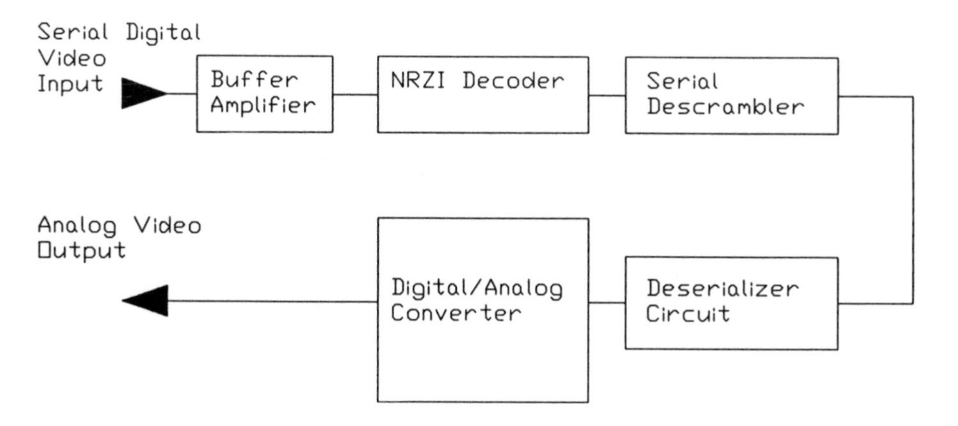

Figure 4.12 Basic block diagram of a serial digital receiver/decoder.

interference can shift the zero-crossing time of data pulses. When the interference extends over more pulses, the distortion and resulting jitter become worse. As the cable loss distortion increases, it becomes impossible for the system to recover the digital data.

The rounding and losses that occur in a coaxial cable are illustrated in Figure 4.14. Waveform (a) shows the recovered data stream output of a short cable, with all pulses at full amplitude and exhibiting good zero crossings. Some overshoot is evident on the leading edges, but this usually does not present a problem for digital detection hardware. For the 100 m cable shown in (b), notice how single bit-interval pulses do not reach the final peak amplitude before they change direction. All pulses still pass through zero, but the width and timing of the pulses are distorted and will cause

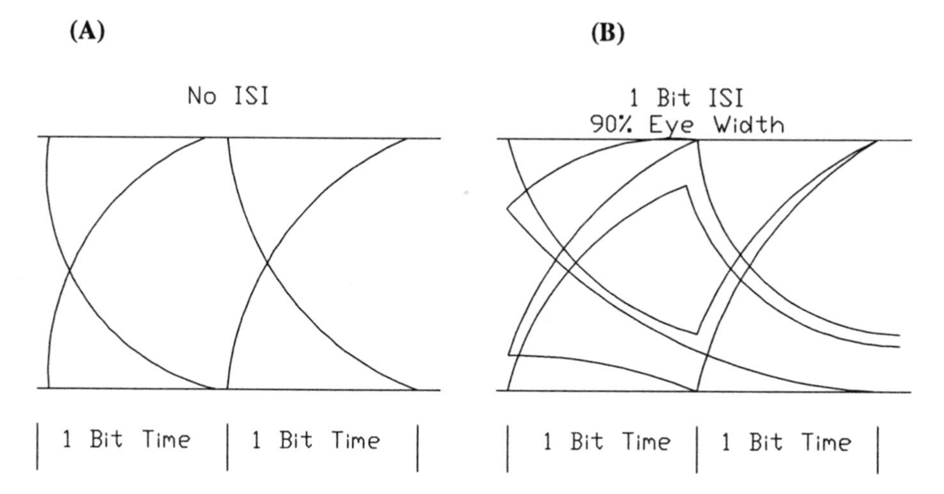

Figure 4.13 Comparison of intersymbol interference: (a) no ISI; (b) ISI present.

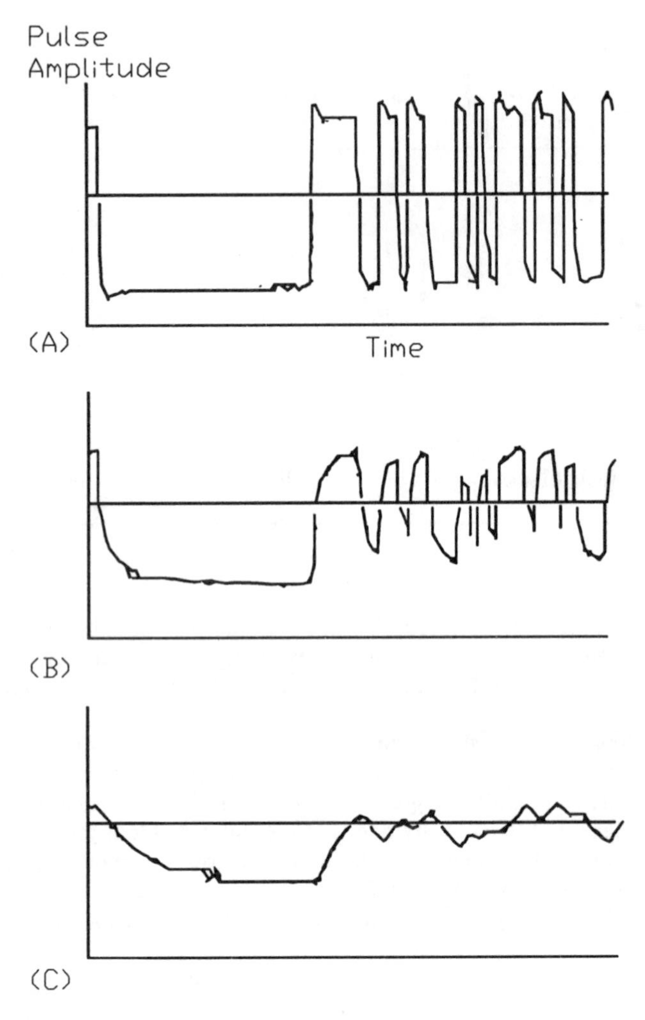

Figure 4.14 Data pattern distortion as a function of coaxial cable length: (a) data pattern through a short cable; (b) data pattern through a 100 m cable; (c) data pattern through a 300 m cable.

---

displacement of the edges at the output of a bit-slicer or comparator. This timing displacement will appear as jitter. Waveform (c) shows the output of a 300 m cable. Notice the additional rounding of pulses and attenuation. Some of the single bit-interval pulses do not even cross zero, because longer pulses have preceded them. Data will be lost unless proper equalization is applied. With proper equalization, data can be recovered completely. It is important that the equalization extend from the period of the long pulse strings to that of single pulses. Proper equalization will eliminate ISI and minimize the jitter that it causes.

Improper cable equalization can make matters worse, however, by increasing ISI. Cascading several equalizers with cable and amplifier sections in series will cause an accumulation of jitter. But if the equalization is properly adjusted, the increase will be minimal. If a DA or other line receiver has a phase-locked system to recover data, it can tolerate larger errors in equalization, as long as the resulting jitter is within the capability of the system to lock up and properly sample the digital data. PLLs must be carefully designed and adjusted, because they can be a source of jitter, particularly if the loop bandwidth and damping have not been properly set. The resulting jitter will then accumulate as the PLLs are cascaded.

Because the shape of waveform of the serial digital signal is predictable, it is possible to automatically equalize for normal coaxial cable losses. For example, the system can be designed to adjust the equalizer for fast rise times with no overshoot, or it can be designed to achieve a given ratio of high-frequency component amplitude to low-frequency component amplitude. Other schemes for automatically adjusting equalization are also possible.

4.4.6 Stress-Testing a Serial System

Simply monitoring error rate will not tell how far a serial link is from its error threshold. As discussed previously, the error level is practically zero above a certain channel S/N. Error monitoring will identify when the system has reached the threshold. Although this represents an important piece of information, the link is effectively broken at this point. For example, an experimental 143 Mb/s system showed no errors at 370 m, occasional errors at 380 m, and large numbers of errors at 390 m[3].

The link margin or headroom may be estimated by measuring how much stress must be added to the system before the link enters threshold. One common method of accomplishing this is to add additional loss to the channel. This can be done by adding coax until the error rate becomes observable on a picture monitor or *error detection and handling* (EDH) indicator. The results can be quantified in "meters of coax" or (preferably) in decibel loss at half the clock frequency. It is easy to transfer the latter method between different types of coax. For example, Belden 8281 coax has approximately 7.5 dB of loss per 100 m at 71.5 MHz (half of 143 Mb/s). Thus, a link that entered threshold with 50 additional meters of 8281 would have 3.7 dB of margin.

Other stresses might include varying the transmitted pulse amplitude, adding noise and jitter, and using bit sequences that stress equalization and clock recovery circuits.[4]

Stress-testing is highly recommended when a system is first installed. Logging the results and repeating the tests at intervals will pinpoint deteriorating performance.

4.4.7 Monitoring Topology

Figure 4.15 illustrates two suggested monitoring connections. In (a), a serial source connects to a single input receiver, such as a video processor or VTR. The input signal loops through the waveform monitor just before terminating at the receiver. This allows content monitoring of the incoming signal and inspection of waveform patterns. The receiver's regenerated output loops through the second input of the monitor, and

3 Belden 8281 coax
4 *Sony SBX1602A Serial Interface/Transmission Decoder Specification*, pp. 22-23, Sony Corporation, 1991.

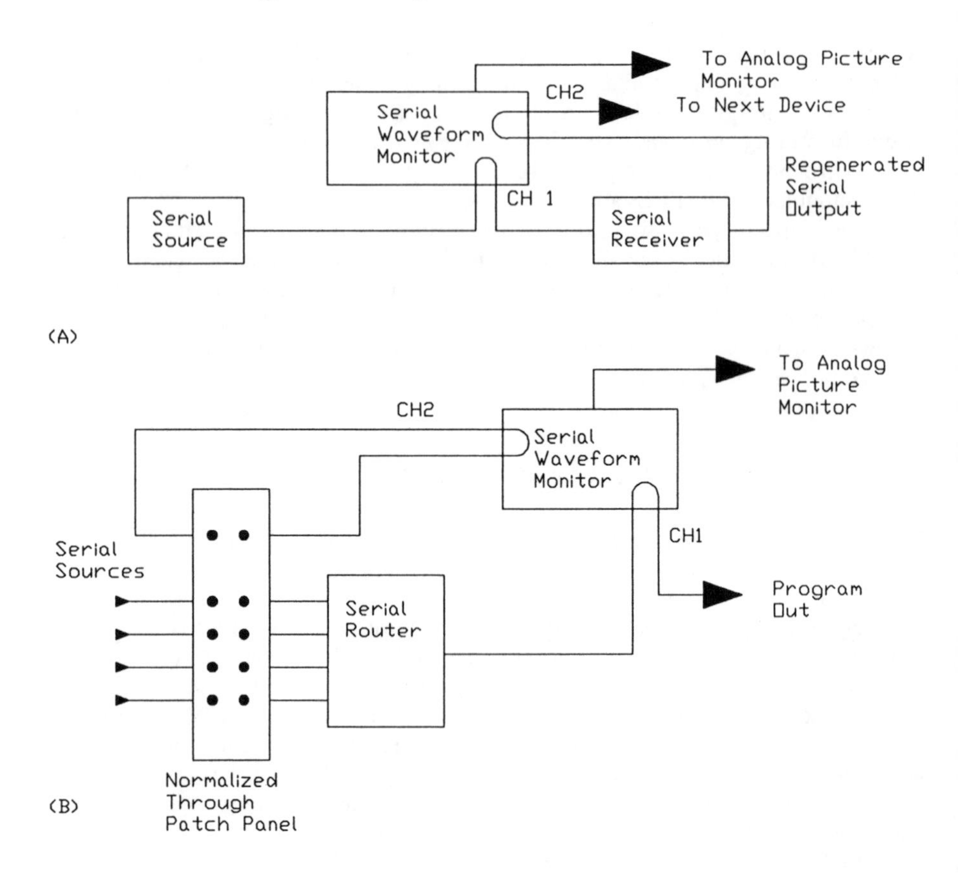

Figure 4.15 Two suggested monitoring connections: (a) the serial signal loops through the waveform monitor before termination at the receiver; (b) any router input can be monitored via the patch panel.

---

then either terminates or is sent to another device. An EDH-equipped monitor can be used to measure the serial receiver error rate. This is possible because the regenerated serial output is derived from the equalized and detected input, and will generally contain the same errors that are experienced by the receiver.

Figure 4.15 (b) shows a serial router with a single output. The channel 1 connection allows content and error-rate monitoring. The second input can be connected in series with any of the input signals via the patch panel.

4.5 Bibliography

Benson, K. B., and Jerry Whitaker, eds.: *Television Engineering Handbook*, rev. ed., McGraw-Hill, New York, 1991.

CHAPTER 5

INTERCONNECTING AUDIO SYSTEMS

5.1 Introduction

System interfacing problems have existed since the first audio engineer tried to wire two pieces of equipment together. Hardware that works flawlessly alone can fail miserably when connected.

In past years, audio engineers routinely used transformers to correct for differences in ground potentials among equipment, and to eliminate electrical noise picked up in cabling. These transformers helped to usher in the age of the balanced and floating interface. However, they also added significant cost, weight, and distortion to the equipment.

As the performance of electronic equipment improved, transformer shortcomings became more noticeable. The large size of transformers also made them inconvenient for use in the ever-shrinking chassis of transistorized equipment. As solid-state technology became more common, designers began to look for ways to eliminate transformers. The result was electronically balanced (active) inputs and outputs (I/O).

The move away from transformers has presented audio engineers with several difficult problems. First, installation practices that provided acceptable performance with transformer-isolated equipment may not be acceptable for active-balanced I/O equipment. Second, active-balanced I/O circuits are less forgiving of wiring errors and accidental short-circuits than transformers. Some active-balanced output stages will self-destruct if their output lines are shorted. Third, attention to ground loops is more important in an active-balanced I/O system than in a transformer system. Active-balanced input circuits can provide excellent noise rejection, but only to a point. Beyond a certain threshold, the noise rejection of an active-balanced input circuit deteriorates rapidly.

Some professional studio designers and system engineers will only use hardware equipped with transformer-isolated inputs and outputs. Premium-quality transformers are available today that overcome many of the shortcomings of older designs.

5.1.1 Decibels

Audio signals span a wide range of levels. The sound pressure of a rock-and-roll band is about 1 million times that of rustling leaves. This range is too wide to be accommodated on a linear scale. The decibel is a logarithmic unit that compresses this wide range down to a more easily handled range. Order-of-magnitude (factor-of-10) changes result in equal increments on a decibel scale. Furthermore, the human ear perceives changes in amplitude on a logarithmic basis, making measurements with the decibel scale reflect audibility more accurately.

A decibel may be defined as the logarithmic ratio of two power measurements or as the logarithmic ratio of two voltage measurements. The following equations define the decibel for power and voltage.

$$dB = 20 \log (E_1/E_2)$$

$$dB = 10 \log (P_1/P_2)$$

There is no difference between decibel values from power measurements and from voltage measurements if the impedances are equal. In both equations, the denominator variable is usually a stated reference. It is illustrated with an example in Figure 5.1. Whether the decibel value is computed from the power-based equation or from the voltage-based equation, the result is the same.

A doubling of voltage will yield a value of 6.02 dB, and a doubling of power will yield 3.01 dB. This is true because doubling voltage results in a factor-of-4 increase in power. Table 5.1 shows the decibel values for some common voltage and power ratios.

Audio engineers often express the decibel value of a signal relative to some standard reference, rather than another signal. The reference for decibel measurements may be pre-defined as a power level, as in dBm (decibels above 1 mW), or it may be a voltage reference. When measuring dBm or any power-based decibel value, the reference impedance must be specified or understood. For example, 0 dBm (600 Ω) would be the correct way to specify a level. Both 600 and 150 Ω are common reference impedances in audio work.

The decibel equations assume that the circuit being measured is terminated in the reference impedance used in the decibel calculation. However, most voltmeters are high-impedance devices and are calibrated in decibels relative to the voltage required to reach 1 mW in the reference impedance. This voltage is 0.775 V in the 600 Ω case. Termination of the line in 600 Ω is left to the user. If the line is not terminated, it is not correct to speak of a dBm measurement. The case of decibels in an unloaded line is referred to as dBu to denote that it is referenced to a 0.775 V level without regard to impedance.

Another common decibel reference in voltage measurements is 1 V. When using this reference, measurements are presented as dBV.

It is often desirable to specify levels in terms of a reference transmission level somewhere in the system under test. These measurements are designated dBr, where the reference point or level must be separately conveyed.

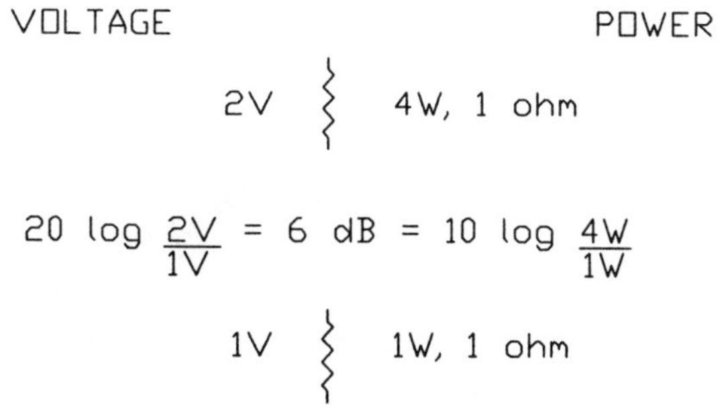

VOLTAGE POWER

2V 4W, 1 ohm

$$20 \log \frac{2V}{1V} = 6 \text{ dB} = 10 \log \frac{4W}{1W}$$

1V 1W, 1 ohm

Figure 5.1 Illustration of the equivalence of voltage and power decibels.

5.2 Audio System Interconnection Considerations

Common-Mode Rejection Ratio (CMRR) is the measure of how well an input circuit rejects ground noise. The concept is illustrated in Figure 5.2. The input signal to a differential amplifier is applied between the plus and minus amplifier inputs. The stage will have a certain gain for this signal condition, called the *differential gain*. Because the ground-noise voltage appears on the plus and minus inputs simultaneously, it is common to both inputs.

The amplifier subtracts the two inputs, yielding only the difference between the voltages at the input terminals at the output of the stage. The gain under this condition should be zero, but in practice, it is not. CMRR is the ratio of these two gains (the differential gain and the common-mode gain) in decibels. The larger the number, the

Table 5.1 Common decibel values and conversion ratios.

dB Value	Voltage ratio	Power ratio
-40	0.01	0.0001
-20	0.1	0.01
-10	0.3163	0.1
-6	0.501	0.251
-3	0.707	0.501
-2	0.794	0.631
-1	0.891	0.794
0	1	1
+1	1.122	1.259
+2	1.259	1.586
+3	1.412	1.995
+6	1.995	3.981
+10	3.162	10
+20	10	100

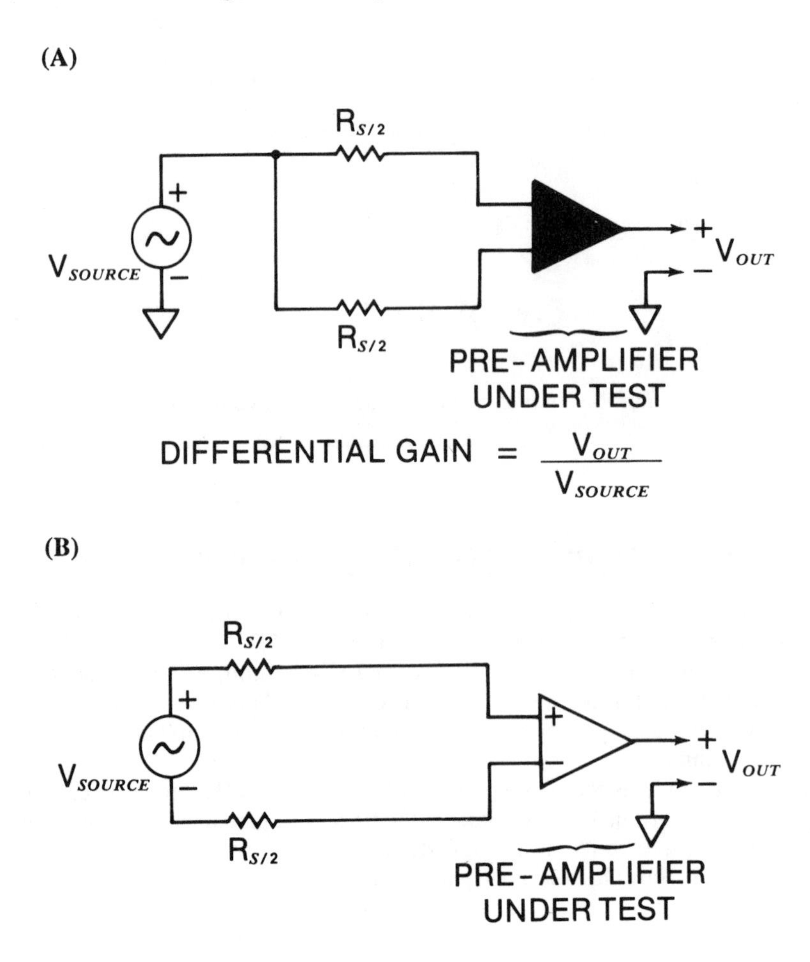

Figure 5.2 The concept of common-mode rejection ratio (CMRR) for an active-balanced input circuit: (a) differential gain measurement; (b) calculating CMRR.

better. For example, a 60 dB CMRR means that a ground signal common to the two inputs will have 60 dB less gain than the desired differential signal. If the ground noise is already 40 dB below the desired signal level, the output noise will be 100 dB below the desired signal level. If, however, the noise is already part of the differential signal, the CMRR will do nothing to improve it.

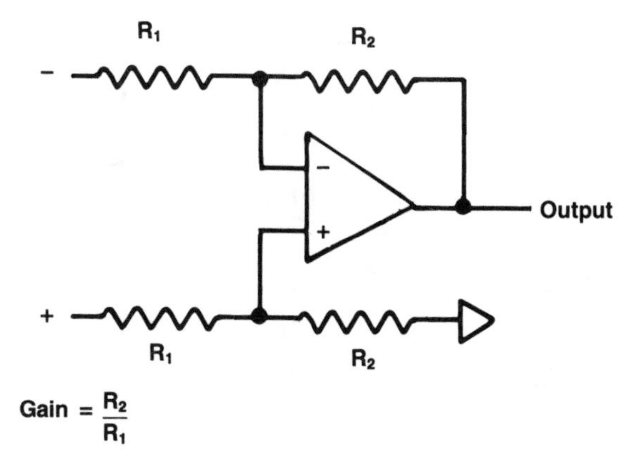

Figure 5.3 The simplest and least expensive active-balanced input op-amp circuit. Performance depends on resistor matching and the balance of the source impedance.

5.2.1 Active-Balanced Input Circuit

Active-balanced I/O circuits are the basis for nearly all professional audio interconnections (except for speaker connections). A wide variety of circuit designs have been devised for active-balanced inputs. All have the common goal of providing high CMRR and adequate gain for subsequent stages. All also are built around a few basic principles.

Figure 5.3 shows the simplest and least expensive approach, using a single operational amplifier (op-amp). For a unity gain stage, all of the resistors are the same value. This circuit presents an input impedance to the line that is different for the two input sides. The positive input impedance will be twice that of the negative input. The CMRR is dependent on the matching of the four resistors and the balance of the source impedance. The noise performance of this circuit, which usually is limited by the resistors, is a tradeoff between low loading of the line and low noise.

Another approach, shown in Figure 5.4, uses a buffering op-amp stage for the positive input. The positive signal is inverted by the op-amp, then added to the negative input of the second inverting amplifier stage. Any common-mode signal on the positive input (which has been inverted) will cancel when it is added to the negative input signal. Both inputs have the same impedance. Practical resistor matching limits the CMRR to about 50 dB. With the addition of an adjustment potentiometer, it is possible to achieve 80 dB CMRR, but component aging will degrade this over time.

Adding a pair of buffer amplifiers before the summing stage results in an instrumentation-grade circuit, as shown in Figure 5.5. The input impedance is increased substantially, and any source impedance effects are eliminated. More noise is introduced by the added op-amp, but the resistor noise usually can be decreased by reducing impedances, causing a net improvement (reduction) in system noise.

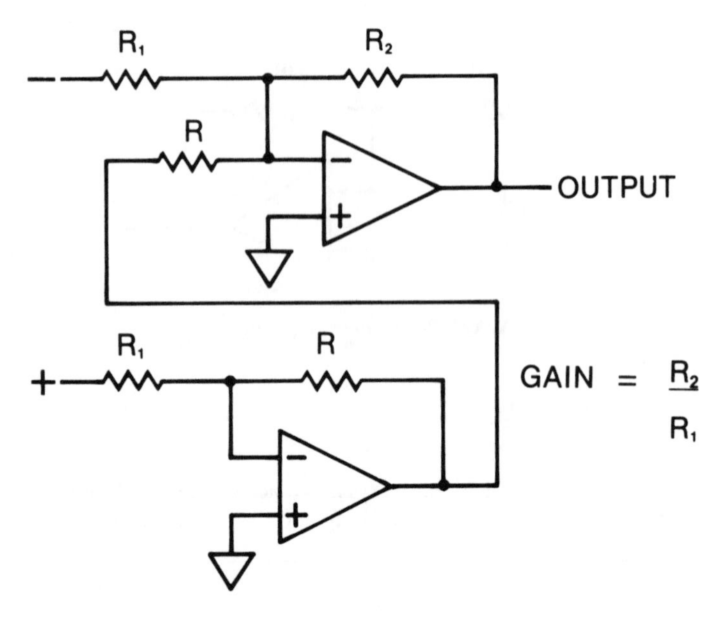

Figure 5.4 An active-balanced input circuit using two op-amps, one to invert the positive input terminal and the other to buffer the difference signal. Without adjustments, this circuit will provide about 50 dB CMRR.

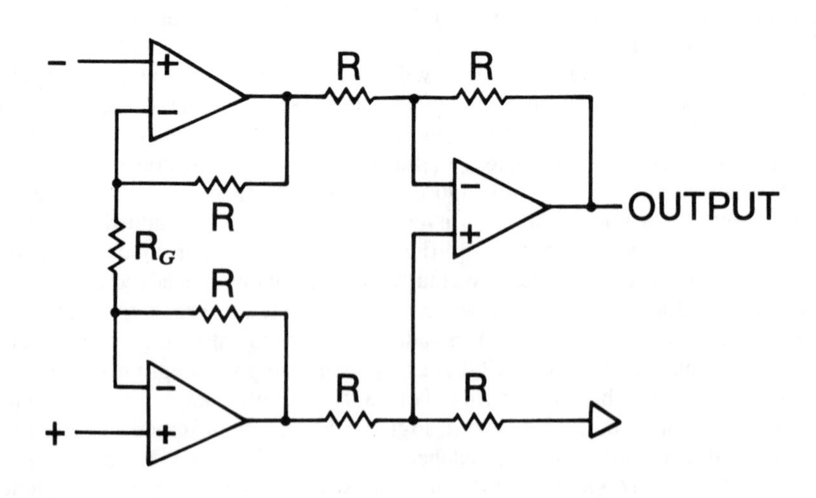

Figure 5.5 An active-balanced input using three op-amps to form an instrumentation-grade circuit. The input signals are buffered and then applied to a differential amplifier.

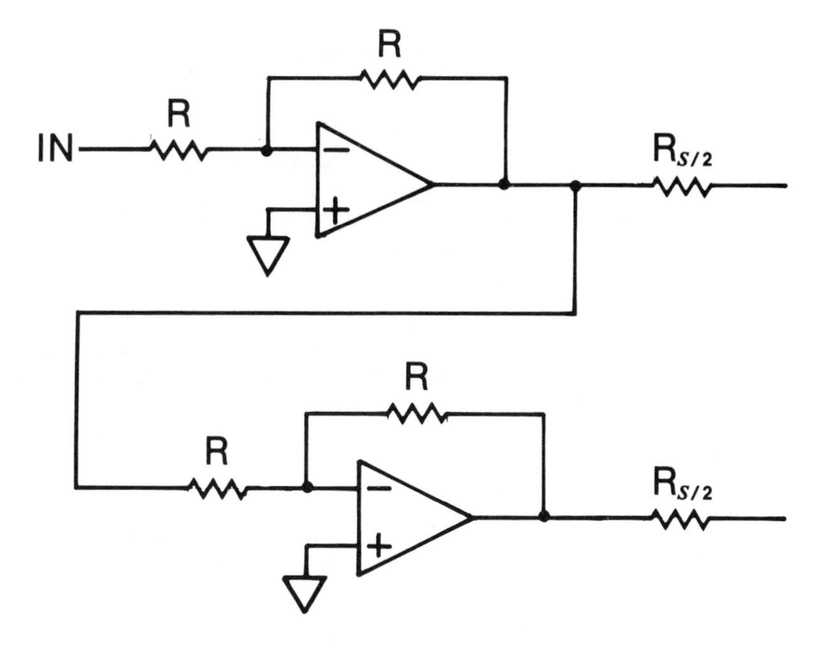

Figure 5.6 A basic active-balanced output circuit. This configuration works well when driving a single balanced load.

5.2.2 Active-Balanced Output Circuit

Early active-balanced output circuits used the approach shown in Figure 5.6. The signal is buffered to provide one phase of the balanced output. This signal then is inverted with another op-amp to provide the other phase of the output signal. The outputs are taken through two resistors, each of which represents half of the desired source impedance. Because the load is driven from the two outputs, the maximum output voltage is double that of an unbalanced stage.

The circuit shown in Figure 5.6 works reasonably well if the load is always balanced, but it suffers from two problems when the load is not balanced. If the negative output is shorted to ground by an unbalanced load connection, the first op-amp is likely to distort. This produces a distorted signal at the input to the other op-amp. Even if the circuit is arranged so that the second op-amp is grounded by an unbalanced load, the distorted output current will probably show up in the output from coupling through grounds or circuit-board traces. Equipment that uses this type of balanced stage often provides a second set of output jacks that are wired to only one amplifier for unbalanced applications.

The second problem with the circuit in Figure 5.6 is that the output does not float. If any voltage difference, such as power-line hum, exists between the local ground and the ground at the device receiving the signal, it will appear as an addition to the signal. The only ground-noise rejection will be from the CMRR of the input stage at the receive end.

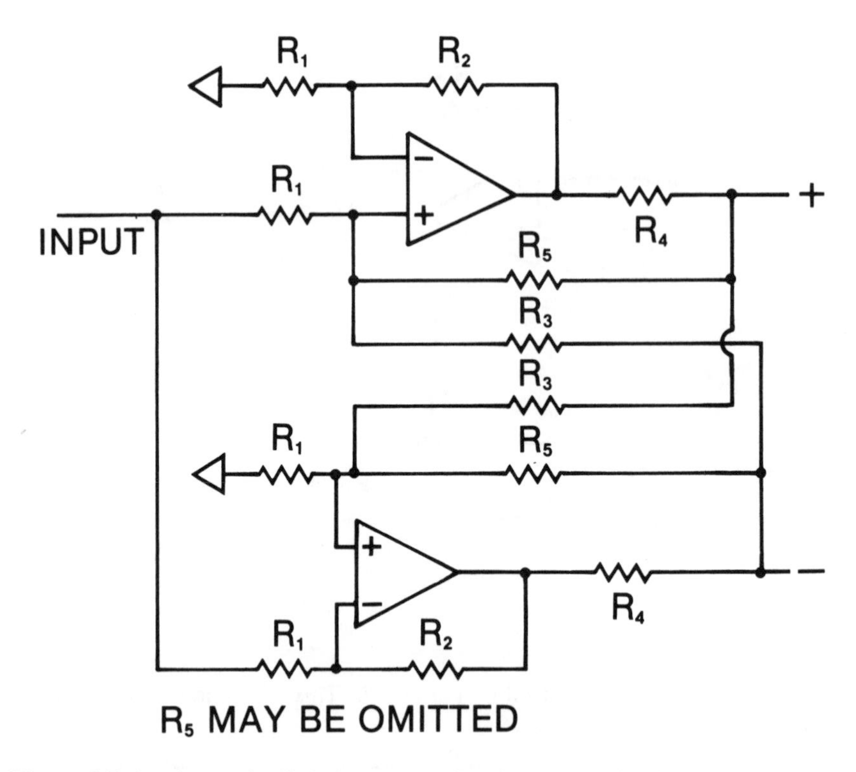

Figure 5.7 An electronically balanced and floating output circuit. A stage such as this will perform well even when driving unbalanced loads.

The preferred output stage is the electronically balanced and floating design, shown in Figure 5.7. The circuit consists of two op-amps that are cross-coupled with positive and negative feedback. The output of each amplifier is dependent on the input signal and the signal present at the output of the other amplifier. This type of design may have gain or loss, depending on the selection of resistor values. The output impedance is set by appropriate selection of resistor values. Some resistance is needed from the output terminal to ground to keep the output voltage from floating to one of the power-supply rails. Care must be taken to properly compensate the devices. Otherwise, stability problems may result.

5.3 Guidelines for Audio Interconnection

Susceptibility to RFI is a common problem with active-balanced circuits. Strong radio signals often can be rectified by nonlinearities in the input operational amplifiers or transistors. Although wideband, low-distortion circuits are less prone to this problem, they are not immune to it. Therefore, signals outside the range of the active circuits must be filtered before they are inadvertently amplified or demodulated.

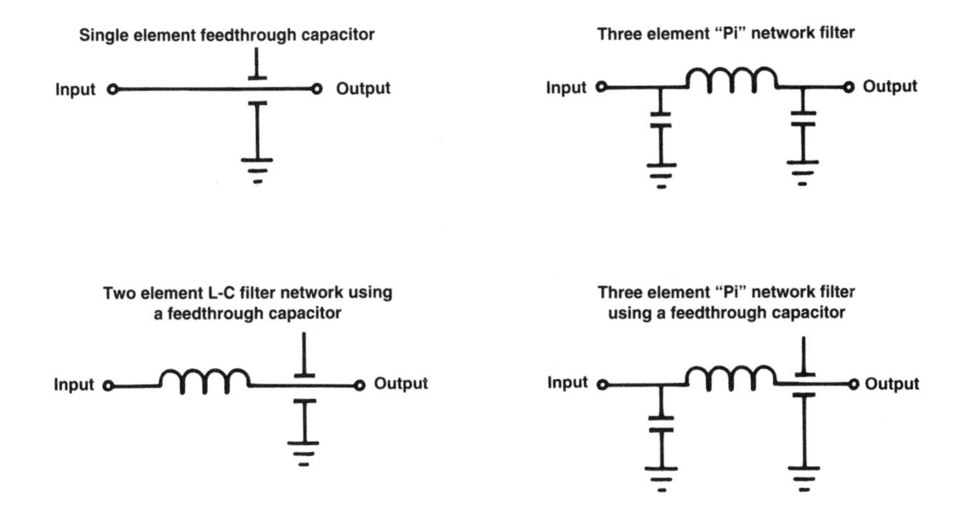

Figure 5.8 Common RF interference filtering networks. These filters, combined with tight mechanical assembly, will significantly reduce the possibility of noise-caused equipment problems.

To reduce RFI problems, most manufacturers add small series resistors and capacitors to ground at the input terminals. Inductors also may be added, but they are susceptible to external magnetic fields. If package shielding is inadequate, the inductors may actually pick up as much noise as they are supposed to filter out. Figure 5.8 shows several common RFI filtering networks.

Successful equipment interfacing is a function of how well the I/O circuits were designed and the care taken by the equipment installer. Mixing of balanced and unbalanced inputs and outputs can be done if care is taken in planning where signals go and how wiring is to be performed. Remember that the ground potential of one device may not be the same as the ground potential of another.

5.3.1 Wiring Guidelines

In designing and building a broadcast or production facility, theory and practice often go in divergent directions. A system design may look elegant on paper, yet it can turn out to be a nightmare when all of the hardware is installed and turned on. Strict adherence to some basic wiring conventions will help prevent equipment interfacing problems.

The microphone is the starting point for most production work. Microphones are considered to be floating sources. Use only 2-conductor shielded cable for all microphone wiring. Connect the microphone shield only to the microphone case and input chassis, as illustrated in Figure 5.9. If a grounded-shell XLR connector is used, connect the shell pin to the ground (shield) wire. The shell ground provides complete RFI shielding at the point where two cables join. This allows users to serially connect several microphone cables and maintain complete shielding. Be certain, however, that

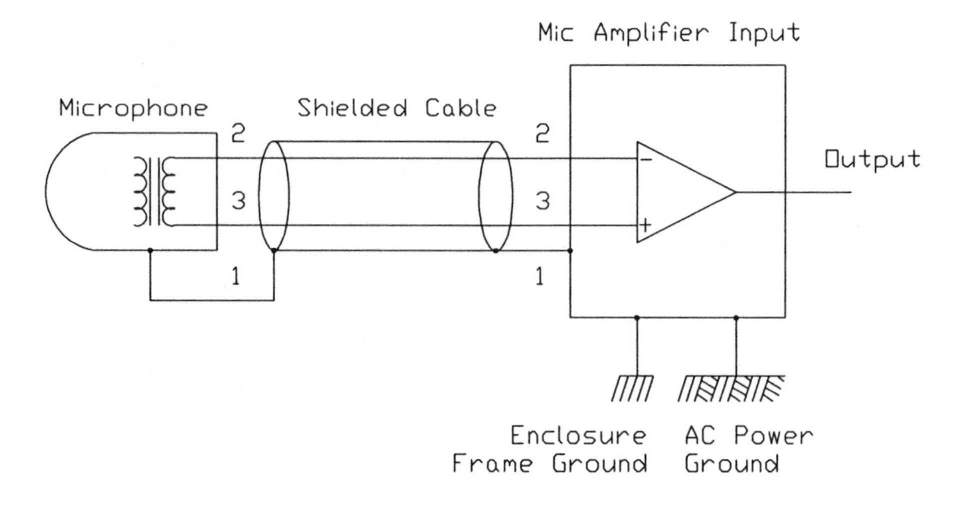

Figure 5.9 Interconnection arrangement for a floating source, such as a microphone. Note that the microphone shield is bonded to the microphone case. It also should be bonded to the microphone pre-amp chassis.

the XLR shells are not allowed to come in contact with any metallic structures or fixtures.

For audio equipment other than microphones, use only 2-conductor shielded cable, and tie the shield of each cable to the signal ground point at the respective source equipment chassis. Do not connect the other end of the shield. Dress the cable so that the shield connects to ground at only one point. Do not, under any circumstances, use the shield of an audio cable to provide safety (ac) grounding. Instead, build a separate ground system to tie the equipment together, as outlined in Section 2.4.

5.3.2 Equipment Grounding

It is a common misconception that the only way for an audio system to be free from hum and buzz is to secure a good earth ground. Although this will certainly help, it is only one part of the picture. The term *grounding*, when applied to audio, often refers loosely to the interconnection of individual chassis with the earth. It is better, however, to view grounding as an entire system. One ground system mistake can seriously impair the performance of an otherwise perfect arrangement.

Grounding schemes can range from simple to complex, but any system serves three primary purposes:

- Provides for operator safety.
- Protects electronic equipment from damage caused by transient disturbances.
- Diverts stray electromagnetic energy away from sensitive audio equipment.

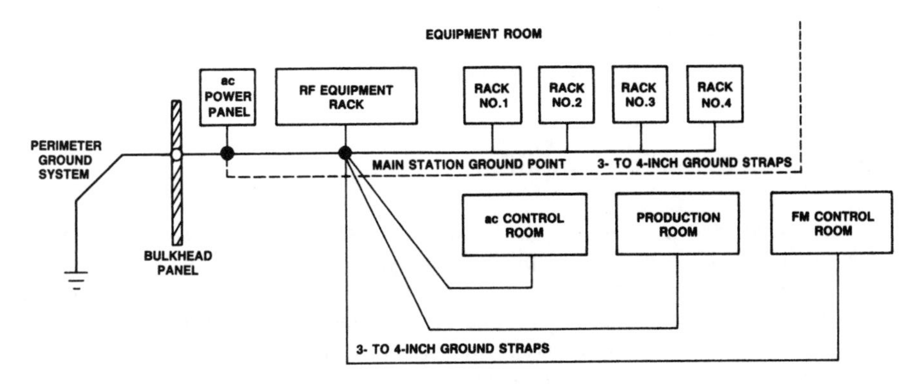

Figure 5.10 A typical grounding arrangement for a broadcast or audio production facility. The *main facility ground point* is the reference from which all grounding is done.

Most engineers view grounding mainly as a method to protect equipment from damage or malfunction. However, the most important element is operator safety. The 120 Vac line current that powers most audio equipment can be dangerous—even deadly—if improperly handled. Grounding of equipment and structures provides protection against wiring errors or faults that could endanger human life.

Many different approaches can be taken to designing a studio facility ground system. But some conventions should always be followed to ensure a low-resistance (and low-inductance) ground that will perform as required. Proper grounding is important whether or not the facility is located in an RF field. Figure 5.10 shows the

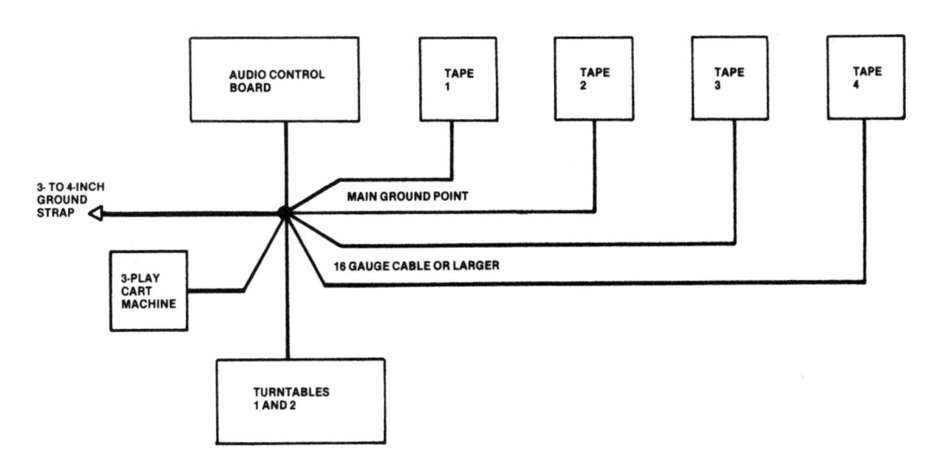

Figure 5.11 A typical grounding arrangement for individual equipment rooms at a broadcast facility. The ground strap from the main station ground point establishes a *local ground point* in each room to which all source and control equipment is bonded. The interconnected equipment rooms constitute a star-of-stars grounding scheme.

recommended grounding arrangement for a typical broadcast facility. This basic star system can be expanded to a star-of-stars system, as illustrated in Figure 5.11.

5.3.3 Distribution Amplifier

It is often necessary in an audio system to connect a single source to several different loads. Although this can be accomplished simply by tying the inputs in parallel, it is a poor solution. And, although the parallel arrangement is inexpensive, it leaves the audio system vulnerable to problems caused by changing levels, ground loops, and wiring faults. With a parallel hookup, shorting one of the input lines will result in a complete loss of signal to all of the other inputs.

Audio distribution amplifier (DA) systems have been developed to solve these problems. DAs are available in stand-alone and modular configurations. Stand-alone DAs are appropriate for applications where a small number of sources must be distributed. Modular DAs are better suited to multi-source distribution. Generally, if more than five channels must be distributed, and they can all be fed conveniently from the same physical location, the modular approach is best.

DAs are available with either fixed or adjustable output levels. If all of the devices being fed require approximately the same input level, then a fixed output DA may be adequate. If, on the other hand, the required output levels vary, specify a DA that offers individual output adjustments.

Most DAs incorporate active-balanced inputs and outputs. The output stage may use one of two basic approaches:

- Resistive divider network, illustrated in Figure 5.12
- Individual driver amplifiers, illustrated in Figure 5.13

Each approach has specific benefits to users. Some engineers like the redundancy provided by having an amplifier for each output. Such a design prevents a single component failure from crashing the entire system. The cost for a multiple-output DA system, however, is higher than for a comparable resistive divider circuit.

5.4 Cable for Audio Applications

A number of cable types and sizes are available for use in audio circuits. Virtually all professional and industrial applications use 2-pair shielded cable. There are three basic types of cable in use today:

- Microphone cable
- Line-level cable
- Speaker cable

Multi-pair audio cables are also available that consist of any number of individually shielded 2-pair audio cables. Multi-pair cables afford the benefits of individually shielded 2-pair audio lines, plus the installation ease of a single cable. Construction of a multi-pair audio cable is illustrated in Figure 5.14.

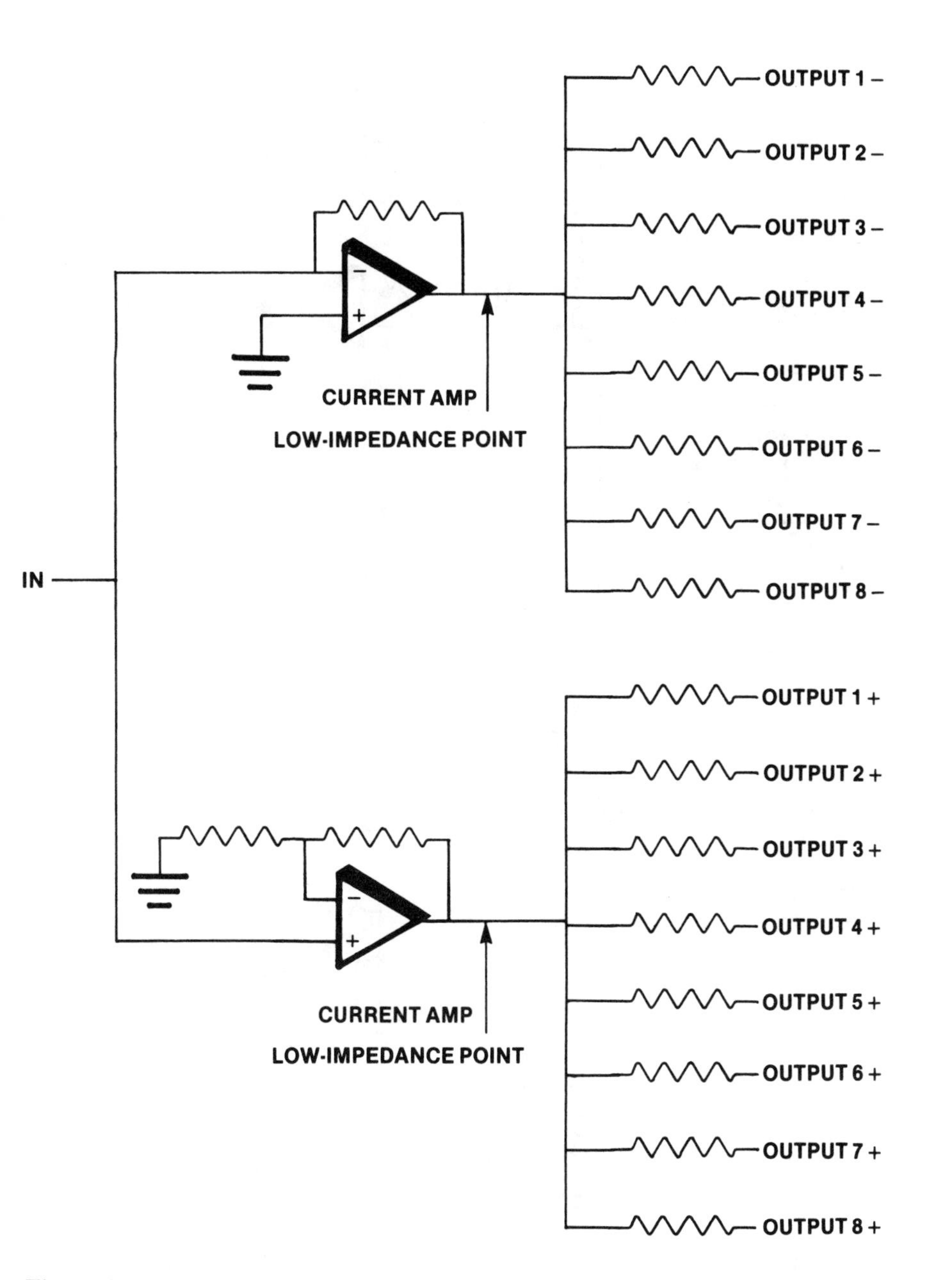

Figure 5.12 A DA output stage using resistive dividers. This system relies on a single high-current driver stage to feed all outputs.

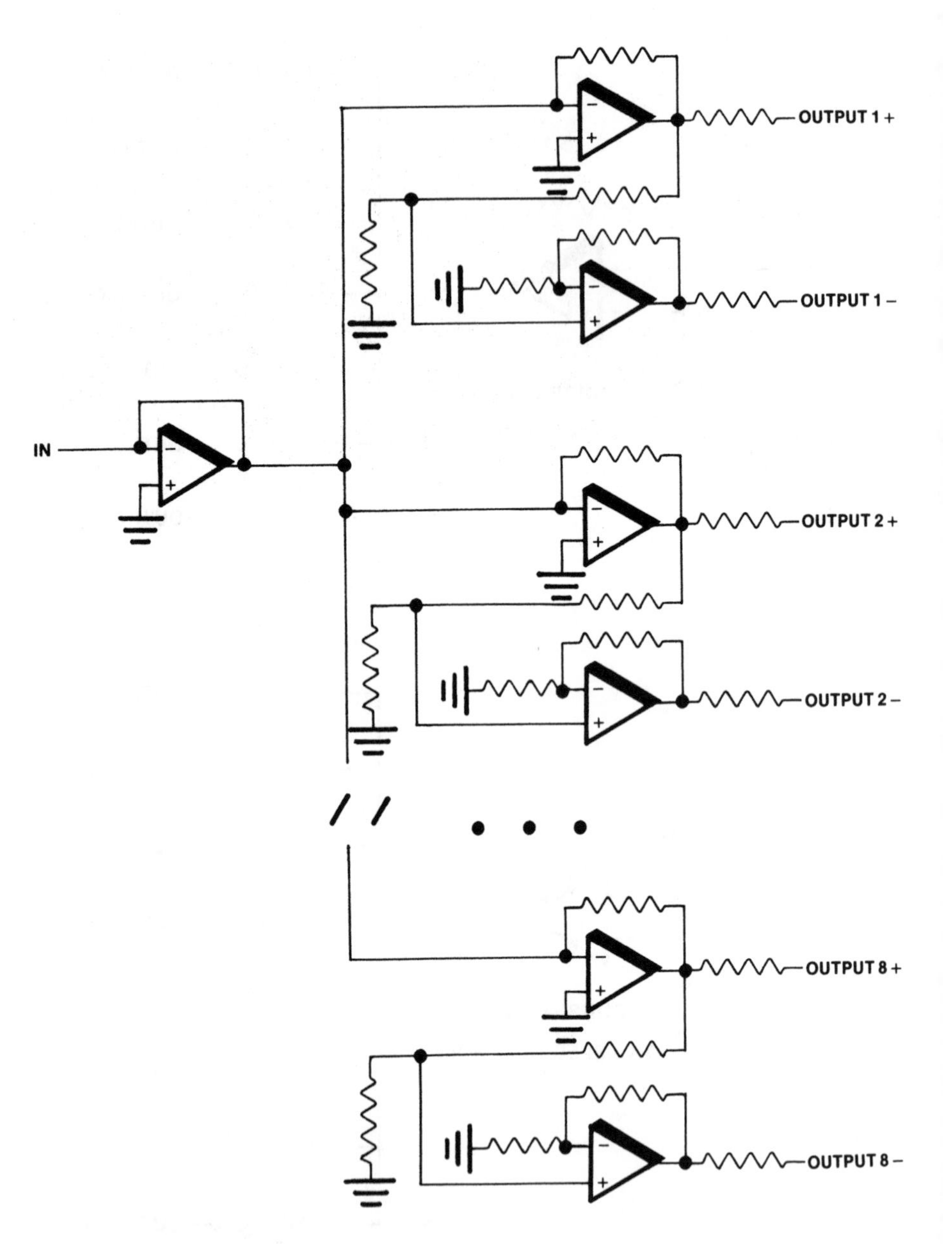

Figure 5.13 A DA output stage using individual driver amplifiers. With this arrangement, the failure of one driver will not cause failure of the entire system.

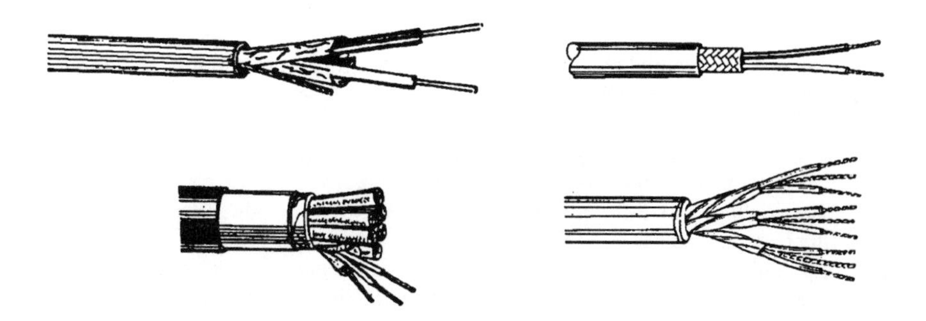

Figure 5.14 Physical construction of single- and multi-pair audio cable with individually shielded pairs.

5.4.1 Separating Signal Levels

Good engineering practice dictates that different signal levels be grouped and separated. It is common to separate cables into the following categories:

- ac power
- Speaker lines
- Line-level audio
- Microphone-level audio
- Video lines
- Control and data lines

Separate each type of cable as much as is practical, given the physical layout of the facility. When different categories of cables must cross, specify that they do so at a right angle. Avoid long parallel runs of cable from different level classifications.

5.4.2 Microphone Cable

Microphone cables are used to connect microphones to mixers or pre-amplifiers. They are generally required to be flexible and durable to facilitate placement or carrying by the user. Microphone cable can be used to carry line-level signals in the range of +4 dBm to +24 dBm; however, the very-low-level signal generated by a microphone may be on the order of -120 dBm to -70 dBm. Because the low-level microphone signal must be amplified by a large amount in the microphone pre-amp and subsequent amplifier stages, even the smallest noise signals picked up by the cable can become audible. Interference can be picked up by microphone cable from external sources through two primary mechanisms:

- Electrostatic coupling
- Electromagnetic coupling or induction

Because of the high gain levels required, *microphonic* noise may also represent a problem. It can be caused by capacitance changes that come from mechanical flexing, vibration, or dimensional changes in the cable, resulting from variations in temperature. The cable itself generates the unwanted signal. Long cables are more vulnerable to these sources.

The longer the cable, the greater its susceptibility to potential sources of noise. Noise, however, is not the only problem to be overcome with microphone cable. All cable has a characteristic impedance. This is the result of inductive and capacitive coupling between the various conductors in the cable. This impedance, combined with the input and output circuits to which the cable is connected, can act as a filter that can degrade the high-frequency response of the audio system. With runs of 100 m (328 ft) or more, microphone cable quality is especially critical.

In professional applications, particularly with microphones or portable sound systems, where cables are subject to continuous handling and flexing, durability is another important factor. Although two cables may seem to deliver similar results when new, they may deteriorate at different rates under actual field conditions. The nature of the insulation, shielding, conductor metallurgy, strain-relief fibers, friction reducers, twisting pitch, and other fabrication details all affect the practical life of a cable.

5.4.3 Electrostatic Noise

Electrostatic noise can be generated by a number of sources:

- Sparks at the armatures of motors or generators
- Gas-discharge lighting (neon and fluorescent) fixtures
- Portable electronic devices and appliances

Electrostatic noise can invade a microphone cable (or other sound system cables and components) by means of capacitive coupling. Electrostatic shielding, such as a metallic braided jacket, a swerved (spiral-wrapped) jacket, or a foil tape, can reduce electrostatic noise, provided the shield offers a low-resistance path to ground.

A circuit generating or carrying electrostatic noise acts as one plate of a capacitor. The microphone cable can act as the other plate of the capacitor. A portion of the noise source voltage will, therefore, be electrostatically (capacitively) coupled into the microphone cable. The nature of capacitive reactance is such that higher frequencies are more readily admitted into the mic cable. Moreover, the higher the impedance of the microphone circuit, the greater the inducted noise voltage. This is one reason low-impedance mics are preferred over comparable high-impedance units.

Wrapping the signal-carrying conductors of the mic cable with a grounded, electrically conductive screen (shield) offers a low-resistance path to ground. This electrostatic shielding provides protection against the noise that would otherwise be induced by electrostatic coupling. The effectiveness of the shield depends on the *percentage of coverage*. The percentage of coverage is a measure of how much space there is within the shield structure for electrostatic and electromagnetic noise to leak into the signal-carrying conductors.

(A)

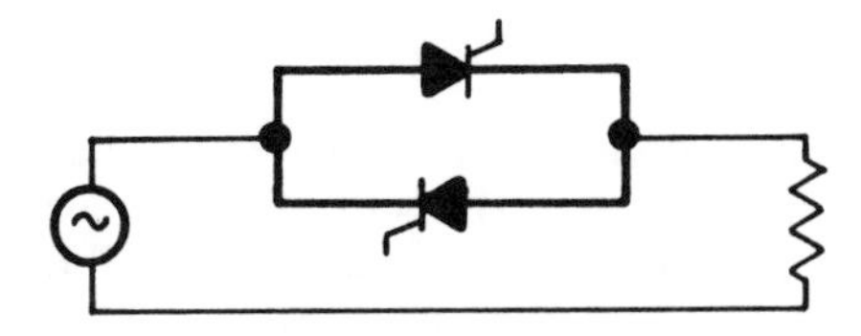

(B)

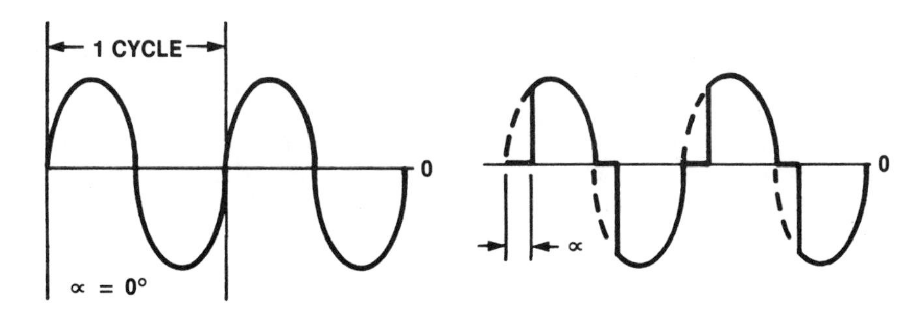

Figure 5.15 Inverse parallel thyristor ac power control: (a) circuit diagram; (b) voltage and current waveforms for full and reduced thyristor control angles. The waveforms apply to a purely resistive load.

5.4.4 Electromagnetic Noise

Electromagnetic noise may be generated by a number of electronic devices, including the following:

- Electric motors
- Fluorescent lighting ballasts
- Silicon-controlled rectifier dimmers

Electromagnetic noise can invade a sound system or cable by means of inductive coupling. Conventional electrostatic shielding offers no protection. Instead, solid conduit (iron or steel) or simply physical distance is required to minimize electromagnetic noise.

The magnetic fields generated by various sources cut across the conductors of a microphone (or other audio signal) cable. Because these fields alternately build and collapse, they induce a corresponding alternating noise voltage in the cable. The induced voltage is affected by the following:

- Power line frequency—the higher the frequency, the greater the problem.

- Current flowing in the source—the greater the current, the greater the induced noise.
- Proximity of the interfering source to the cable.
- The length of cable exposed to the noise source.

The ac power line waveform in most parts of the world is 50 or 60 Hz, but this can become contaminated by a rich harmonic spectrum. The harmonics are generated by various sources, most notably by the clipped waveforms emitted by SCR dimmers. SCR dimmers are a major source of noise problems because they generate high harmonics at some settings, and because these higher frequencies more readily couple into audio circuits.

Although SCR dimmers are a major factor contributing to higher-order power line harmonics, they are not the only problem source of electromagnetic noise. Frequent offenders include saturated power transformer cores and reactive fluorescent lamp ballasts. The noise caused by these sources includes 60 Hz hum and also considerable energy at 120, 240, and 480 Hz. If the power utility service is three-phase, it is also possible to obtain harmonics at 180, 300, 360 Hz, and so forth. Still, it is predominantly low-frequency energy that is heard as "hum," rather than the higher order harmonic energy (as from SCRs), heard as "buzz." The method of power control used by SCR devices is illustrated in Figure 5.15. The sharp turn-on point of an SCR creates a wide spectrum of noise unless the switching device is properly filtered and shielded.

5.4.5 Shield Considerations

A microphone cable must be capable of withstanding repeated flexing under heavy commercial use. It must also maintain mechanical integrity and shield effectiveness under a variety of operating conditions. Some cables are preferred over others, even though laboratory bench tests of shielding effectiveness may not show a clear advantage. Foil tape shields are known to be intolerant of flexing, and such cables are primarily used in fixed installations (within equipment racks or conduit).

Spiral-wrapped shielding is much less tolerant to flexing than braided shielding. Repeated flexing can cause the conductors in the shielding to break. Although spiral-wrapped cable appears to give better electrostatic noise immunity than braided cable in initial lab tests, it is likely to give worse results in actual field conditions. Figure 5.16 illustrates some of the more popular types of cable shield.

5.5 Audio Interconnection Components

A large number of individual elements go into the makeup of an audio system. Passive and active devices are both used to bring distant parts of a system together to perform a specific task. Some of the primary components include the following:

- Patch panels
- Connectors
- Terminal blocks

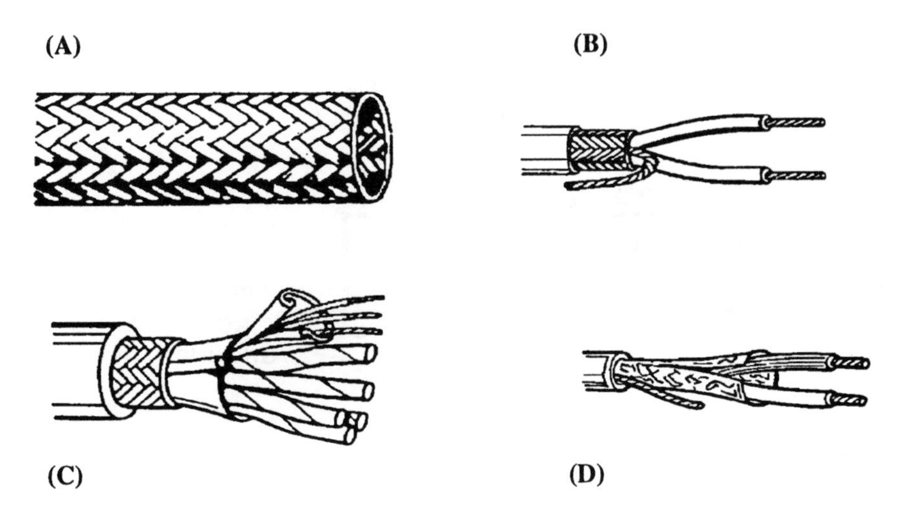

Figure 5.16 Common types of audio cable shielding: (a) basic braid material; (b) 2-conductor braided cable with a drain ground wire; (c) multi-pair cable using foil wrapped pairs enclosed in a braid; (d) 2-conductor cable using foil shield and a drain wire.

5.5.1 Audio Patch Panel

Designing patchbays into an audio system can provide a number of benefits to the users. Without patchbays, interconnections between the devices in a system are fixed, permanent, and difficult to access. If troubleshooting or set-up adjustments are required, it is usually necessary to access the connections on the equipment. With patchbays, access to equipment inputs and/or outputs is readily available.

Most patchbays are configured with two horizontal rows of vertically aligned 3-conductor *tip-ring-sleeve* (TRS) jacks. The jacks on the top row are connected to equipment outputs or sources. The bottom row of jacks is connected to equipment inputs or loads.

If the signal path changes frequently, patchcords are used to complete the circuit between the desired jacks in the patchbay. When the system circuit path is relatively permanent, switching jacks can be used. These jacks incorporate TRS terminals, which are an integral part of the contacts that mate with the patchcord plug. When a pair of jacks is *normalled* (connected) to each other, the signal is routed through the pair. When a patchcord is inserted in either of the two normalled jacks, the circuit path is broken and the signal is routed to the patchcord. In 150 Ω and 600 Ω circuits, a terminating resistor can also be switched across the circuit opened by the insertion of the patch connector.

If durability is required, use the standard-size connector. If space or weight is limited, use the smaller bantam connector. The smaller-size connectors have become more popular as system complexity has increased.

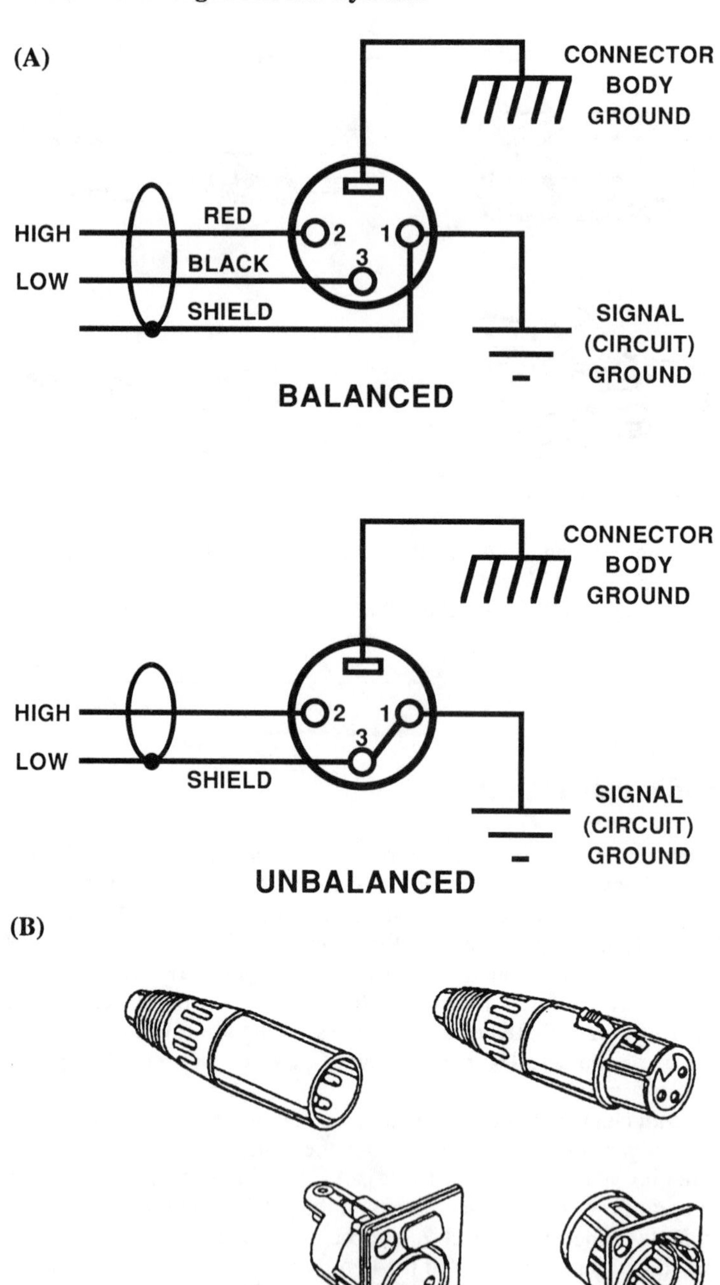

Figure 5.17 XLR connector for audio: (a) standard wiring for balanced and unbalanced circuits; (b) illustration of common XLR jacks and plugs.

5.5.2 Audio Connectors

A variety of connector types are available for use with audio frequency signals. Connectors that have become standardized in the industry for professional use include XLR and phone. Phono (or RCA) and miniphone connectors are used on industrial-grade and consumer equipment. Quality, construction, and workmanship of connectors vary. Quality and reliability are determined largely by the materials used. This is especially true for the contact and contact spring elements used in connectors. These elements, in turn, determine the cost of the connector. The method of fastening component parts together also affects the durability and cost of the product.

5.5.2.A *XLR Connector*

XLR connectors are the standard connector used for professional audio. This connector uses three pins. When using a shielded-pair cable, the shield is connected to pin 1, the high side is connected to pin 2, and the low side is connected to pin 3. (See Figure 5.17.)

Phone plugs are used for quick-connect applications, such as headphones, microphones (for consumer or industrial applications), and speakers. Mono and stereo versions are available. The mono version has two elements: tip and sleeve. The tip is the high side, and the sleeve is the low or ground side. The stereo phone connector has three elements: tip, ring, and sleeve. The sleeve is used as the common and the left and right channels are connected to the tip and ring, respectively. This is illustrated in Figure 5.18. Phone connectors are used in audio patch panels. Using shielded-pair cable, the shield is connected to the sleeve, the high side to the tip, and the low side to the ring.

5.5.3 Terminal Blocks

Three basic types of terminals are available on terminal blocks:

- Solder
- Twist-on, including barrier strips
- Push-on, including *insulation displacement* types

Solder-type terminals provide the most reliable connection but require more work to connect and make changes. Push-on-type connectors provide the easiest and fastest connections. Figure 5.19 shows the basic design of an insulation displacement terminal block.

In a facility designed around a central distribution scheme, all inputs and outputs are brought together at one location for interconnection by way of patch panels or a signal router. Jacks from the individual patch panels are wired to terminals on terminal blocks. All interconnections to equipment are also brought to terminal blocks at the central location. Interconnection between equipment and patch panels is made by connecting short jumpers from point to point on the terminal blocks. Changes are simply a matter of removing one end of a particular wire and connecting it to a new set of terminals on the same or another terminal block.

When designing a central equipment interconnection point, incorporate wire forms or hangers to support the cables entering and leaving the terminal point. This approach will reduce or eliminate the need for cable ties or other cable bundling in the distribution area. A typical wire form installation is shown in Figure 5.20.

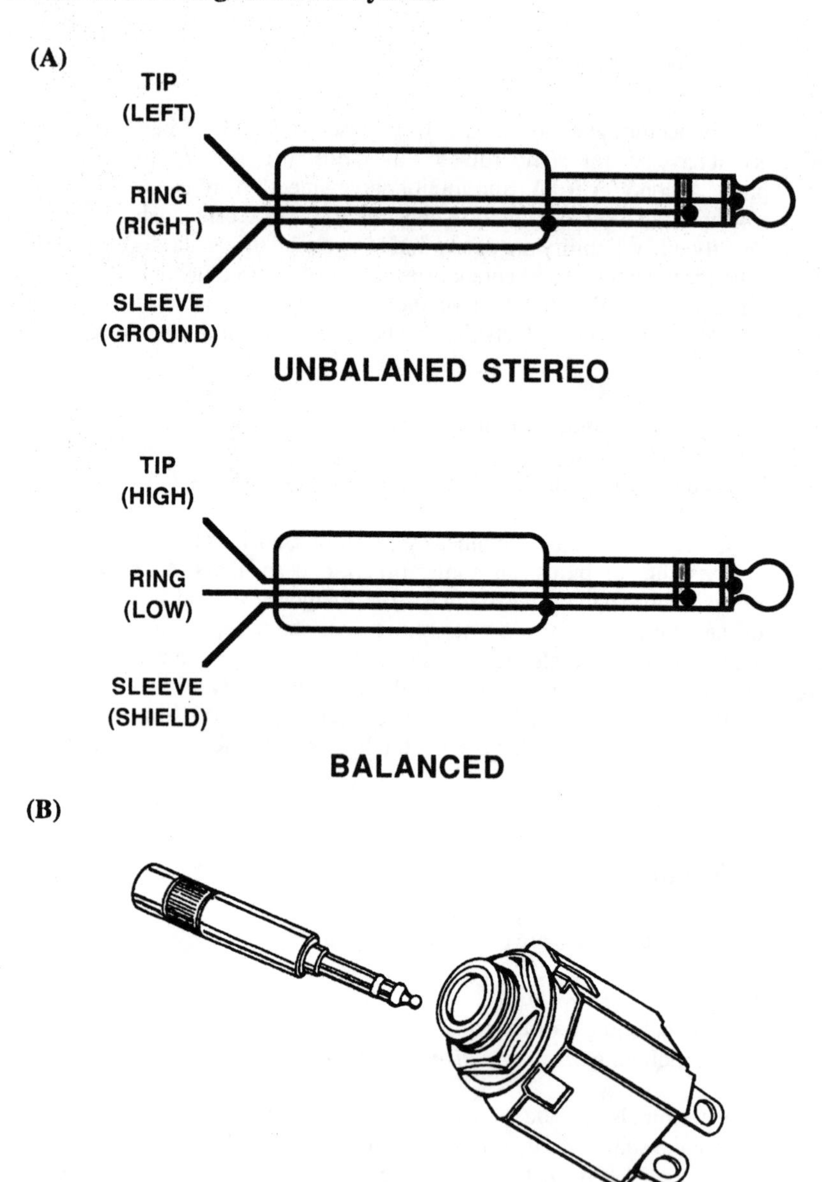

Figure 5.18 1/4 in stereo phone connector: (a) standard wiring diagram; (b) illustration of a jack and mating plug.

5.6 Bibliography

Benson, K. B., and J. Whitaker: *Television and Audio Handbook for Engineers and Technicians*, McGraw-Hill, New York, 1989.

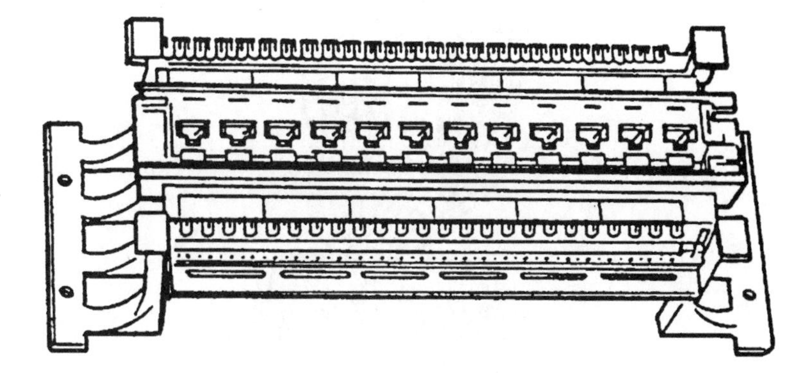

Figure 5.19 A standard insulation displacement terminal block similar to the type used by telephone companies in the United States.

(A)

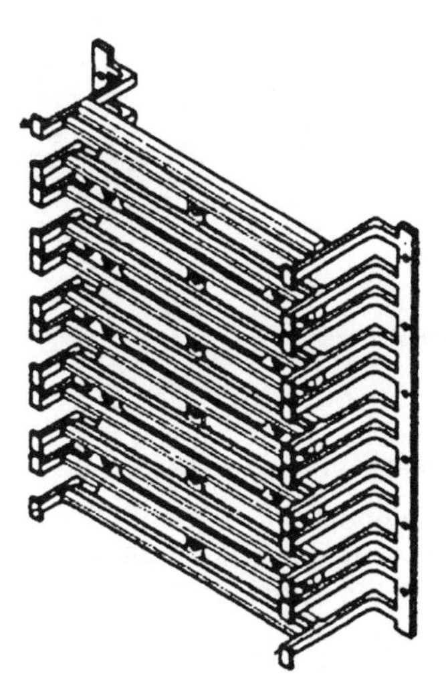

(B) next page

Figure 5.20 Use of wire forms to organize a terminal center: (a) basic wiring block; (b) blocks formed into a wire center.

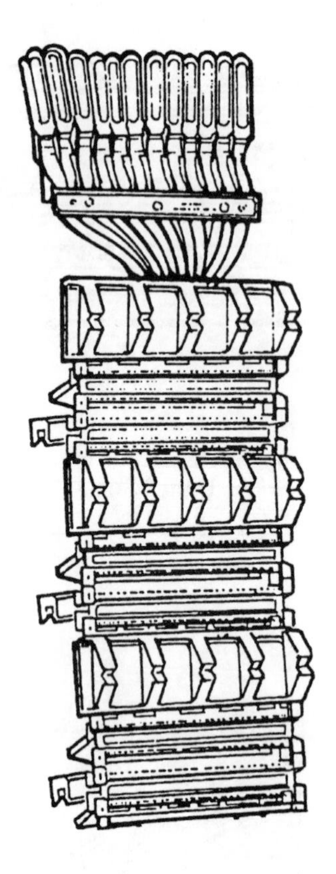

Figure 5.20(b)

Cabot, Richard, "Active-Balanced Inputs and Outputs," *Broadcast Engineering* magazine, Intertec Publishing, Overland Park, KS, July 1988.

Davis, Gary, and Ralph Jones: *Sound Reinforcement Handbook*, Yamaha Music Corporation, Hal Leonard Publishing, Milwaukee, 1987.

Lanphere, John: "Establishing a Clean Ground," *Sound & Video Contractor* magazine, Intertec Publishing, Overland Park, KS., August 1987.

Mullinack, Howard G.: "Grounding for Safety and Performance," *Broadcast Engineering* magazine, Intertec Publishing, Overland Park, KS., October 1986.

Whitaker, Jerry: *AC Power Systems Handbook*, CRC Press, Boca Raton, FL, 1991.

CHAPTER 6

INTERCONNECTING DATA SYSTEMS

6.1 Introduction

Getting computer equipment to work as a stand-alone system is one thing; getting computers to talk with each other in a networked system is quite another. In today's business environment, communication is a vital link to success in any enterprise. For many operations, *local area networks* (LANs) provide an efficient way to exchange information. LAN technology answers many business problems by providing the means for efficient communication between systems. For the system engineer, the introduction of LANs has brought the need for a different approach to design and maintenance. Because one of the major assets of a network is machine-to-machine communication, interface questions can become complex. The system engineer must have an understanding of the machines in the network and the interaction between machines as well.

6.2 Computer-Peripheral Interface

The three most common interface methods currently used to connect small (personal) computers with peripheral hardware and instruments are:

- RS-232, a serial format widely supported for bidirectional data transfer at low to moderate rates.
- Centronics parallel, a parallel format supported by most printer manufacturers for one-way data distribution. Table 6.1 details the connector assignments for the standard parallel interface.
- IEEE-488, also referred to as the *general purpose interface bus* (GPIB) or the *Hewlett-Packard interface bus* (HPIB).

Table 6.1 Standard parallel (Centronics) connector pin assignments.

Pin	Signal
1	-Strobe (Input)
2	Data 1 (Input)
3	Data 2 (Input)
4	Data 3 (Input)
5	Data 4 (Input)
6	Data 5 (Input)
7	Data 6 (Input)
8	Data 7 (Input)
9	Data 8 (Input)
10	-Acknowledge (Output)
11	Busy (Output)
12	Paper Error (Output)
13	Select (Output)
14	NC
15	NC
16	0 VDC
17	Chassis GND
18	+5 Vdc (output)
19	Strobe Return (GND)
20	Data 1 (GND)
21	Data 2 (GND)
22	Data 3 (GND)
23	Data 4 (GND)
24	Data 5 (GND)
25	Data 6 (GND)
26	Data 7 (GND)
27	Data 8 (GND)
28	Acknowledge Return (GND)
29	Busy Return (GND)
30	Signal GND
31	-Input Prime (Input)
32	-Nfault (Output)
33	Auxout I (Output)
34	NC
35	Auxout 2 (Output)
36	NC

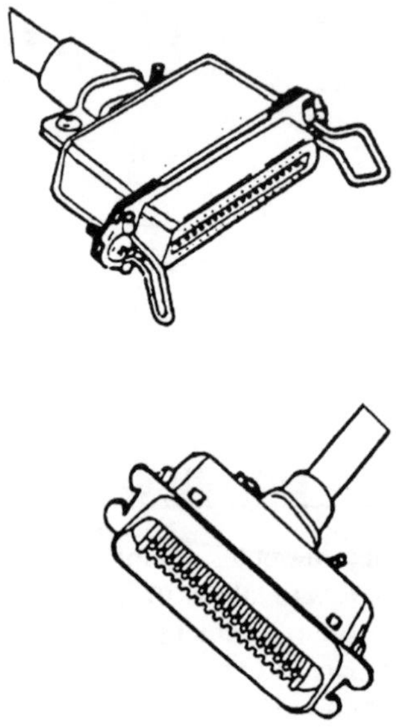

RS-232 is the standard serial interface used on most computers. The three interface systems each have their own advantages and applications. All provide for connection of a computer to one or more peripherals. RS-232 and GPIB are bidirectional, which allows the computer to either send information or receive it from the outside world.

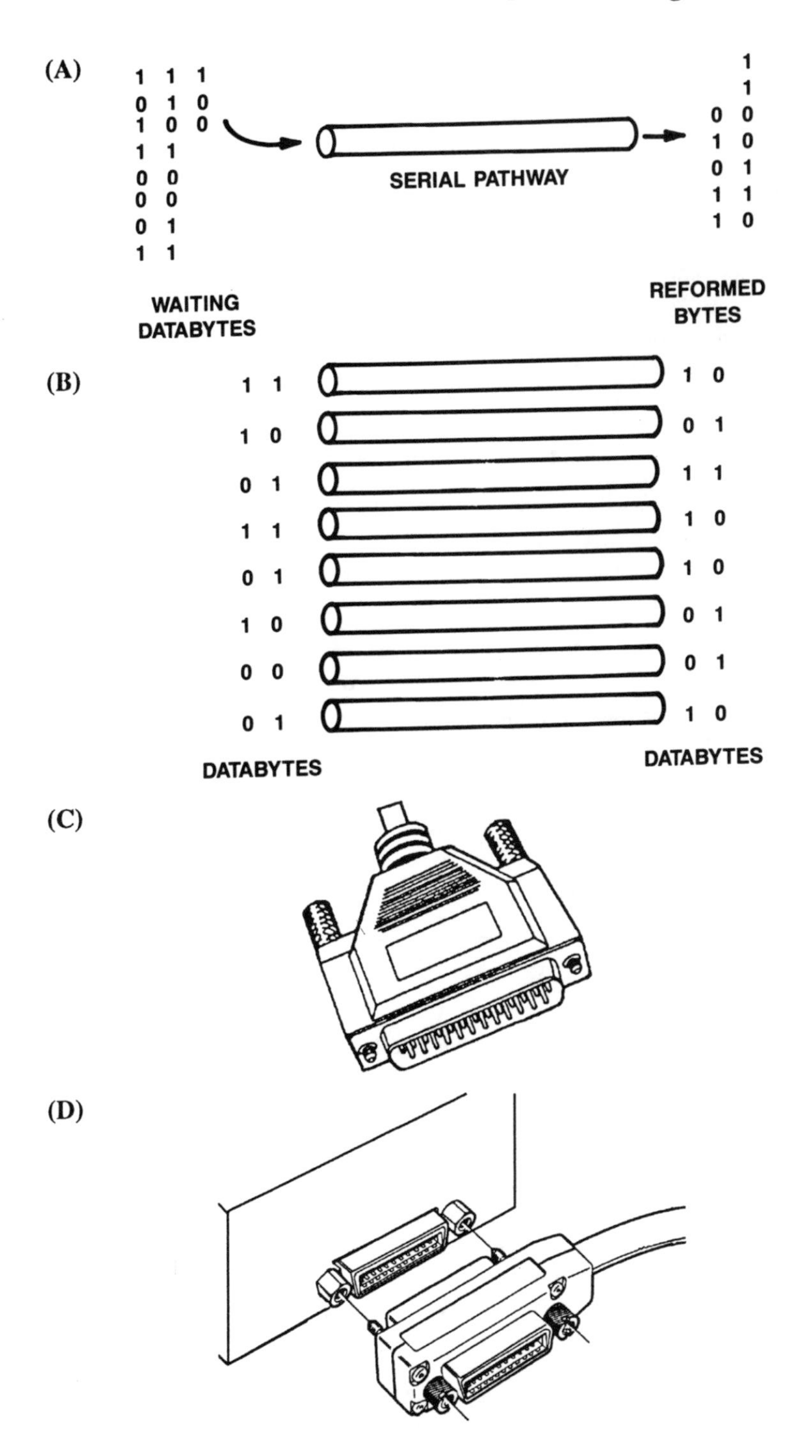

Figure 6.1 Comparison of data formats: (a) serial RS-232 protocol; (b) parallel GPIB protocol. Comparison of connector types: (c) RS-232; (d) GPIB.

Table 6.2 Comparison of the RS-232 and GPIB standards.

Advantages	**RS-232**	**GPIB**
Bidirectional data transfer	X	X
Works with long cables	X	
Sends data by phone	X	
Included in most computers	X	
Inexpensive cables and connectors	X	
Controls one to 15 units		X
Fast data transfer		X
Standard to most test equipment		X
Automatically adjusts speed		X
Plug-together compatibility		X
One standard connector		X
Advanced software available	X	X
Disadvantages		
Only controls one unit	X	
Speed of computer must match controlled unit	X	
Many data formats	X	
Many wiring variations	X	
Several connector styles	X	
Higher cost to add		X
Short cable runs only		X
Expensive multi-conductor cable		X

Most systems provide a Centronics parallel interface and either an RS-232 or GPIB interface. Figure 6.1 illustrates the differences between RS-232 and GPIB.

Test instruments using the GPIB interface greatly outnumber those with RS-232. Several thousand test instruments are available with GPIB interfacing as an option. Some plotters and printers also accept a GPIB input.

By comparison, RS-232 is more common than GPIB in computer applications. Printers, plotters, scanners, and modems often use this *standard serial interface*. Test instruments incorporating RS-232 typically are those used for remote sensing, such as RF signal-strength meters or thermometers.

Neither RS-232 nor GPIB is ideal for every application. Each protocol works well in some uses, marginally in others, and poorly in still others. Table 6.2 lists the relative advantages and disadvantages of the two protocols from a test and measurement perspective. Notice that the only advantage common to both is the ability to move data in both directions. Beyond that, the two are very different. The decision of which protocol to use for a particular application is based on what the system needs to do. Because RS-232 is already built into personal computers (PCs), many users want to use it for interface to peripheral hardware. Yet GPIB is the preferred protocol for most test equipment applications.

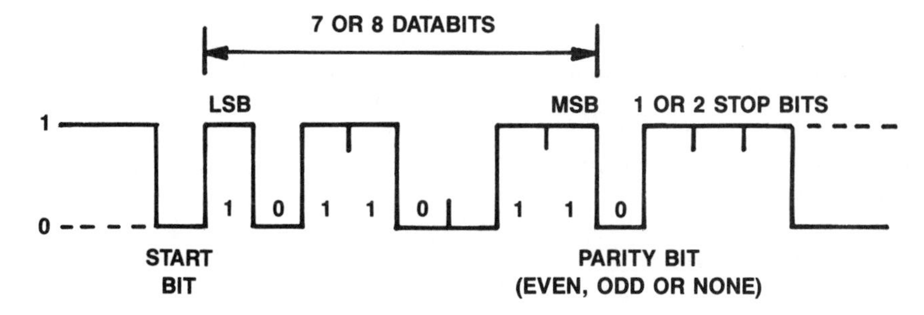

Figure 6.2 Characteristic data format for RS-232.

6.2.1 RS-232 and GPIB

The greatest advantage of RS-232 over GPIB is that it can easily send signals over longer distances. It can directly transmit data about 1,000 ft in one run of cable. A *line extender* permits greater distances. Many mainframe computers use RS-232 to send data to printers located in different locations of a business. Inexpensive twisted-pair cables interconnect the computer and printers. If data must be sent over longer distances, a modem can be used to convert the RS-232 signals into a form that can be fed over a standard telephone line. GPIB signals must first be converted to RS-232 if a modem is needed.

By contrast, GPIB's biggest advantage is that it can work with several instruments simultaneously. This capability is essential when an automated test requires more than one item to be under computer control. For example, a manufacturer might use GPIB to automate several pieces of test equipment at the end of a production line. Up to 15 different units can be connected simultaneously when GPIB is used.

The reason for these differences lies in the way the signals are fed to and from the computer. RS-232 is a bidirectional serial system. GPIB is a parallel format. Eight separate wires carry the GPIB data into or out of the computer, allowing an entire byte to move at one time. If all things were equal, this would make GPIB eight times faster than RS-232. GPIB can actually transfer data about 260 times faster than the fastest RS-232 data rate because of other electrical differences in GPIB. The parallel structure of GPIB permits external instruments to be addressed one at a time or in groups, allowing interconnection of several units. RS-232 requires mechanical or electrical signal switching to work with multiple instruments.

6.2.2 Availability of Interfaces

RS-232 interface ports are either included as part of a PC or easily added with a low-cost accessory board. RS-232 will interface with many printers. Third-party manufacturers make accessories to add GPIB to IBM (and IBM-compatible) and Apple/Macintosh computers. (IBM is a registered trademark of International Business Machines. Apple and Macintosh are registered trademarks of Apple Computers.)

Specialized computers are also available that offer GPIB as their main input/output port, intended for application as instrument or machine controllers.

Because RS-232 is common on computers but GPIB is often needed for instrument automation, several manufacturers make protocol converters that translate RS-232 signals to GPIB. These converters allow the benefits of both communications protocols to be employed. RS-232 and GPIB are both based on industry-wide standards, but only GPIB is a *true standard.* RS-232 has numerous variations, making direct connection more difficult.

6.2.2.A *RS-232 Formats*

The RS-232 standard specifies voltage levels and polarity so that one RS-232 feeds another directly. However, the many variations in RS-232 make it difficult to work with. Most initial problems result from variations in data transfer rates, data formats, and electrical connectors. Once an RS-232 link is up and running, it usually works well.

RS-232 encompasses 15 different data transfer rates. The rate of data transfer is measured in *baud.* One baud is the transfer of one data bit per second. A 300 baud device transfers data at 300 b/s. It takes about 10 bits (seven or eight data bits and two or three control bits) to form one character (byte). Data transfer occurs, therefore, at 1/10 the baud rate. A baud rate of 300 yields about 30 characters per second; the fastest RS-232 baud rate of 19,200 sends data at approximately 1,900 characters per second.

The computer and the external device must use the same baud rate to communicate. If the two rates are different, each character is garbled and all data are lost. Most RS-232 devices have configuration switches for matching the baud rate to the computer system. Aside from speed variations, there are nearly a dozen data format variations. Data bytes can either be seven or eight bits long. RS-232 adds stop bits and parity bits, which help ensure accurate data transfer. (See Figure 6.2.) There can be one or two stop bits, and parity can be none, even, odd, mark, or space. The number of stop and parity bits must also match for data to move from one device to the other. Again, switches on the unit let the user match a peripheral to the computer system.

RS-232 also uses at least four different physical connectors with five, eight, nine, or 25 pins. The 25-pin version is the most common, but it has dozens of different wiring variations. Luckily, the main data and ground pins are always the same. Table 6.3 shows the four pins that are always the same. Table 6.4 lists the most commonly used pin assignments for RS-232. A nine- to 25-pin interface is shown in Figure 6.3.

The data-out and data-in pins interchange, depending on whether the computer or peripheral end of the cable is referenced. The computer's *out* pin must feed the *in* pin of the external device, and vice versa. There is a 50-50 chance that an RS-232 cable will connect the inputs and the outputs correctly. Some devices have switches or jumpers to let the user exchange the wiring of the in and out pins. If there is no way to internally switch the pins, one of the following methods may be used:

- Custom-wired cables
- An adapter plug that reverses the connectors (a null-modem adapter, modem eliminator, cable switcher, or line reverser)

The null-modem typically provides a female connector on one side and a male connector on the other, with the data wires exchanged between them; pin 2 on one side connects to pin 3 on the other, and vice versa. The null-modem also exchanges the pins used for handshaking functions (see Section 6.2.3).

Table 6.3 The four standard pins of the RS-232 connector.

Function	Pin
Safety ground	1
Data out	2 or 3
Data in	3 or 2
Data ground	7

Table 6.4 Pin assignments for the RS-232 format commonly found in hardware intended for PC applications.

Pin number	Description
1	Protective ground (shield)
2	Transmitted data
3	Received data
4	Request to send
5	Clear to send
6	Data set ready
7	Signal ground
20	Data terminal ready

6.2.2.B *GPIB Format*

All GPIB connections and signals are the same. Any GPIB device may be connected to any GPIB computer. Because of its standard format, GPIB does not require settings for baud rate, stop bits, and parity bits. The format uses the same pins for all data going to or coming from the computer. Also, the system automatically adjusts the data transfer speed. GPIB can transfer data at any speed, from less than one character per second to 500,000 characters per second, making it up to 260 times faster than an RS-232 system operating at 19,200 baud. Few instruments supply data that quickly, but the system is capable of this speed without modification.

The GPIB connector, cable, and signals are always the same. A standard 24-pin connector hooks one piece to another. Each connector has a male and a female connector, allowing them to be stacked on top of each other, as illustrated in Figure 6.1(d). Some systems are built in a *star* arrangement, with each cable terminating in a single point; others are configured as a *daisy chain*, looping from one instrument to the next. Mixed stacking and chaining are possible for any connection scheme required. There is a limit of about 2 m per connection, however, which restricts GPIB to short runs. Longer runs cause capacitive loading, which may distort high-speed data.

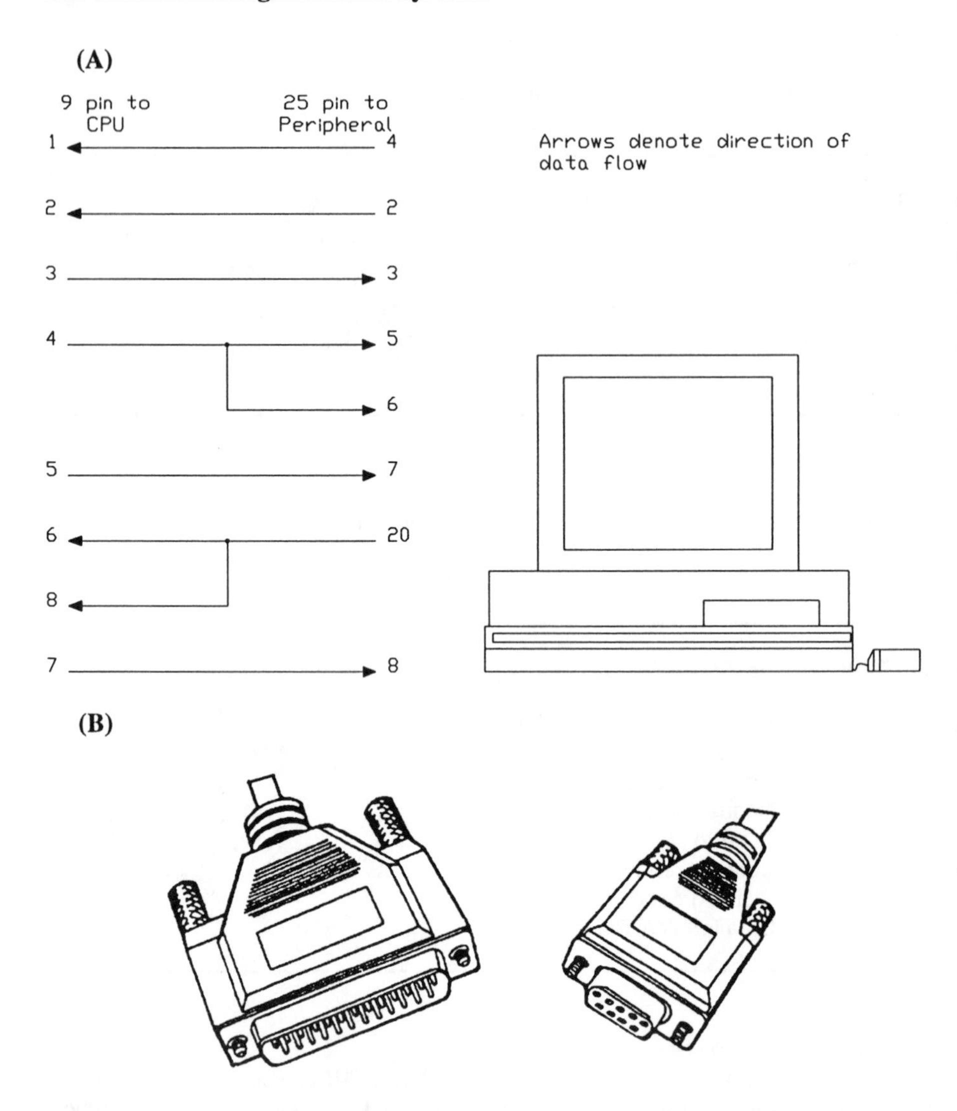

Figure 6.3 The 9-pin to 25-pin RS-232 adapter: (a) wiring configuration commonly used for personal computer equipment; (b) illustration of common connector design.

6.2.3 Handshaking

Handshaking is the method by which data transfer is controlled between the computer and the external device. Figure 6.4 illustrates the concept. GPIB takes care of handshaking with a single, standard method. There is nothing to consider in the design of a system; it either meets GPIB standards and works, or it does not. RS-232 uses hardware and software handshaking, with variations on each.

Transferring data by telephone requires a method of handshaking that can be encoded for transmission over a standard telephone line. The only method that meets

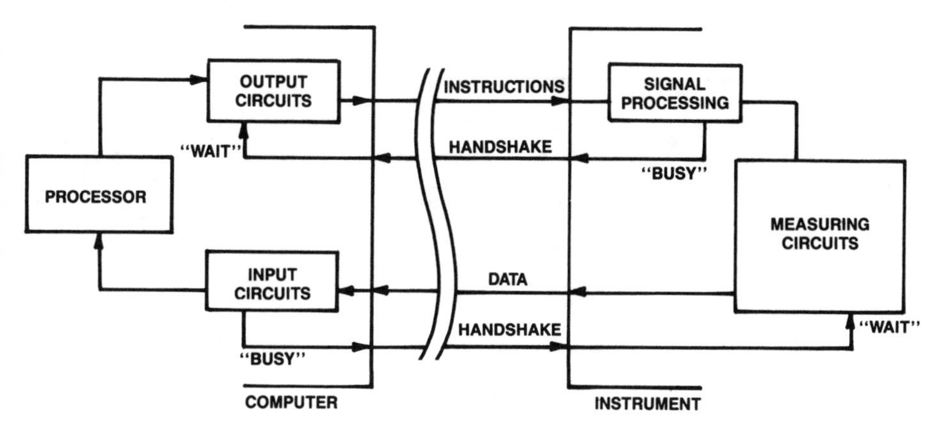

Figure 6.4 Basic principles of handshaking. Handshaking signals halt data flow if the device at either end of the cable is busy. Both RS-232 and IEEE-488 use handshaking.

---

this requirement is software handshaking. Indicators to start and stop data transfer are sent as special characters with the other data. The most commonly used software handshake is the *X-ON/X-OFF* (also called the *DC1/DC3*) system. Sending a special character (Control-S) to the device transmitting data causes it to stop. Sending another special character (Control-Q) causes it to start again.

The *ETX/ACK* protocol is another handshaking format. It is a complex control system that eliminates the need for the sending device to constantly monitor the return line for a busy signal. ETX/ACK is often used when data is sent between mainframe computers. The data are grouped into blocks with the same number of characters. Control characters embedded with the other data mark the end of blocks. The receiving device stores entire blocks of data until they can be processed.

RS-232 systems that do not involve a modem can use either software or hardware handshaking. With hardware handshaking, two additional wires connect the computer to the external device. One signals the computer to stop sending characters if the remote device is busy; the other serves the same function in the opposite direction. Several pins are used for hardware handshaking. The most common are pins 6 and 20. Other systems, however, might use pin 11 or 19; still others use pins 4 and 5 for different handshaking purposes.

A breakout box is a special tester with lights that show which pins are active when an RS-232 system is connected. This tester often saves time configuring a device or system when first connected to a host computer. After the proper connections have been determined, the breakout box is removed and a cable of the proper configuration is connected in its place.

Table 6.5 RS-232-C serial communications standard interchange circuits by category.

				GND	Data		Control		Timing	
Pin no.	Interchange circuit	CCITT equivalent	Description		From DCE	To DCE	From DCE	To DCE	From DCE	To DCE
1	AA	101	Protective ground	X						
7	AB	102	Signal ground/common return	X						
2	BA	103	Transmitted data			X				
3	BB	104	Received data		X					
4	CA	105	Request to send					X		
5	CB	106	Clear to send				X			
6	CC	107	Data set ready				X			
20	DC	108.2	Data terminal ready					X		
22	CE	125	Ring indicator				X			
8	CF	109	Received line signal detector				X			
21	CG	110	Signal quality detector				X			
23	CH	111	Data signal rate selector (DTE)					X		
23	CI	112	Data signal rate selector (DCE)				X			
24	DA	113	Transmitter element signal timing (DTE)							X
15	DB	114	Transmitter element signal timing (DCE)						X	
17	DD	115	Receiver signal element timing (DCE)						X	
14	SBA	116	Secondary transmitted data			X				
16	SBB	117	Secondary received data		X					
19	SCA	120	Secondary request to send					X		
13	SCB	121	Secondary clear to send				X			
12	SCF	122	Secondary received line signal detector				X			

6.2.4 Software Considerations

Most peripheral hardware manufacturers provide software that permits the user to customize an interconnected system to meet the required task. The software, in effect, writes software. The user enters the codes needed by each peripheral, selects the functions to be performed, and tells the computer where to store the resulting data. The program compiles the final software after the user answers key configuration questions. Automatic generation of programming greatly reduces the need to have experienced programmers on staff. Once installed in the computer, programming is often as simple as fitting graphic symbols together on the screen.

6.2.5 Communicating via RS-232-C

As discussed in previous sections, many variations of RS-232 are currently in use. Table 6.5 lists the interchange circuits of the RS-232-C format by category. A subset of RS-232-C has found wide acceptance in the majority of applications. Common pin and signal assignments on pins 1 through 8, 20, and 22 are found in most systems. The RS-232-C standard allows signals to be redefined for specific applications, providing maximum flexibility. This feature also, however, makes system interconnection difficult. Most problems are encountered during setup of a new system, or expansion of an existing system.

A null-modem adapter is commonly used to permit computer-to-computer communications. Two possible null-modem configurations are shown in Figure 6.5.

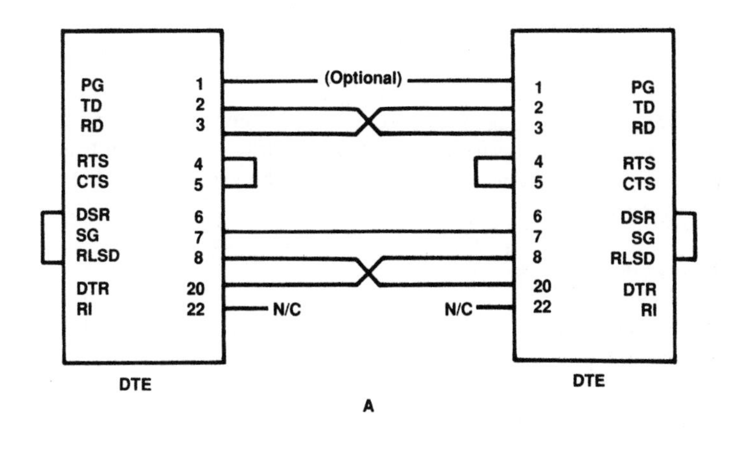

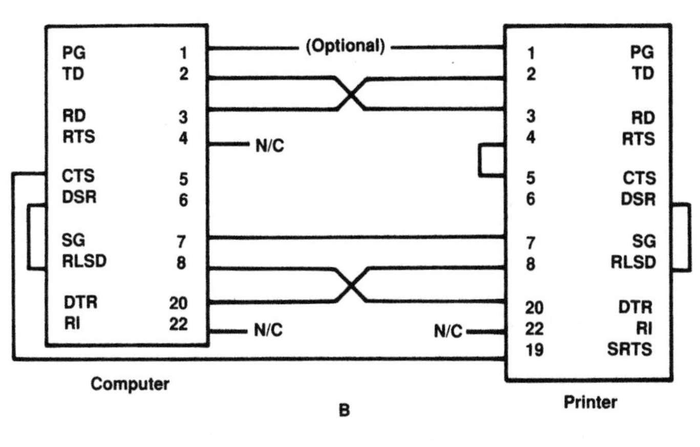

Figure 6.5 Null-modem configurations: (a) basic interface; (b) hardware handshaking interface; (c) illustration of typical null-modem module.

6.3 Local Area Network Operation[1]

The *open system interconnections* (OSI) model is the most broadly accepted explanation of LAN transmissions in an open system. The reference model was developed by the International Organization for Standardization (ISO) to define a framework for computer communication. The OSI model divides the process of data transmission into the following steps:

- Physical layer
- Data-link layer
- Network layer
- Transport layer
- Session layer
- Presentation layer
- Application layer

An overview of the OSI model is illustrated in Figure 6.6.

6.3.1 Physical Layer

Layer 1 of the OSI model is responsible for carrying an electrical current through the computer hardware to perform an exchange of information. The physical layer is defined by the following parameters:

- Bit transmission rate.
- Type of transmission medium (twisted-pair, coaxial cable, or fiber-optic cable).
- Electrical specifications, including voltage- or current-based, and balanced or unbalanced.
- Type of connectors used (normally RJ-45 or DB-9).

Many different implementations exist at the physical layer.

6.3.1.A *Installation Considerations*

Layer 1 can exhibit error messages as a result of overusage. For example, if a file server is being burdened with requests from workstations, the results may show up in error statistics that reflect the server's inability to handle all incoming requests. An overabundance of *response timeouts* may also be noted in this situation. A response timeout (in this context) is a message sent back to the workstation stating that the waiting period allotted for a response from the file server has passed without action from the server.

Error messages of this sort, which can be gathered by any number of commercially available software diagnostic utilities, can indicate an overburdened file server or a hardware flaw within the system. Intermittent response timeout errors can be caused

---

1 Portions of this section were adapted from: Michael W. Dahlgren, "Servicing Local Area Networks," *Broadcast Engineering* magazine, Intertec Publishing, Overland Park, KS, November 1989.

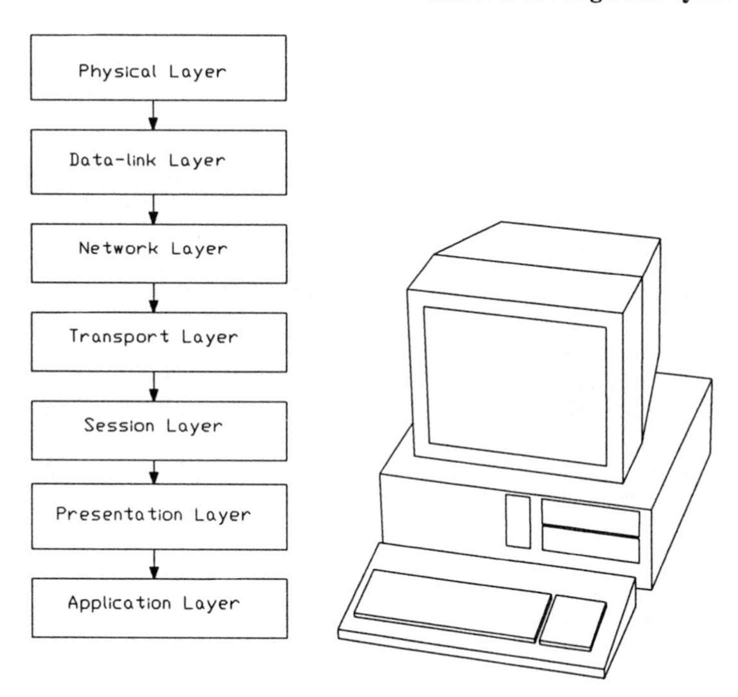

Figure 6.6 The OSI LAN model.

by a corrupted *network interface card* (NIC) in the server. A steady flow of timeout errors throughout all nodes on the network may indicate the need for another server or bridge. Hardware problems are among the easiest to locate. In simple configurations, where something has suddenly gone wrong, the physical and data-link layers are usually the first suspects.

6.3.2 Data-Link Layer

Layer 2 of the OSI model, the data-link layer, describes hardware that enables data transmission (NICs and cabling systems). This layer integrates data packets into messages for transmission and checks them for integrity. Sometimes layer 2 will also send an "arrived safely" or "did not arrive correctly" message back to the transport layer (layer 4), which monitors this communications layer. The data-link layer must define the frame (or package) of bits that is transmitted down the network cable. Incorporated within the frame are several important fields:

- Addresses of source and destination workstations.
- Data to be transmitted between workstations.
- Error control information, such as a *cyclic redundancy check* (CRC), which assures the integrity of the data.

The data-link layer must also define the method by which the network cable is accessed, because only one workstation may transmit at a time on a baseband LAN. The two predominant schemes are:

- *Token passing*, used with the *ARCnet* and token-ring networks.
- *Carrier sense multiple access with collision detection* (CSMA/CD), used with *Ethernet* and *starLAN* networks.

At the data-link layer, the true identity of the LAN begins to emerge.

6.3.2.A *Installation Considerations*

Because most functions of the data-link layer (in a PC-based system) take place in integrated circuits on NICs, software analysis is generally not required in the event of an installation problem. As mentioned previously, when something happens on the network, the data-link layer is among the first to be suspect. Because of the complexities of linking multiple topologies, cabling systems, and operating systems, the following failure modes may be experienced:

- RF disturbance. Transmitters, ac power controllers, and other computers can all generate energy that may interfere with data transmitted on the cable. RF interference (RFI) is usually the single biggest problem in a broadband network. This problem can manifest itself through excessive checksum errors and/or garbled data.
- Excessive cable run. Problems related to the data-link layer may result from long cable runs. Ethernet runs can stretch to 1,000 ft, depending on the cable. A typical token-ring system can stretch 600 ft, with the same qualification. The need for additional distance can be accommodated by placing a bridge, gateway, active hub, equalizer, or amplifier on the line.

The data-link layer usually includes some type of routing hardware with one or more of the following:

- Active hub
- Passive hub
- Multiple access units (for token-ring, starLAN, and ARCnet networks

6.3.3 Network Layer

Layer 3 of the OSI model guarantees the delivery of transmissions as requested by the upper layers of the OSI. The network layer establishes the physical path between the two communicating endpoints through the *communications subnet*, the common name for the physical, data-link, and network layers taken collectively. As such, layer 3 functions (routing, switching, and network congestion control) are critical. From the viewpoint of a single LAN, the network layer is not required. Only one route—the cable—exists. Internetwork connections are a different story, however, because multiple routes are possible. The *internet protocol* (IP) and *internet packet exchange* (IPX) are two examples of layer 3 protocols.

6.3.3.A *Installation Considerations*

The network layer confirms that signals get to their designated targets, and then translates logical addresses into physical addresses. The physical address determines where the incoming transmission is stored. Lost data errors can usually be traced back to the network layer, in most cases incriminating the network operating system. The network layer is also responsible for statistical tracking and communications with other environments, including gateways. Layer 3 decides which route is the best to take, given the needs of the transmission. If *router tables* are being corrupted or excessive time is required to route from one network to another, an operating system error on the network layer may be involved.

6.3.4 Transport Layer

Layer 4, the transport layer, acts as an interface between the bottom three and the upper three layers, ensuring that the proper connections are maintained. It does the same work as the network layer, only on a local level. The network operating system driver performs transport layer tasks.

6.3.4.A *Installation Considerations*

Connection flaws between computers on a network can sometimes be attributed to the shell driver. The transport layer may be able to save transmissions that were en route in the case of a system crash, or reroute a transmission to its destination in case of a primary route failure. The transport layer also monitors transmissions, checking to make sure that packets arriving at the destination node are consistent with the *build specifications* given to the sending node in layer 2. The data-link layer in the sending node builds a series of packets according to specifications sent down from higher levels, then transmits the packets to a *destination node*. The transport layer monitors these packets to ensure that they arrive according to specifications indicated in the original build order. If they do not, the transport layer calls for a retransmission. Some operating systems refer to this technique as a *sequenced packet exchange* (SPX) transmission, meaning that the operating system guarantees delivery of the packet.

6.3.5 Session Layer

Layer 5 is responsible for turning communications on and off between communicating parties. Unlike other levels, the session layer can receive instructions from the application layer through the network basic input/output operation system (netBIOS), skipping the layer directly above it. The netBIOS protocol allows applications to "talk" across the network. The session layer establishes the session, or logical connection, between communicating host processors. Name-to-address translation is another important function; most communicating processors are known by a common name, rather than a numerical address.

6.3.5.A *Installation Considerations*

Multi-vendor problems can often arise in the session layer. Failures relating to gateway access usually fall into layer 5 for the OSI model, and are often related to compatibility issues.

6.3.6 Presentation Layer

Layer 6 translates application layer commands into syntax that is understood throughout the network. It also translates incoming transmissions for layer 7. The presentation layer masks other devices and software functions, allowing a workstation to emulate a 3270 terminal through an emulation card and software. Reverse video, blinking cursors, and graphics also fall into the domain of the presentation layer. Layer 6 software controls printers and plotters, and may handle encryption and special file formatting. Data compression, encryption, and ASCII translations are examples of presentation layer functions.

6.3.6.A *Installation Considerations*

Failures in the presentation layer are often the result of products that are not compatible with the operating system, an interface card, a resident protocol, or another application. *Terminate-and-stay-resident* (TSR) programs are particularly troublesome.

6.3.7 Application Layer

At the top of the seven-layer stack is the application layer. It is responsible for providing protocols that facilitate user applications. Print spooling, file sharing, and E-mail are components of the application layer, which translates local application requests into network application requests. Layer 7 provides the first layer of communications into other open systems on the network.

6.3.7.A *Installation Considerations*

Failures at the application layer usually center on software quality and compatibility issues. The program for a complex network may include latent faults that will manifest only when a specific set of conditions are present. The compatibility of the network software with other programs, particularly TSR utilities, is another source of potential complications.

6.3.8 Ethernet

Ethernet is a LAN standard that breaks the data stream into packets for transmission. An address header is added to each packet to identify its destination and source. So-called *thick Ethernet* uses large-diameter coaxial cable as the transmission medium. The cable is tapped by inserting a bus transceiver onto the cable, which penetrates the cable's shield and makes contact with the center conductor and the shield. A cable connected to the transceiver has a 15-pin D-type connector on the other end for connection to the computer or peripheral device that is to be tied to the network.

Thin Ethernet uses standard-diameter 50 Ω coaxial cable as the transmission medium. A BNC connector is used at the computer. A BNC T connector is installed at this connector. The coaxial cable is looped between the devices on the network. Another distribution method uses a *hub box* (a multi-port null-modem), which acts as a junction box through which several devices can be interconnected via their 15 conductor cables.

6.4 Transmission System Options

A variety of options beyond the traditional local serial interface are available for linking intelligent devices. The evolution of *wide area network* (WAN) technology has permitted efficient two-way transmission of data between distant computer systems. High-speed facilities are now cost-effective and available from the telephone company (telco) central office to the customer premises. Private communications companies also provide interconnection services.

LANs have proliferated and are being integrated with WANs through bridges and gateways. Interconnections via fiber-optic cable are common. Further extensions of the basic LAN include the following:

- *Campus area network* (CAN)—designed for communications within an industrial or educational campus.
- *Metropolitan area network* (MAN)—designed for communications among different facilities within a certain metropolitan area. MANs generally operate over common-carrier-owned switched networks installed in and over public rights of way.
- *Regional area network* (RAN)—interconnecting MANs within a unified geographicaal area, generally installed and owned by interexchange carriers (IECs).
- *Wide area network* (WAN)—communications systems operating over large geographic areas. Common carrier networks interconnect MANs and RANs within a contiguous land mass, generally within a country's political boundaries.
- *Global area network* (GAN)—networks interconnecting WANs, both across national borders and ocean floors, including between continents.

These network systems can carry a wide variety of multiplexed analog and/or digital signal transmissions on a single piece of coax or fiber. (Fiber-optic technology is discussed in Chapter 8.)

6.4.1 System Design Alternatives

The signal form at the input and/or output interface of a large cable or fiber system may be either analog or digital, and the number of independent electrical signals transmitted may be one or many. Independent electrical signals may be combined into one signal for optical transmission by virtually unlimited combinations of electrical analog frequency division multiplexing (using analog AM and/or FM carriers) and digital bit stream multiplexing. *Frequency division multiplexing* involves the integration of two or more discrete signals into one complex electrical signal.

With the current availability of fiber-optic transmission lines, fiber interconnection of data networks is the preferred route for new systems. Three primary multiplexing schemes are used for fiber transmission:

- *Frequency division multiplexing* (FDM)
- *Time division multiplexing* (TDM)
- *Wave(length) division multiplexing* (WDM)

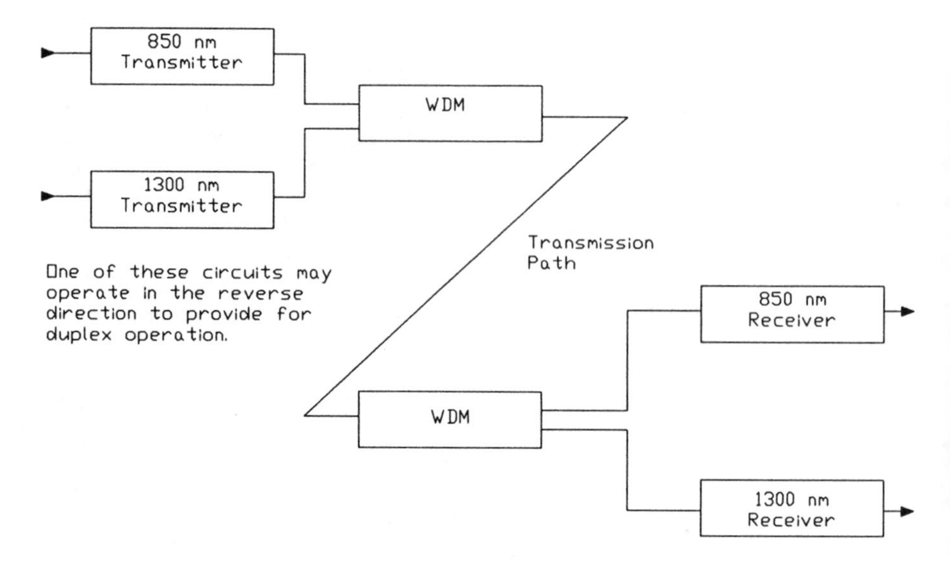

Figure 6.7 Basic operation of a wave division multiplexer. This type of passive assembly is created by fusing optical-fiber pigtails.

6.4.1.A *Frequency Division Multiplexing*

The FDM technique of summing multiple AM or FM carriers is widely used in coaxial cable distribution. Unfortunately, nonlinearity of optical devices operated in the intensity-modulation mode results in substantial—and often unacceptable—noise and intermodulation distortion in the delivered signal channels. Wide and selective spacing of carriers ameliorates this problem to some degree.

6.4.1.B *Time Division Multiplexing*

TDM involves sampling the input signals at a high rate, converting the samples to high-speed digital codes, and interleaving the codes into pre-determined time slots. The principles of digital TDM are straightforward. Specific-length bit groups in a high-speed digital bit stream are repetitively allocated to carry the digital representations of individual analog signals and/or the outputs of separate digital devices.

6.4.1.C *Wave(length) Division Multiplexing*

This multiplexing technique, illustrated in Figure 6.7, reduces the number of optical fibers required to meet a specific transmission requirement. Two or more complete and independent fiber-transmission systems operating at different optical wavelengths can be transported over a single fiber by combining them in a passive optical multiplexer. This device is an assembly in which pigtails from multiple optical transmitters are fused together and spliced into the transporting fiber. Demultiplexing the optical signals at the receiver end of the circuit is accomplished in an opposite-oriented passive optical multiplexer. The pigtails are coupled into photodetectors through wavelength-selective optical filters.

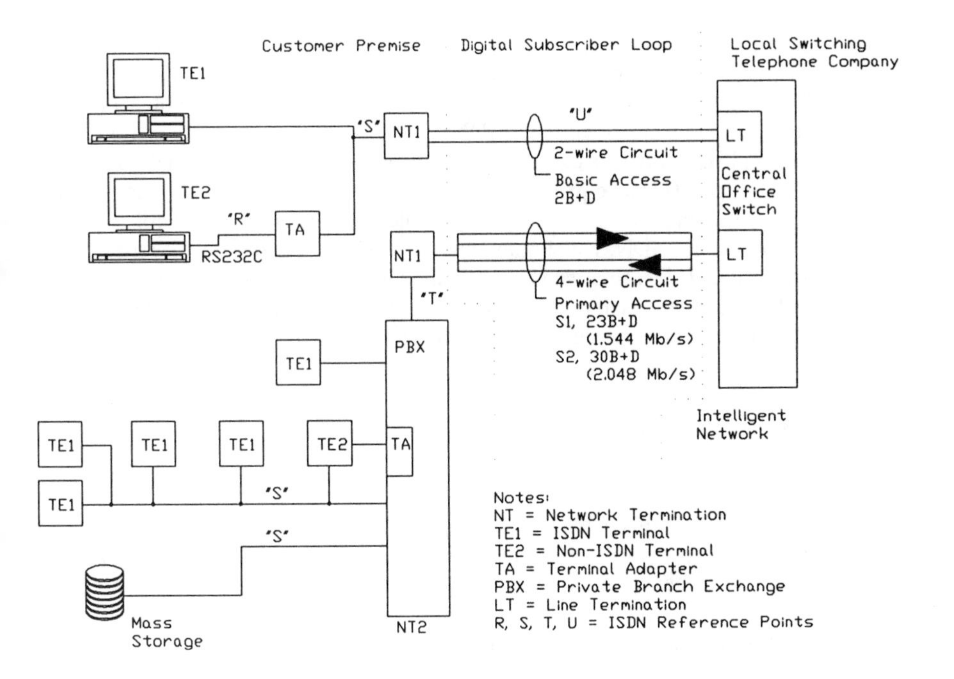

Figure 6.8 Basic operating structure of an integrated services digital network.

6.4.2 Integrated Services Digital Network

An integrated services digital network (ISDN) operates under the basic principle of out-of-band signaling. Signaling is carried on one channel, and the actual data is carrier on two or more separate channels, often referred to as *clear channels*. The *demand* (D) channel is used to establish the necessary connections on the *bearer* (B) channels, where data, voice, or video images may be exchanged. Most of today's ISDN systems can be categorized into one of two types:

- *Basic rate ISDN*, which consists of two B channels and one D channel on a single physical interface.
- *Primary rate ISDN*, which is defined as 23 B channels and one D channel, in North America and Japan, and 30 B channels and one D channel in Europe.

Figure 6.8 illustrates the basic principles of an ISDN. Note that there are several *reference points* for ISDN that denote the change from one functional block to another. For example, *network termination* (NT) and *terminal endpoint* (TE) represent demarcation points that may or may not constitute a physical interface that is accessible for connection.

6.4.2.A *Performance Measurement*

Most performance measurement and testing of an ISDN system are performed on the D channel, because it is where information is transmitted to set up and disconnect the traffic. The B channels are clear 64 kb/s links that transmit the voice or data using methods and protocols not specific to ISDN.

Primary rate ISDN traffic is usually carried over lines that were originally installed to carry T1/E1 data. Because primary rate data runs at the same data rates used by T1/E1 circuits, the threat of media (physical cable system) problems is not nearly as great as for basic rate ISDN, which may operate over low-bandwidth circuits.

Measurement of ISDN lines can be performed with a media tester common to any cabling technology. Excessive signal loss is a point of particular concern. Other potential problems include excessive cable runs and improper cable termination.

6.5 Network Installation and Service

Fiber optics is the medium of choice for computer communications networks. Fiber systems offer the numerous benefits, including immunity to EMI, small size, and zero capacitive loading. The cost of fiber interface hardware, however, is currently high, relative to copper. Fiber installation posses several distinct challenges, including:

- Fiber is difficult to splice, requiring the use of special equipment that aligns the glass filament and ensures that the ends of the fiber meet perfectly to eliminate deflection.
- Network drops require an optic splitter. A drop cable cannot be tapped-in, as with coax.
- Equipment standards relating to fiber have yet to be formally resolved by industry.

6.5.1 Acceptance Testing

Acceptance testing of equipment during the installation process is basic to establishing any type of network. Testing involves verifying the performance of network components in the operating environment according to user requirements. Acceptance testing may occur during installation, while troubleshooting, or after a repair or modification. Successful acceptance testing requires test equipment that can simulate the operating environment. For terminal testing, the network must be simulated.

Information generated on one system should be capable of being processed or displayed on another, even if manufactured by different companies. The process of ensuring compatibility among systems is referred to as *interoperability testing*. Interoperability testing during product development increases the likelihood of interoperability with complementary equipment but does not guarantee it; there are simply too many devices and systems to test. Also, the standards may change. Field interoperability testing should occur during network installation, while troubleshooting, and in verifying performance after repair or modification.

6.5.2 Test Instruments

Data communication test instruments can be classified into four general categories:

- Interface tester—a powered or nonpowered breakout-box device.
- Equipment tester—an instrument that monitors the status of the physical interface between two network components, such as a terminal and a modem, or a PBX and a channel service unit (CSU). Equipment testers interact with a specific network component, such as a printer, terminal, or front-end processor, to check functionality and performance. Such instruments can range from simple testers with pre-programmed messages to complex PC-based emulators.
- Link tester—an instrument that checks the transmission integrity of the channel between two network components. *Bit error rate test* (BERT) sets are used for checking digital channels. *Transmission impairment measurement* (TIM) sets are used for checking analog lines that carry digital data.
- Protocol analyzer—an instrument that monitors data on the network. A protocol analyzer also can simulate network components when substitution troubleshooting is required.

Instrument criteria for interoperability testing are similar to those for acceptance testing. The gear's ability to capture faithfully every detail of the exchange of information between interoperating systems is a primary consideration. The ability to simulate the target environment (terminal equipment, PBX, or switch) to the component being tested is another key capability.

Physical, functional, electrical, and procedural problems related to the interface between two devices constitute a large percentage of OSI network installation problems.

6.5.2.A *Integrated Services Digital Network (ISDN)*

Systems built around the ISDN protocol present special instrumentation requirements. With ISDN, two or more physical channels are time-division multiplexed onto a single interface. Therefore, the instrument must be capable of extracting (demultiplexing) the channel to be measured. It is also necessary to compare the demand channel (D channel) with one of the bearer channels (B channel). Measurement of *pulse code modulation* (PCM) interfaces involve a similar consideration—the ability to extract data from a multiplexed data stream. For applications requiring simulation, this function is commonly referred to as "drop and insert."

6.5.2.B *PC-Based Test Instruments*

An add-on board and the appropriate software can permit an off-the-shelf computer to perform sophisticated protocol analysis. PC-based instruments provide a number of features:

- Trigger flexibility
- Mixed-mode (transmit/receive) monitoring
- Data communication equipment (DCE) and data terminal equipment (DTE) emulation
- BERT functions
- Clock signal outputs
- User-defined messages of up to 512 bytes

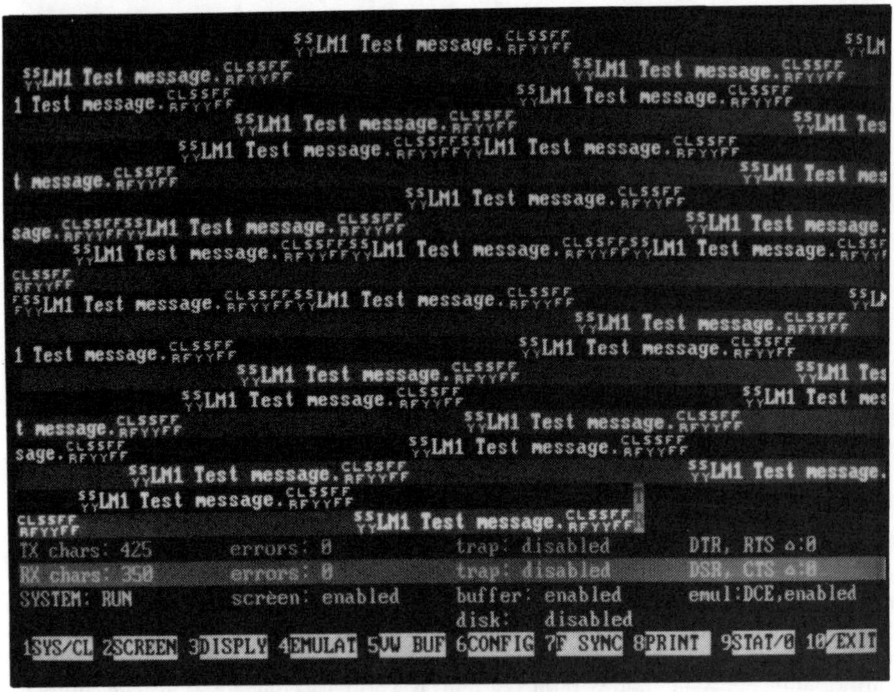

Figure 6.9 Display page from a PC-based protocol analyzer.

- Remote operation capability with appropriate control software
- Time-stamps for all recorded events

PC-based systems can display data in real time or store it directly on disk for later examination. Graphics capabilities provide a more understandable display of the data present on the network. Figure 6.9 shows the display screen of a protocol analyzer operating on a PC platform.

6.5.3 Isolating System Faults

Design and installation of a LAN can be a formidable challenge, particularly in a multi-vendor environment. The quality of a network service is most often measured in terms of network availability.

The most common problems that occur on LANs are faults in the cabling system. An efficient fault-management strategy begins with effective tools and techniques for locating problems within the physical LAN cabling system. Faults in the LAN transmission media may result from one of several sources:

- Cable and/or connector open-circuit
- Cable and/or connector short-circuit
- Improper or mismatched cable types

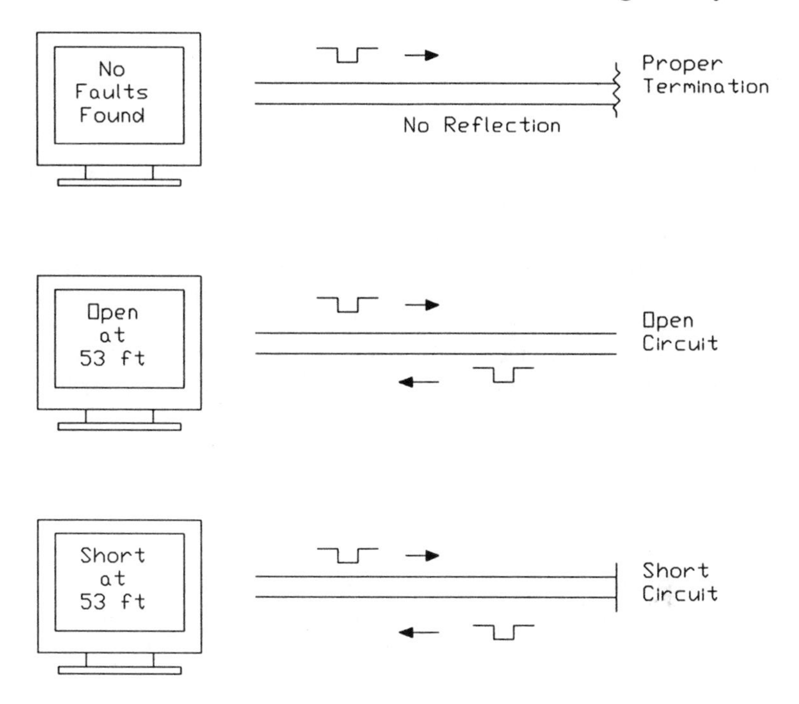

Figure 6.10 Measurement of data cables using a time domain reflectometer.

- Improper termination devices
- Electrical noise in the environment
- Initial installation wiring errors, resulting in excessive attenuation and cross-talk

The most obvious physical fault is an open- or short-circuit in the LAN cable or conductor. Open-circuits can occur when physical stress separates conductors and breaks the electrical connection. They also can result from inadvertent disconnection of a cable. Short-circuits are typically the result of physical stress, faulty insulation, or improperly installed connectors.

The presence of a short- or open-circuit normally can be detected easily using a dc continuity check. However, it may be difficult to pinpoint the location of the failure. A time domain reflectometer (TDR) can be used to identify the exact location of a cable fault. The TDR transmits a pulse and measures the relative time of arrival and polarity of any reflected pulse. Figure 6.10 illustrates the concept. If there are no discontinuities on the LAN and the lines are properly terminated, there will be no significant reflected pulse. The presence of a short- or open-circuit, however, will cause a reflection, which will be detected by the instrument.

Advanced TDR instruments automatically interpret the characteristics of the reflected pulse and report the nature and location of the fault. In order to interpret correctly the reflected pulse on a conventional TDR, the velocity of propagation of the cable type being measured must be known. More sophisticated instruments provide this data on demand and automatically calculate the location of the fault. Discontinuities other than complete open- or short-circuits can also be detected and located with the help of a LAN *cable scanner* TDR.

6.5.3.A *Crosstalk*

Data traffic on an active twisted-pair LAN can leak or radiate into another active pair when the two lines are routed close together. This problem often occurs among pairs carried within the same cable or bundle. Crosstalk will result in an interfering signal being coupled onto one or more of the other twisted pairs (in a twisted-pair system). Depending on the degree of coupling, crosstalk can cause significant data errors.

Crosstalk effects are usually the most troublesome at the transceiver, where the transmitted signal is the strongest and the received signal is the weakest. For this reason, crosstalk is measured at the *near end*. This measurement is referred to as *near-end crosstalk* (NEXT).

A common cause of excessive crosstalk is the result of a *split pair* in a multi-pair cable. If a failure in one wire of a pair occurs, the cable installer or technician will sometimes attempt to substitute another wire in the same bundle to complete the circuit. A split pair results when the differential signal travels on two wires that are each twisted with other wires, but are not paired with each other.

Failure to maintain twisting uniformity of the pairs throughout terminal blocks in the system is another common cause of crosstalk. Even a short length of untwisted wire can result in significant crosstalk, particularly at high data rates.

Crosstalk performance is especially important when qualifying existing twisted-pair wiring systems for carrying LAN traffic. Wiring systems that are acceptable for voice transmission may be unacceptable for high-speed data.

6.6 Selecting Cable for Digital Signals

Cable for the transmission of digital signals is selected on the basis of its electrical performance: the ability to transmit the required number of pulses at a specified bit rate over a specified distance, and its conformance to appropriate industry or government standards. A wide variety of data cables are available from manufacturers. Figure 6.11 illustrates some of the more common types. The type of cable chosen for an application is determined by the following:

- Type of network involved. Different network designs require different types of cable.
- Distance to be traveled. Long runs require low-loss cable.
- Physical environment. Local and national safety codes require specific types of cable for certain indoor applications. Outdoor applications require a cable suitable for burial or exposure to the elements.
- Termination required. The choice of cable type may be limited by the required connector termination on one or both ends.

6.6.1 Data Patch Panel

The growth of LANs has led to the development of a variety of interconnection racks and patch panels. Figure 6.12 shows two common types. Select data patch panels that offer many cycles of repeated insertion and removal. If a patch panel is designed using

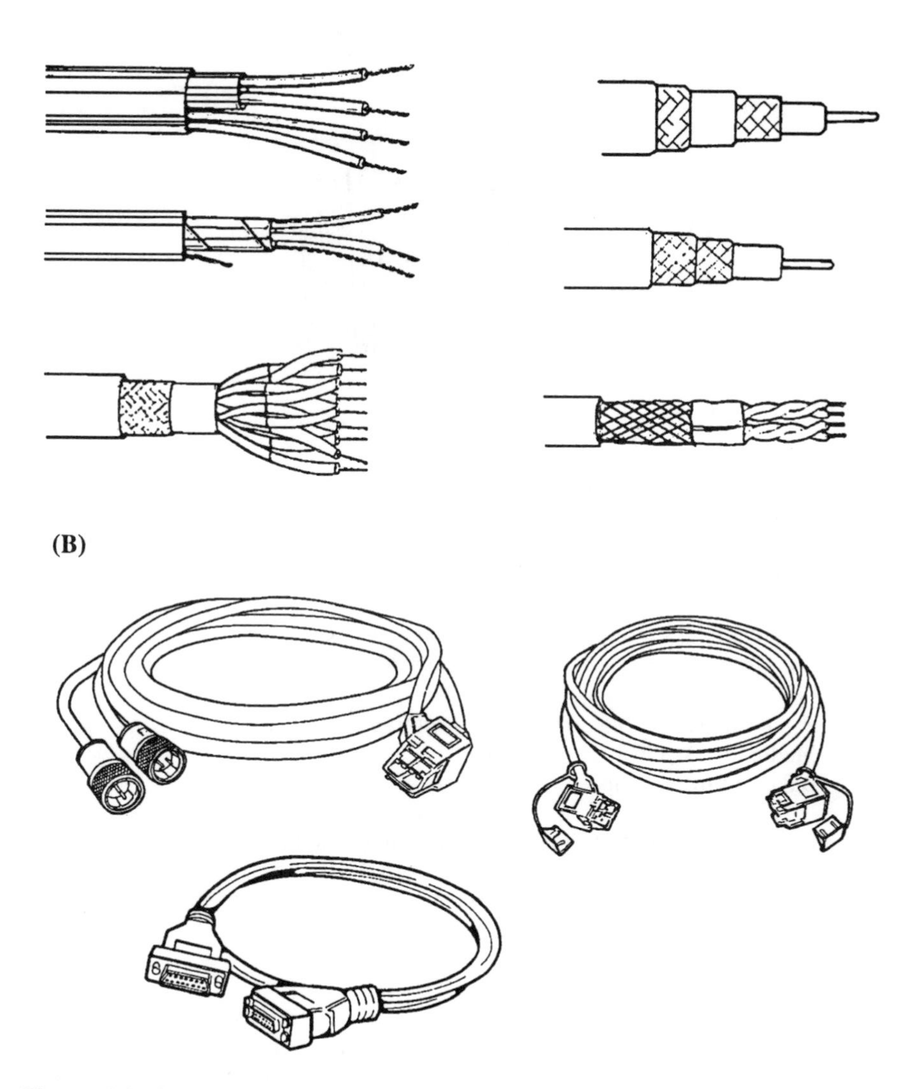

Figure 6.11 Common types of data cable: (a) shielded pair, multi-pair shielded, and coax; (b) data cable with various terminations.

---

normal D-subminiature-type connectors, premature problems may result from contact wear. Instead, use components specifically designed for network interconnection. Such components include the following:

- Twisted-pair network patch panels
- Coax-based network patch panels
- Fiber-based network patch panels

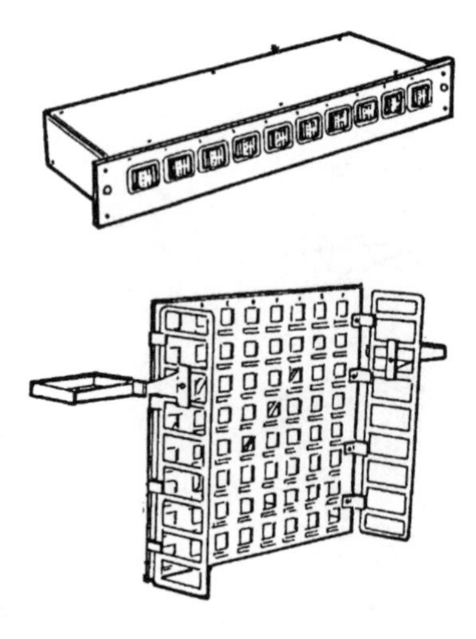

Figure 6.12 Data network patch-panel hardware.

- Modular feed-through (normalled) patch panels
- Pre-assembled patch cables of various lengths
- Pre-assembled "Y" patch cables
- Patch cables offering different connectors on each end
- Media filter cables
- Balanced-to-unbalanced (balun) cable assemblies

Connector termination options for patch hardware include the following:

- Insulation displacement (punch-block) for twisted-pair cable
- Screw terminal (for twisted-pair)
- BNC connectors for coax
- Fiber-termination hardware

Although the cost of pre-assembled network patch panels and patch cables is higher than purchasing the individual components and then assembling them, most system engineers should specify factory-assembled hardware. Reliability is greater with pre-assembled elements, and installation is considerably faster.

6.7 Bibliography

Ardire, Phil, "Bit Error Rate Testing," *Microservice Management* magazine, Intertec Publishing, Overland Park, KS, February 1990.

Benzel, Susan E., "Service and ISDN Testing: What You Already Know May Surprise You," *Microservice Management* magazine, Intertec Publishing, Overland Park, KS, November 1991.

Blog, Thomas, "Solving Data Comm Problems With a PC," *Microservice Management* magazine, Intertec Publishing, Overland Park, KS, July 1987.

Buckland, Phillip, "Solving Data Comm Problems," *Microservice Management* magazine, Intertec Publishing, Overland Park, KS, September 1988.

Carey, Gregory, "Automated Test Instruments," *Broadcast Engineering* magazine, Intertec Publishing, Overland Park, KS, November 1989.

Dahlgren, Michael W., "Servicing Local Area Networks," *Broadcast Engineering* magazine, Intertec Publishing, Overland Park, KS, November 1989.

Gorman, Ron, "Transmission Line Testing," *Microservice Management* magazine, Intertec Publishing, Overland Park, KS, November 1986.

International Organization for Standardization, "Information Processing Systems—Open Systems Interconnection—Basic Reference Model," ISO 7498, 1984.

Miller, Mark A., "Servicing Local Area Networks," *Microservice Management* magazine, Intertec Publishing, Overland Park, KS, February 1990.

Miller, Mark A.: *LAN Troubleshooting Handbook*, M&T Books, Redwood City, CA, 1990.

Pozzi, Michael A., "Learning Intelligent Troubleshooting," *Microservice Management* magazine, Intertec Publishing, Overland Park, KS, February 1992.

Whitaker, Jerry: *Maintaining Electronic Systems*, CRC Press, Boca Raton, FL, 1991.

Wilkin, Donald, "Trends in Data Communications and Test Equipment," *Microservice Management* magazine, Intertec Publishing, Overland Park, KS, December 1991.

CHAPTER 7

INTERCONNECTING RF SYSTEMS

7.1 Introduction[1]

The mechanical and electrical characteristics of the transmission line, waveguide, and associated hardware that carry power from a power source—usually a transmitter—to the load—usually an antenna—are critical to proper operation of any RF system. Mechanical considerations determine the ability of the components to withstand temperature extremes, lightning, rain, and wind. In other words, they determine the overall reliability of the system. A number of different types of hardware are available. The approach taken depends on the power level, frequency, length of the run from the transmitter to the antenna, and the installation method preferred by the system engineer.

7.2 Coaxial Transmission Line

Two types of coaxial transmission line are commonly used today: *rigid* line and corrugated (*semi-flexible*) line. Rigid coaxial cable is constructed of heavy-wall copper tubes with Teflon or ceramic spacers. (Teflon is a registered trademark of Du Pont.) Rigid line provides electrical performance approaching an ideal transmission line, including:

- High power-handling capability

1 Portions of this chapter were adapted from: Jerry Whitaker: *Radio Frequency Transmission Systems: Design and Operation*, McGraw-Hill, New York, 1991. Used with permission.

(A)

(B)

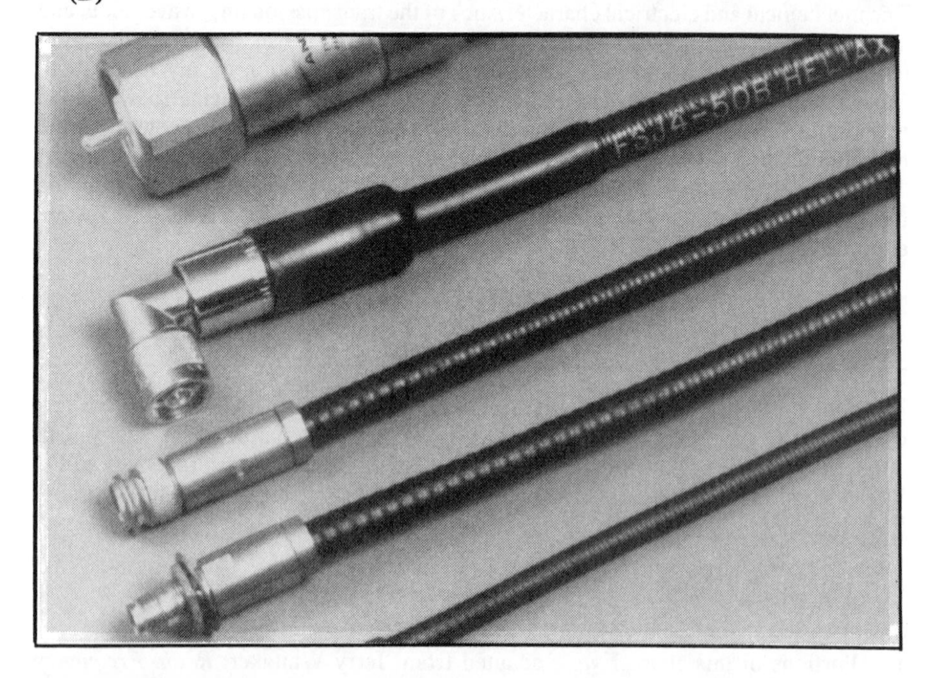

Figure 7.1 Semi-flexible coaxial cable: (a) a section of cable showing the basic construction; (b) cable with various terminations. (*Courtesy of Andrew Corp.*)

- Low loss
- Low VSWR (*voltage standing wave ratio*)

The primary alternative to rigid coax is semi-flexible transmission line made of corrugated outer and inner conductor tubes with a spiral polyethylene (or Teflon) insulator. The internal construction of a semi-flexible line is illustrated in Figure 7.1. Semi-flexible line has three primary benefits:

- It is manufactured in a continuous length, rather than the 20 ft sections typically used for rigid line.
- Because of the corrugated construction, the line may be shaped as required for routing from the transmitter to the antenna.
- The corrugated construction permits differential expansion of the outer and inner conductors.
- Each size of line has a minimum bending radius. For most installations, the nature of flexible corrugated line permits the use of a single piece of cable from the transmitter to the antenna, with no elbows or other transition elements. This speeds installation and provides for a more reliable system.

7.2.1 Velocity of Propagation

A signal traveling in free space is unimpeded; it has a free-space velocity equal to the speed of light. In a transmission line, capacitance and inductance slow the signal as it propagates along the line. The amount that the signal is slowed is represented as a percentage of the free-space velocity. This quantity is called the *relative velocity of propagation* and is described by the equation:

$$V_p = \frac{1}{\sqrt{L \times C}}$$

where:
L = inductance in H/ft
C = capacitance in F/ft

and

$$V_r = \frac{V_p}{C} \times 100\%$$

where:
V_p = velocity of propagation
C = 9.842×10^8 ft/s (free-space velocity)
V_r = velocity of propagation as a percentage of free-space velocity

7.2.1.A *Transverse Electromagnetic Mode*

The principal mode of propagation in a coaxial line is the *transverse electromagnetic mode* (TEM). This mode will not propagate in a waveguide. That is why coaxial lines can propagate a broad band of frequencies efficiently. The cut-off frequency for a coaxial transmission line is determined by the line dimensions. Above cut-off, modes other than TEM can exist, and the transmission properties are no longer defined. The cut-off frequency is equivalent to:

$$F_C = \frac{7.5 \times V_r}{D_i + D_o}$$

where:
F_C = cut-off frequency in GHz
V_r = velocity (percent)
D_i = inner diameter of outer conductor in inches
D_o = outer diameter of inner conductor in inches

At dc, current in a conductor flows with uniform density over the cross section of the conductor. At high frequencies, the current is displaced to the conductor surface. The effective cross section of the conductor decreases and the conductor resistance increases. For RF signals, current only flows in a thin "skin" of the conductor (the *skin effect*). (The skin effect is discussed in detail in Section 3.3.3.)

Center conductors are made from copper-clad aluminum or high-purity copper and can be solid, hollow tubular, or corrugated tubular. Solid-center conductors are found on semi-flexible cable with 1/2 in or smaller diameters. Tubular conductors are found in 7/8 in- or larger-diameter cables. Although the tubular center conductor is used primarily to maintain flexibility, it can also be used to pressurize an antenna through the feeder.

7.2.1.B *Dielectric*

Coaxial lines use two types of dielectric construction to isolate the inner conductor from the outer conductor. The first is air dielectric, with the inner conductor supported by a dielectric spacer and the remaining volume filled with air or nitrogen gas. The spacer, which may be constructed of spiral or discrete rings, is typically made of Teflon or polyethylene. Air-dielectric cable offers lower attenuation and higher average power ratings than foam-filled cable, but it requires pressurization to prevent moisture entry.

Foam-dielectric cables are ideal for use as feeders, with antennas that do not require pressurization. The center conductor is surrounded completely by foam-dielectric material, resulting in a high dielectric breakdown level. The dielectric materials are polyethylene-based formulations, which contain anti-oxidants to reduce dielectric deterioration at high temperatures.

7.2.2 Electrical Considerations

VSWR, attenuation, and power-handling capability are key electrical factors in the application of coaxial cable. High VSWR can cause power loss, voltage breakdown, and thermal degradation of the line. High attenuation means less power delivered to the antenna, higher power consumption at the transmitter, and increased heating of the transmission line.

VSWR is a common measure of the quality of a coaxial cable. High VSWR indicates nonuniformities in the cable that can be caused by variations in the dielectric core diameter, variations in the outer conductor, poor concentricity of the inner conductor, or a nonhomogeneous or periodic dielectric core. Although these may

Table 7.1 Representative specifications for various types of flexible air-dielectric coaxial cable.

Cable size (in)	Maximum frequency (MHz)	Veloci-ty (%)	Peak power (kW)	Average Power 1 MHz (kW)	Average Power 100 MHz (kW)	Attenuation* 1 MHz (dB)	Attenuation* 100 MHz (dB)
1 5/8	2.7	92.1	145	145	14.4	0.020	0.207
3	1.64	93.3	320	320	37	0.013	0.14
4	1.22	92	490	490	56	0.010	0.113
5	0.96	93.1	765	765	73	0.007	0.079

* Attenuation specified in dB/100 ft.

contribute only a small reflection, they can add up to a measurable VSWR at a particular frequency.

Rigid transmission line is typically available in a standard length of 20 ft, and in alternative lengths of 19.5 ft and 19.75 ft. The shorter lines are used to avoid VSWR buildup caused by discontinuities resulting from the physical spacing between line section joints. If the section length selected and the operating frequency have a half-wave correlation, the connector junction discontinuties will add. This effect is known as *flange build-up*. The result can be excessive VSWR. The critical frequency at which a half-wave relationship exists is given by:

$$F_{cr} = \frac{490.4 \times N}{L}$$

where:

F_{cr} = the critical frequency
N = any integer
L = transmission line length in feet

The critical frequency for a chosen line length should not fall closer than ±2 MHz of the pass band at the operating frequency.

Attenuation is related to the construction of the cable and varies with frequency, product dimensions, and dielectric constant. Larger-diameter cable exhibits lower attenuation than smaller-diameter cable of similar construction when operated at the same frequency. Therefore, larger-diameter cables should be used for long runs.

Air-dielectric coax exhibits less attenuation than comparable-size foam-dielectric cable. The attenuation characteristic of a given cable is also affected by standing waves present on the line, resulting from an impedance mismatch. Table 7.1 gives a representative sampling of semi-flexible coaxial cable specifications for a variety of line sizes.

7.2.3 Coaxial Cable Ratings

Selection of a type and size of transmission line is determined by a number of parameters, including power-handling capability, attenuation, and phase stability.

7.2.3.A *Power Rating*

Both *peak* and *average* power ratings are required to fully describe the capabilities of a transmission line. In most applications, the peak power rating limits the low frequency or pulse energy, and the average power rating limits high-frequency applications, as shown in Figure 7.2. Peak power ratings are usually stated for the following conditions:

- VSWR = 1.0:1
- Zero modulation
- One atmosphere absolute dry air pressure at sea level

The peak power rating of a selected cable must be greater than the following expression. It must also satisfy the average power-handling criteria:

$$P_{pk} > P_t \times (1 + M)^2 \times VSWR$$

where:
P_{pk} = cable peak power rating in kW
P_t = transmitter power in kW
M = amplitude modulation percentage expressed decimally (100 percent = 1.0)
$VSWR$ = voltage standing wave ratio

From this equation, notice that 100 percent amplitude modulation will increase the peak power in the transmission line by a factor of four. Further, the peak power in the transmission line increases directly with VSWR.

The peak power rating is limited by the voltage breakdown potential between the inner and outer conductors of the line. The breakdown point is independent of frequency. It varies, however, with the line pressure (for an air-dielectric cable) and the type of pressurizing gas.

The average power rating of a transmission line is limited by the safe long-term operating temperature of the inner conductor and the dielectric. Excessive temperatures on the inner conductor will cause the dielectric material to soften, leading to mechanical instability inside the line.

The primary purpose of pressurization of an air-dielectric cable is to prevent the ingress of moisture. Moisture, if allowed to accumulate in the line, can increase attenuation and reduce the breakdown voltage between the inner and outer conductors. Pressurization with high-density gases can be used to increase the average power and the peak power ratings of a transmission line. For a given line pressure, the increased power rating is more significant for peak power than for average power. High-density gases used for such applications include Freon 116 and sulfur hexafluoride. Figure 7.3 illustrates the effects of pressurization on cable power rating.

An adequate safety factor is necessary for peak and average power ratings. Most transmission lines are tested at two or more times their rated peak power before shipment to the customer. This safety factor is intended as a provision for transmitter transients, lightning-induced effects, and high-voltage excursions resulting from unforeseen operating conditions.

(A)

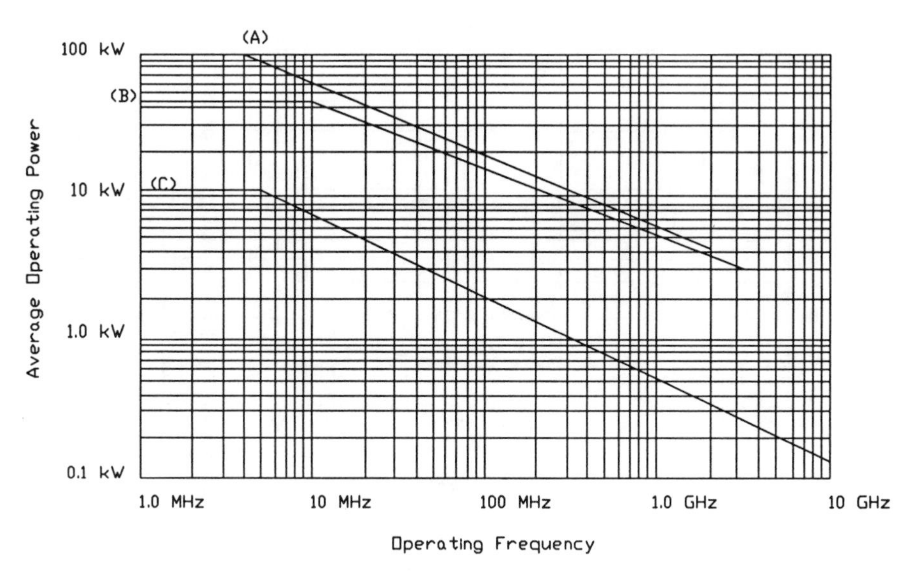

(B)

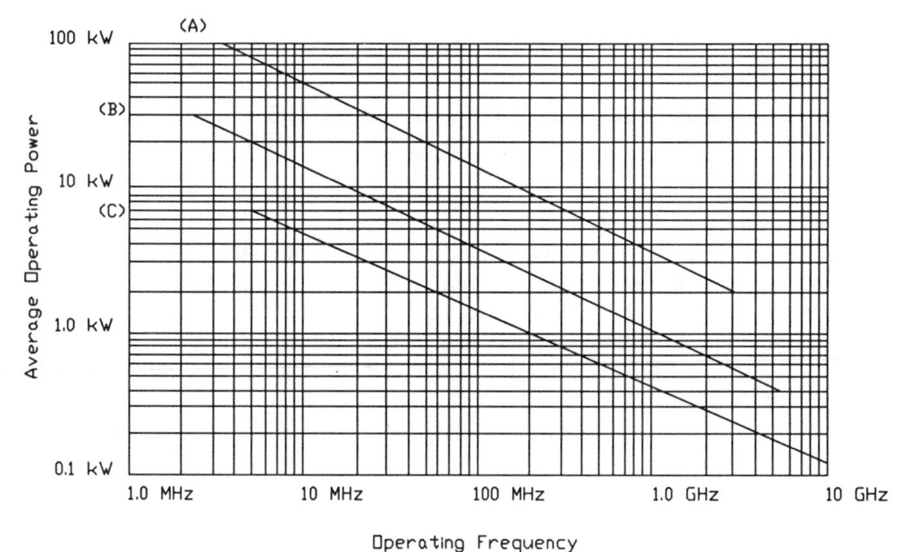

Figure 7.2 Power ratings data for a variety of coaxial transmission lines: (a) 50 Ω line; (b) 75 Ω line.

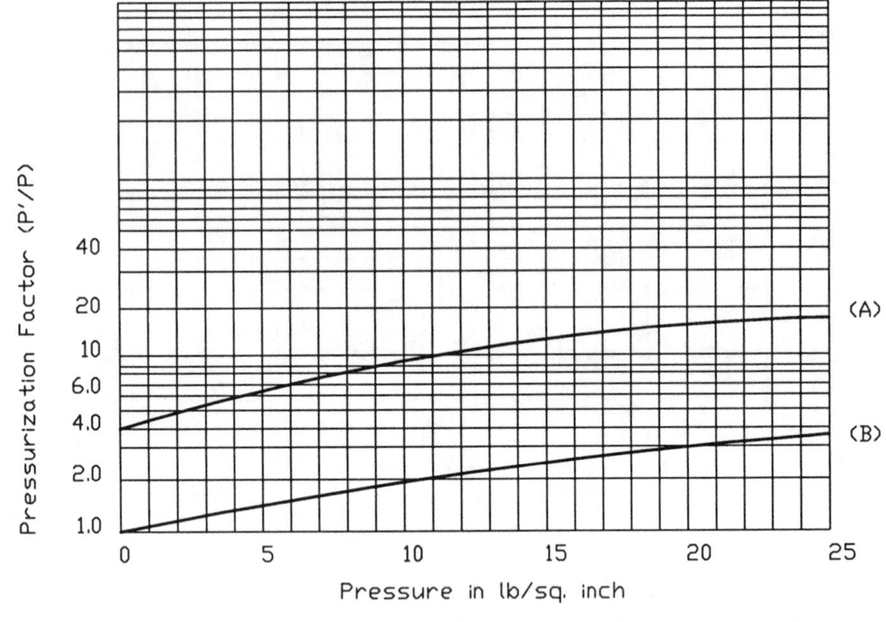

Figure 7.3 Effects of transmission line pressurization on peak power rating. Note that P′ = the rating of the line at the increased pressure, and P = the rating of the line at atmospheric pressure.

7.2.3.B *Connector Effects*

Foam-dielectric cables typically have a greater dielectric strength than air-dielectric cables of similar size. For this reason, foam cables should exhibit higher peak power ratings than air lines. Higher values, however, usually cannot be realized in practice, because the connectors commonly used for foam cables have air spaces at the cable/connector interface that limit the allowable RF voltage to "air cable" values.

The peak power handling capability of a transmission line is the smaller of the values for the cable and the connectors attached to it. Table 7.2 lists the peak power ratings of several common RF connectors at standard conditions (defined in the previous section).

7.2.3.C *Attenuation*

The attenuation characteristics of a transmission line vary as a function of the operating frequency and the size of the line. The relationships are shown in Figure 7.4.

The *efficiency* of a transmission line dictates how much power output by the transmitter actually reaches the antenna. Efficiency is determined by the length of the line and the attenuation per unit length.

Table 7.2 Electrical characteristics of common RF connectors

Connector type	DC test voltage (kW)	Peak power (kW)
SMA	1.0	1.2
BNC, TNC	1.5	2.8
N, UHF	2.0	4.9
GR	3.0	11
HN, 7/16	4.0	20
LC	5.0	31
7/8 EIA, F Flange	6.0	44
1 5/8 EIA	11.0	150
3 1/8 EIA	19.0	44

The attenuation of a coaxial transmission line is defined by the equation:

$$\alpha = 10 \text{ x Log } (P_1/P_2)$$

where:
α = attenuation in dB/100 m
P_1 = input power into a 100 m line terminated with the nominal value of its characteristic impedance
P_2 = power measured at the end of the line

Stated in terms of efficiency:

$$\text{Efficiency } (\%) = 100\ (P_o/P_i)$$

where:
P_i = power delivered to the input of the transmission line
P_o = power delivered to the antenna

The relationship between efficiency and loss in decibels (*insertion loss*) is illustrated in Figure 7.5.

Manufacturer-supplied attenuation curves are typically guaranteed to within ±5 percent. The values given are usually rated at 24°C (75°F) ambient temperature. Attenuation increases slightly with higher temperature or applied power. The effects of ambient temperature on attenuation are illustrated in Figure 7.6.

Loss in connectors is negligible, except for small (SMA and TNC) connectors at frequencies of several gigahertz and higher. Small connectors used at high frequencies typically add 0.1 dB of loss per connector.

When a transmission line is attached to a load, such as an antenna, the VSWR of the load increases the total transmission loss of the system. This effect is small under conditions of low VSWR. Figure 7.7 illustrates the dependence of these two elements.

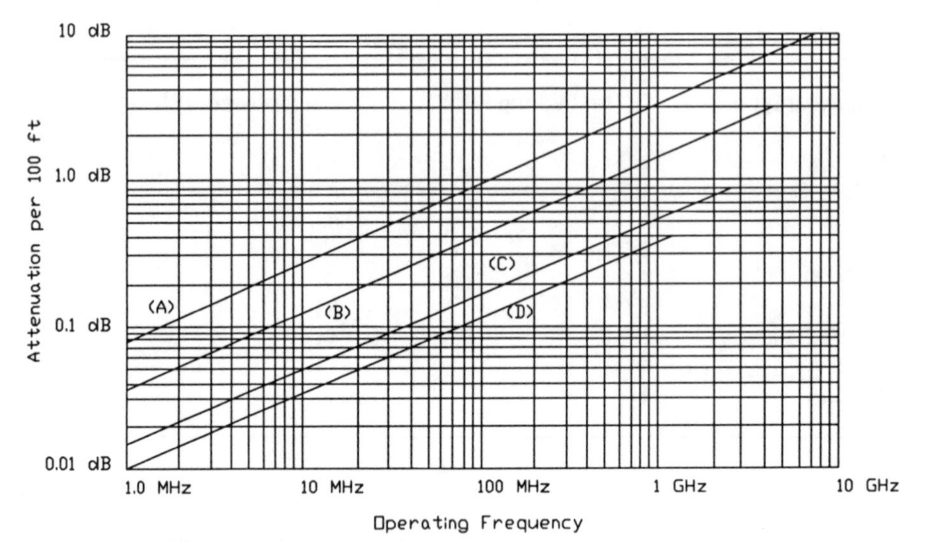

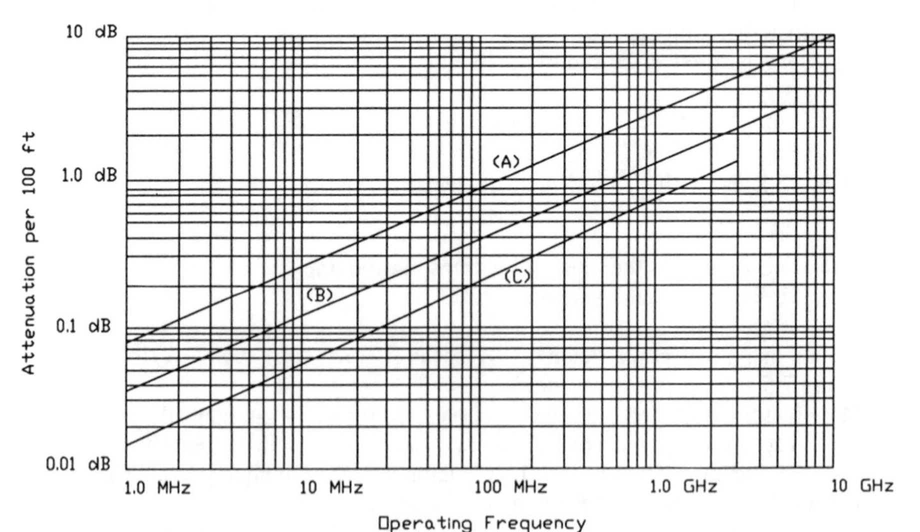

Figure 7.4 Attenuation characteristics for a selection of coaxial cables: (a) 50 Ω line; (b) 75 Ω line.

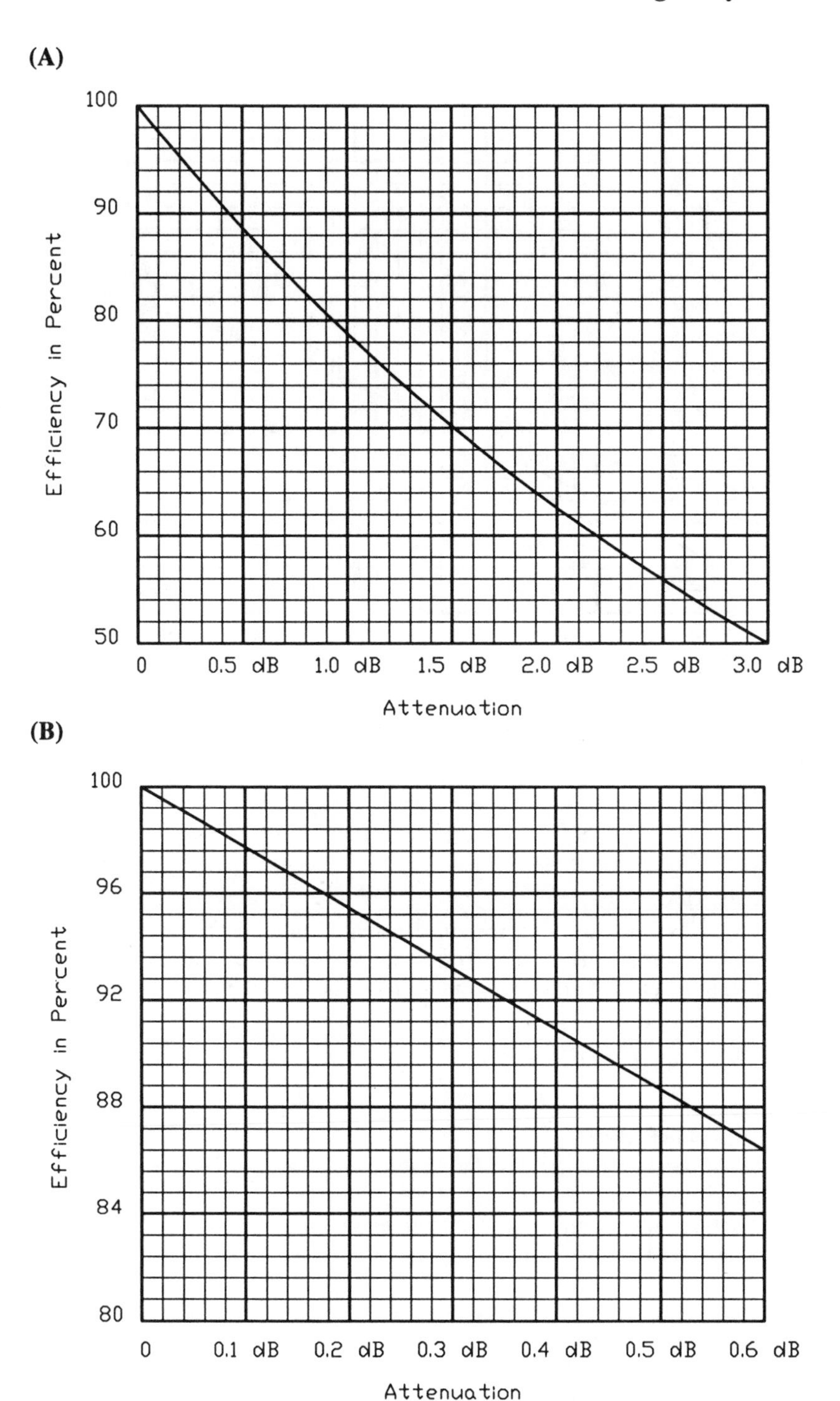

Figure 7.5 Conversion chart showing the relationship between decibel loss and efficiency of a transmission line: (a) high-loss line; (b) low-loss line.

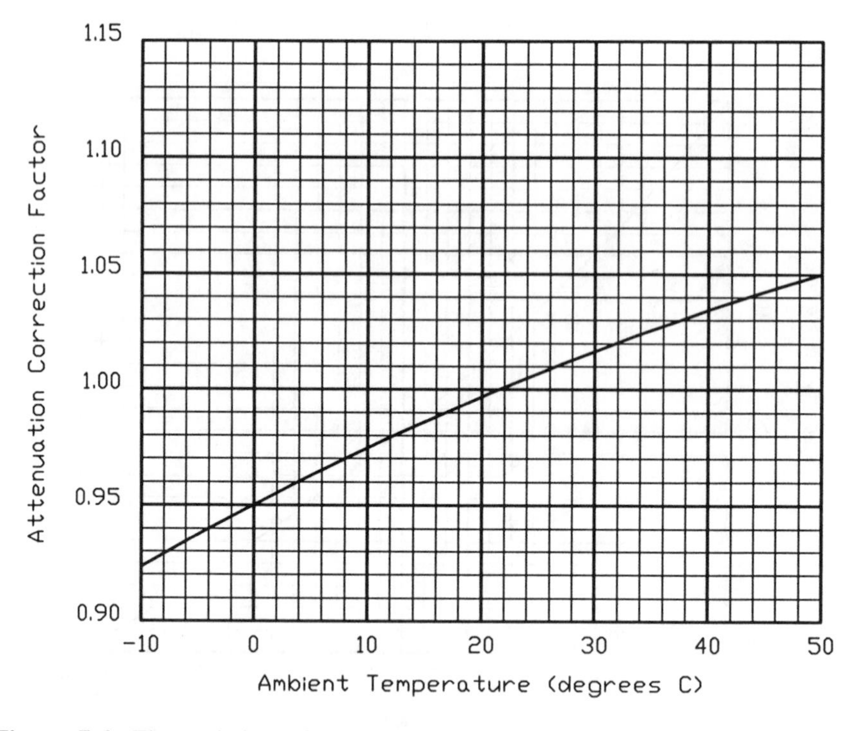

Figure 7.6 The variation of coaxial cable attenuation as a function of ambient temperature.

7.2.3.D *Phase Stability*

A coaxial cable expands as the temperature increases, causing the electrical length of the line to also increase. This factor results in phase changes that are a function of operating temperature. The phase relationship can be described by the equation:

$$\theta = 3.66 \times 10^{-7} \times P \times L \times T \times F$$

where:
θ = phase change in degrees
P = phase temperature coefficient of the cable
L = length of coax in feet
T = temperature range (minimum-to-maximum operating temperature)
F = frequency in MHz

Phase changes that are a function of temperature are important in systems that use multiple transmission lines, such as a directional array fed from a single phasing source. To maintain proper operating parameters, the phase changes of the cables must be minimized. Specially designed coaxial cables offer low phase-temperature characteristics. Two types of coax are commonly available:

- *Phase stabilized* cables, which have undergone extensive temperature cycling until such time as they exhibit their minimum phase temperature coefficient.

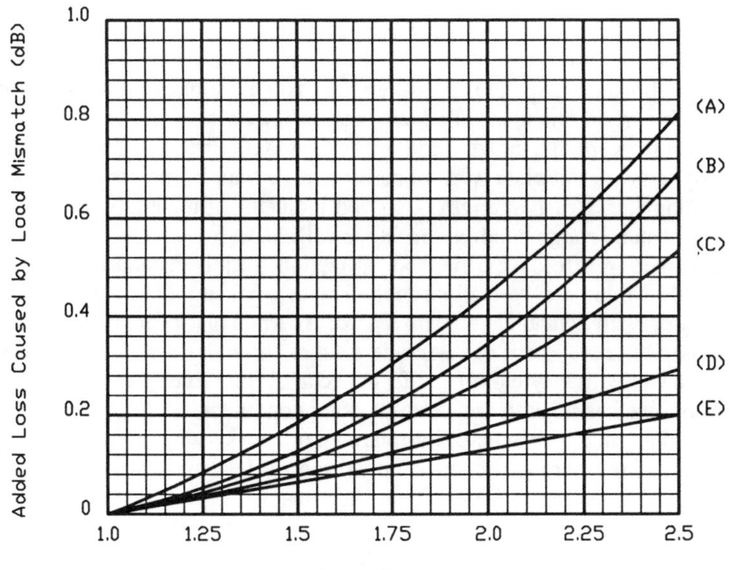

Figure 7.7 The effect of load VSWR on transmission line loss.

- *Phase compensated* cables, in which changes in the electrical length have been minimized through adjustment of the mechanical properties of the dielectric and inner/outer conductors.

7.2.4 Mechanical Parameters

Corrugated copper cables are designed to withstand bending with no change in properties. Low-density foam- and air-dielectric cables generally have a minimum bending radius of 10 times the cable diameter. So-called *super flexible* versions provide a much smaller bending radius.

Rigid transmission line will not tolerate bending. Instead, transition elements (elbows) of various sizes are used. Individual sections of rigid line are secured by multiple bolts around the circumference of a coupling flange.

When a large cable must be used to meet attenuation requirements, short lengths of a smaller cable (jumpers or *pigtails*) may be used on each end for installation ease in low-power systems. The tradeoff is slightly higher attenuation and some additional cost.

The *tensile strength* of a cable is defined as the axial load that may be applied to the line with no more than 0.2 percent permanent deformation after the load is released.

When divided by the weight per foot of cable, this gives an indication of the maximum length of cable that is self-supporting and, therefore, can be installed readily on a tower with a single hoisting grip. This consideration usually applies only to long runs of corrugated line; rigid line is installed one section at a time.

The *crush strength* of a cable is defined as the maximum force per linear inch that may be applied by a flat plate without causing more than a 5 percent deformation of the cable diameter. Crush strength is a good indicator of a cable's ruggedness and ability to withstand rough handling during installation.

Cable jacketing affords mechanical protection during installation and service. Semi-flexible cables are typically supplied with a jacket. This jacket consists of low-density polyethylene blended with 3 percent carbon black for protection from the sun's ultraviolet rays (which can degrade plastics through time). This approach has proved to be effective, yielding a life expectancy of more than 20 years. Rigid transmission line has no covering over the outer conductor.

For indoor applications, where fire-retardant properties are required, cables can be supplied with a fire-retardant jacket, usually listed by Underwriters Laboratories. Note that under the provisions of the National Electrical Code, outside plant cables, such as standard black polyethylene-jacketed coaxial line, may be run for as much as 50 ft inside a building with no additional protection. The line also may be placed in conduit for longer runs.

Low-density foam cables are designed to prevent water from traveling along their lengths, should it enter through damage to a connector or the cable sheath. This is accomplished by mechanically locking the outer conductor to the foam dielectric by annular corrugations. Annular or ring corrugations, unlike helical- or screw-thread-type corrugations, provide a water block at each corrugation. Closed-cell polyethylene dielectric foam is bonded to the inner conductor, completing the moisture seal.

7.2.4.A *Installation Considerations*

A coaxial cable is of no use if it can't be reliably connected to other parts of the transmission system. Connector design, therefore, takes into account several key requirements. The connector interface must provide a weatherproof bond with the cable to prevent water from penetrating the connection. This is ensured by using O-ring seals. The cable-connector interface also must provide a good electrical bond that does not introduce a mismatch and increase VSWR. Good electrical contact between the connector and the cable also ensures that proper RF shielding is maintained.

Windloading is always a concern when a tower is designed. Overall cable diameter and the configuration in which it is hung affect the tower windload. This is especially important when several cable runs are required. If these runs are mounted side by side, they increase the windload. Windload reduction can be achieved through a cluster-mount arrangement, whereby the cables are mounted in a side-by-side circular configuration.

Proper installation of a cable system requires various accessories. The method used to hang a cable system determines immediate and long-term performance. Two basic types of hangers are available for use with flexible cable.

- Wrap-type hanger. This hanger offers quick and easy installation, but there are some drawbacks to its use: Many such hangers are made of nylon and have limited mechanical durability. Others are made of malleable metal and installed by twisting wires together. Such devices have a low resistance to cable slippage, and the clamping force may be inconsistent (determined by the installer).

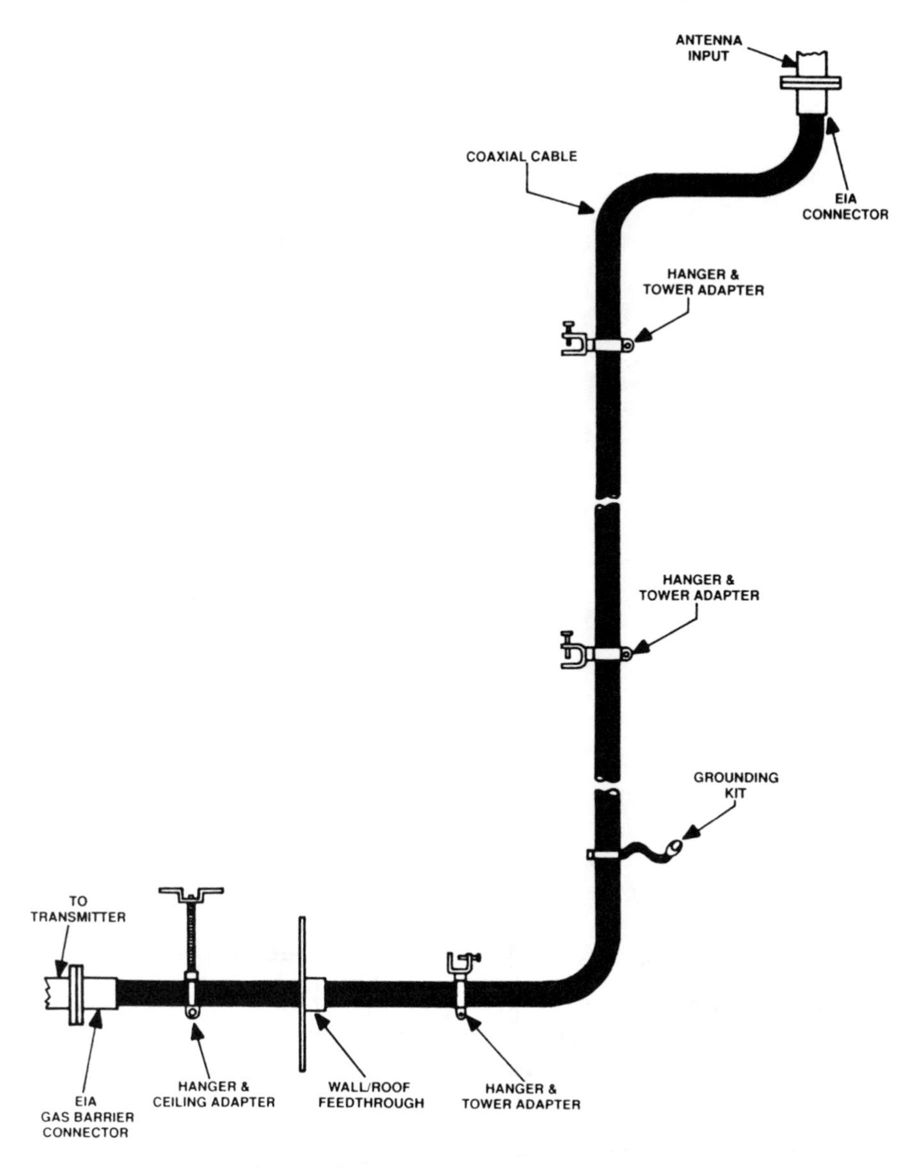

Figure 7.8 Typical installation hardware for a semi-flexible transmission line system.

- Clamp hangers. The clamp hanger is similar to a conduit-type hanger, but the halves are easier to spread apart. Clamp hangers require hardware for mounting to the tower bracket and clamping cable. This type of hanger provides dependable support and should be used wherever possible. The additional time required for installation is usually not significant.

A typical installation using semi-flexible cable is shown in Figure 7.8.

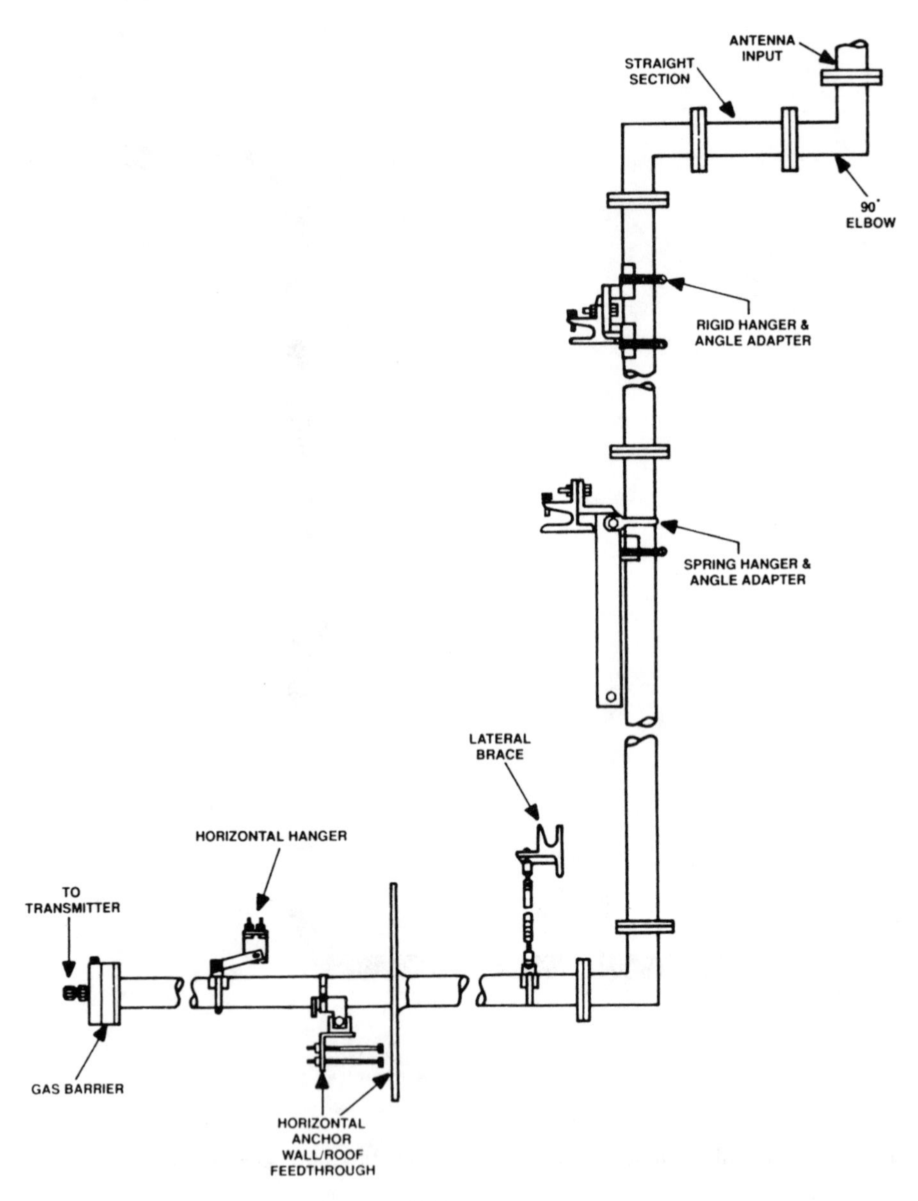

Figure 7.9 Typical installation hardware for a rigid transmission line system.

Installation of rigid line involves more complicated mechanical considerations, as illustrated in Figure 7.9. Common hardware includes:

- *Bullets*—connector elements used to electrically join the inner conductors of transmission line sections.

- *Elbows*—pre-formed copper angle elements used to transition from one direction to another. Elbows are commonly available in 90° and 45° types.
- *Expansion joints*—a flexible section inner conductor element used to accommodate vibration and differential expansion of the inner and outer conductors of a long line.
- *Line clamps*—mounting hardware to secure the line to a tower leg or other supporting structure.
- *Spring hangers*—a flexible version of the line clamp used to compensate for differential thermal expansion of the transmission line and the tower or other supporting structure.

7.3 Waveguide

As the operating frequency of a system reaches into the UHF band, waveguide-based transmission line systems become practical. Waveguide is simplicity itself. There is no inner conductor. RF energy is *launched* into the structure and propagates to the load. Several types of waveguide are available, including rectangular, square, circular, and elliptical. Waveguide offers several advantages over coax. First, unlike coax, waveguide can carry more power as the operating frequency increases. Second, efficiency is significantly better with waveguide at higher frequencies.

Rectangular waveguide is commonly used in high-power transmission systems. Circular waveguide may also be used, especially for applications requiring a cylindrical member, such as a rotating joint for an antenna feed. The physical dimensions of the guide are selected to provide for propagation in the dominant (lowest-order) mode.

Waveguide also has its drawbacks, however. Rectangular or square guide constitutes a large windload surface, which places significant structural demands on the tower. Because of the physical configuration of rectangular and square guide, pressurization is limited, depending on the type of waveguide used (0.5 psi is typical). Excessive pressure can deform the guide shape and result in increased VSWR. Wind also may cause deformation and ensuing VSWR problems. These considerations have led to the development of circular and elliptical waveguide.

7.3.1 Propagation Modes

Propagation modes for waveguide fall into two broad categories:

- *Transverse-electric* (TE) waves
- *Transverse-magnetic* (TM) waves

With TE waves, the electric vector (*E vector*) is perpendicular to the direction of propagation. With TM waves, the magnetic vector (*H vector*) is perpendicular to the direction of propagation. These propagation modes take on integers (from 0 or 1 to infinity) that define field configurations. Only a limited number of these modes can be propagated, depending on the dimensions of the guide and the operating frequency.

Energy cannot propagate in waveguide unless the operating frequency is above the *cut-off frequency*. The cut-off frequency for rectangular guide is:

$$F_c = c/2a$$

where:
F_c = waveguide cut-off frequency
c = 1.179×10^{10} in/s (the velocity of light)
a = the wide dimension of the guide

The cut-off frequency for circular waveguide is:

$$F_c = \frac{c}{2 \times a}$$

where:
a = the radius of the guide

There are four common propagation modes in waveguide:

- $TE_{0,1}$, the principal mode in rectangular waveguide
- $TE_{1,0}$, also used in rectangular waveguide
- $TE_{1,1}$, the principal mode in circular waveguide. $TE_{1,1}$ develops a complex propagation pattern with electric vectors curving inside the guide. This mode exhibits the lowest cut-off frequency of all modes, which allows a smaller guide diameter for a specified operating frequency.
- $TM_{0,1}$, which has a slightly higher cut-off frequency than $TE_{1,1}$ for the same size guide. $TM_{0,1}$ energy, developed as a result of discontinuities in the waveguide such as flanges and transitions, is not coupled out by either dominant or cross-polar transitions. The parasitic energy must be filtered out, or the waveguide diameter picked carefully, to reduce the unwanted mode.

Waveguide will support dual-polarity transmission within a single run of line. A combining element (*dual polarized transition*) is used at the beginning of the run, and a splitter (polarized transition) is used at the end of the line. Square waveguide has found numerous applications in such systems. The $TE_{1,0}$ and $TE_{0,1}$ modes are theoretically capable of propagation without cross coupling, at the same frequency, in lossless waveguide of square cross section. In practice, surface irregularities, manufacturing tolerances, and wall losses give rise to $TE_{1,0}$ and $TE_{0,1}$ mode cross conversion. Because this conversion occurs continually along the waveguide, long guide runs are usually avoided in dual-polarity systems.

7.3.1.A *Efficiency*

Waveguide loses result from the following:

- Power dissipation in the waveguide walls and the dielectric material filling the enclosed space.
- Leakage through the walls and transition connections of the guide.
- Localized power absorption and heating at the connection points.

The operating power of waveguide may be increased through pressurization. Sulfur hexafluoride is commonly used as the pressurizing gas.

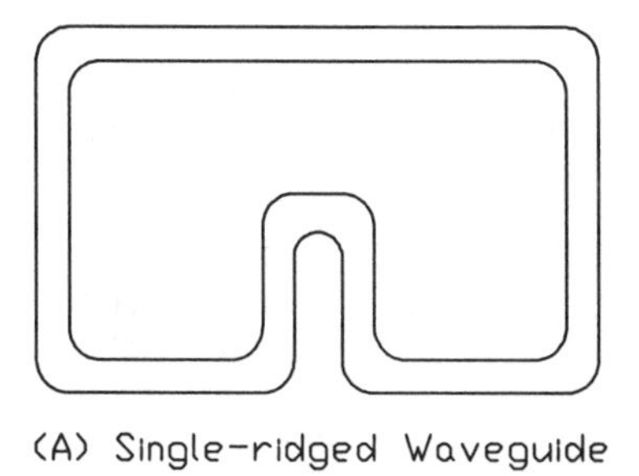

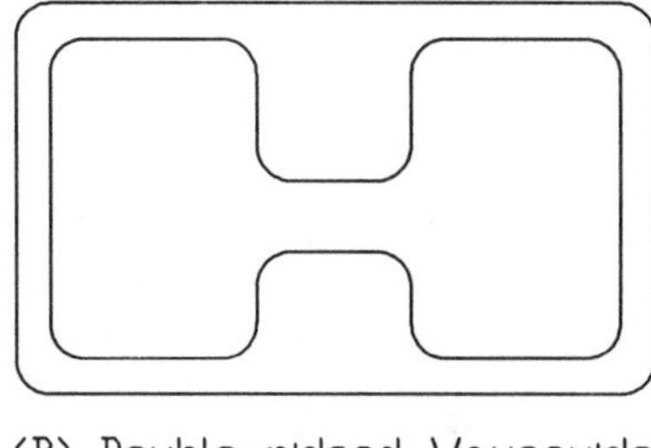

Figure 7.10 Ridged waveguide: (a) single-ridged; (b) double-ridged.

7.3.2 Ridged Waveguide

Rectangular waveguide may be ridged to provide a lower cut-off frequency and thereby permit use over a wider frequency band. One- and two-ridged guides are used, as illustrated in Figure 7.10. Increased bandwidth comes at the expense of increased attenuation, relative to an equivalent section of rectangular guide.

7.3.3 Circular Waveguide

Circular waveguide offers several mechanical benefits over rectangular or square guide. The windload of circular guide is two-thirds that of rectangular waveguide. It also presents lower and more uniform windloading than rectangular waveguide, reducing tower structural requirements.

The same physical properties of circular waveguide that give it good power handling and low attenuation also result in electrical complexities. Circular waveguide has two potentially unwanted modes of propagation, the cross-polarized $TE_{1,1}$ and $TM_{0,1}$ modes.

Circular waveguide, by definition, has no short or long dimension and, consequently, no method to prevent the development of *cross-polar* or *orthogonal* energy. Cross-polar energy is formed by small ellipticities in the waveguide. If the cross-polar energy is not trapped out, the parasitic energy can recombine with the dominant mode energy.

7.3.3.A *Parasitic Energy*

Hollow circular waveguide works as a high-Q resonant cavity for some energy and as a transmission medium for the rest. The parasitic energy present in the cavity formed by the guide will appear as increased VSWR if not disposed of. The polarization in the guide meanders and rotates as it propagates from the source to the load. The end pieces of the guide, typically circular-to-rectangular transitions, are polarization-sensitive (see the top portion of Figure 7.11). If the polarization of the incidental energy is not matched to the transition, energy will be reflected.

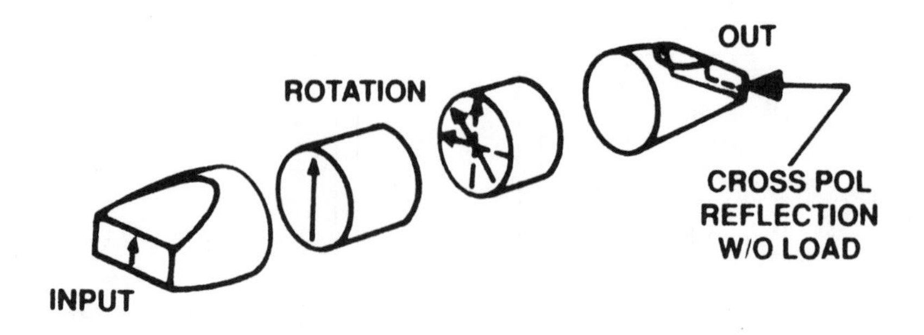

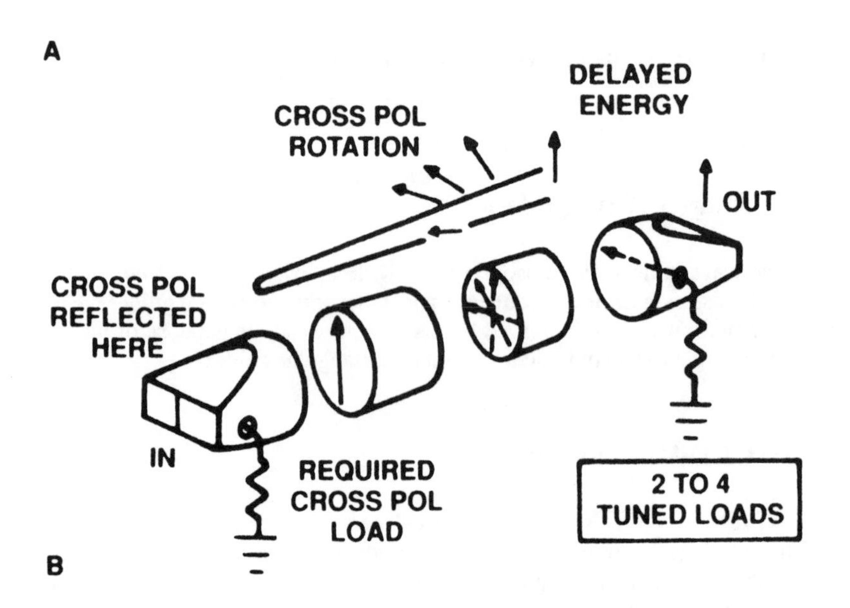

Figure 7.11 The effects of parasitic energy in circular waveguide: (a) trapped cross-polarization energy; (b) delayed transmission of the trapped energy.

Several factors can result in this undesirable polarization. One cause is out-of-round guides that result from nonstandard manufacturing tolerances. In Figure 7.11, the solid lines depict the situation at launching: perfectly circular guide with perpendicular polarization. The dashed lines show how certain ellipticities cause polarization rotation into unwanted states, while others have no effect. A 0.2 percent change in diameter can produce a -40 dB cross-polarization component per wavelength. This is roughly 0.03 in for 18 in of guide length.

Other sources of cross polarization include twisted and bent guides, out-of-roundness, offset flanges, and transitions. Various methods are used to dispose of this energy trapped in the cavity, including absorbing loads placed at the ground and/or antenna level.

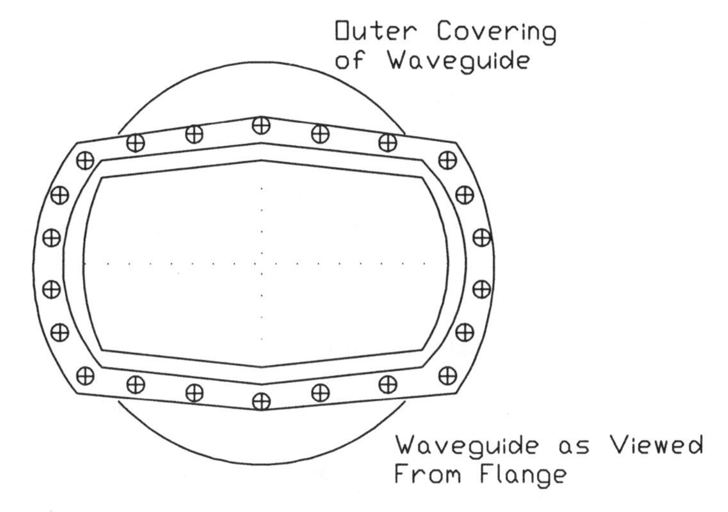

Figure 7.12 Physical construction of doubly truncated waveguide.

7.3.4 Doubly Truncated Waveguide

The design of *doubly truncated waveguide* (DTW) is intended to overcome the problems that may result from parasitic energy in a circular waveguide. As shown in Figure 7.12, DTW consists of an almost elliptical guide inside a circular shell. This guide does not support cross polarization; tuners and absorbing loads are not required. The low windload of hollow circular guide is maintained, except for the flange area.

Each length of waveguide is actually two separate pieces: a doubly truncated center section and a circular outer skin, joined at the flanges on each end. A large hole in the broadwall serves to pressurize the circular outer skin. Equal pressure inside the DTW and inside the circular skin ensures that the guide will not "breathe" or buckle as a result of rapid temperature changes.

DTW exhibits about 3 percent higher windloading than circular waveguide (because of the transition section at the flange joints), and 32 percent lower loading than comparable rectangular waveguide.

7.3.5 Impedance Matching

The efficient flow of power from one type of transmission medium to another requires matching of the field patterns across the boundary to launch the wave into the second medium with a minimum of reflections. Coaxial line is typically matched into rectangular waveguide by extending the center conductor of the coax through the broadwall of the guide, parallel to the electric field lines across the guide. Alternatively, the center conductor can be formed into a loop and oriented to couple the magnetic field to the guide mode.

7.3.5.A *Waveguide Filters*

A section of waveguide beyond cutoff constitutes a simple high-pass reflective filter. Loading elements in the form of posts or stubs may be employed to supply the reactances required for conventional lumped-constant-filter designs.

Absorption filters avoid the reflection of unwanted energy by incorporating lossy material in secondary guides that are coupled through so-called *leaky walls* (small sections of guide beyond cutoff in the passband). Such filters are typically used to suppress harmonic energy.

7.3.6 Installation Considerations

Waveguide system installation is both easier and more difficult than traditional transmission line. There is no inner conductor to align, but alignment pins and more bolts are required per flange. Transition hardware to accommodate loads and coaxial-to-waveguide interfacing is also required. Figure 7.13 shows a typical dual-polarized microwave waveguide system.

Flange reflections can add up in phase at certain frequencies, resulting in high VSWR. The length of the guide must be chosen so that flange reflection buildup does not occur within the operating bandwidth.

Installing DTW is similar to installing circular or rectangular waveguide, as shown in Figure 7.14. With the exception of the rectangular "E" plane sweep at the bottom of the vertical run, the DTW cross section is used in both the horizontal and the vertical runs. The transition from rectangular to DTW on either end of the elbow occurs in the flanges. Constant-force hangers provide lateral support.

7.3.6.A *Flexible Waveguide*

Flexible sections of waveguide are used to join rigid sections or components that cannot be otherwise aligned. Flexible sections also permit controlled physical movement resulting from thermal expansion of the line. Such hardware is available in a variety of forms. Corrugated guide is commonly produced by shaping thin-wall, seamless rectangular tubing. Flexible waveguide can accommodate only a limited amount of mechanical movement. Depending on the type of link, the manufacturer may specify a maximum number of bends.

7.3.6.B *Tuning*

Circular waveguide must be tuned. This process is a two-step procedure. First, the cross-polar $TE_{1,1}$ component is reduced, primarily through *axial ratio compensators* or *mode optimizers*. These devices counteract the net system ellipticity and indirectly minimize cross-polar energy. The cross-polar filters may also be rotated to achieve maximum isolation between the dominant and cross-polar modes. Cross-polar energy manifests itself as a net signal rotation at the end of the waveguide run. A perfect system would have a net rotation of zero.

In the second step, tuning slugs at the top and bottom of the waveguide run are adjusted to reduce the overall system VSWR. Tuning waveguide can be a complicated and time-consuming procedure, but normally, once set, tuning does not drift and must be redone only if major component changes are made.

7.3.6.C *Waveguide Hardware*

Increased use of waveguide has led to the development of waveguide-based hardware for all elements, from the output of the transmitter to the antenna. Waveguide-based

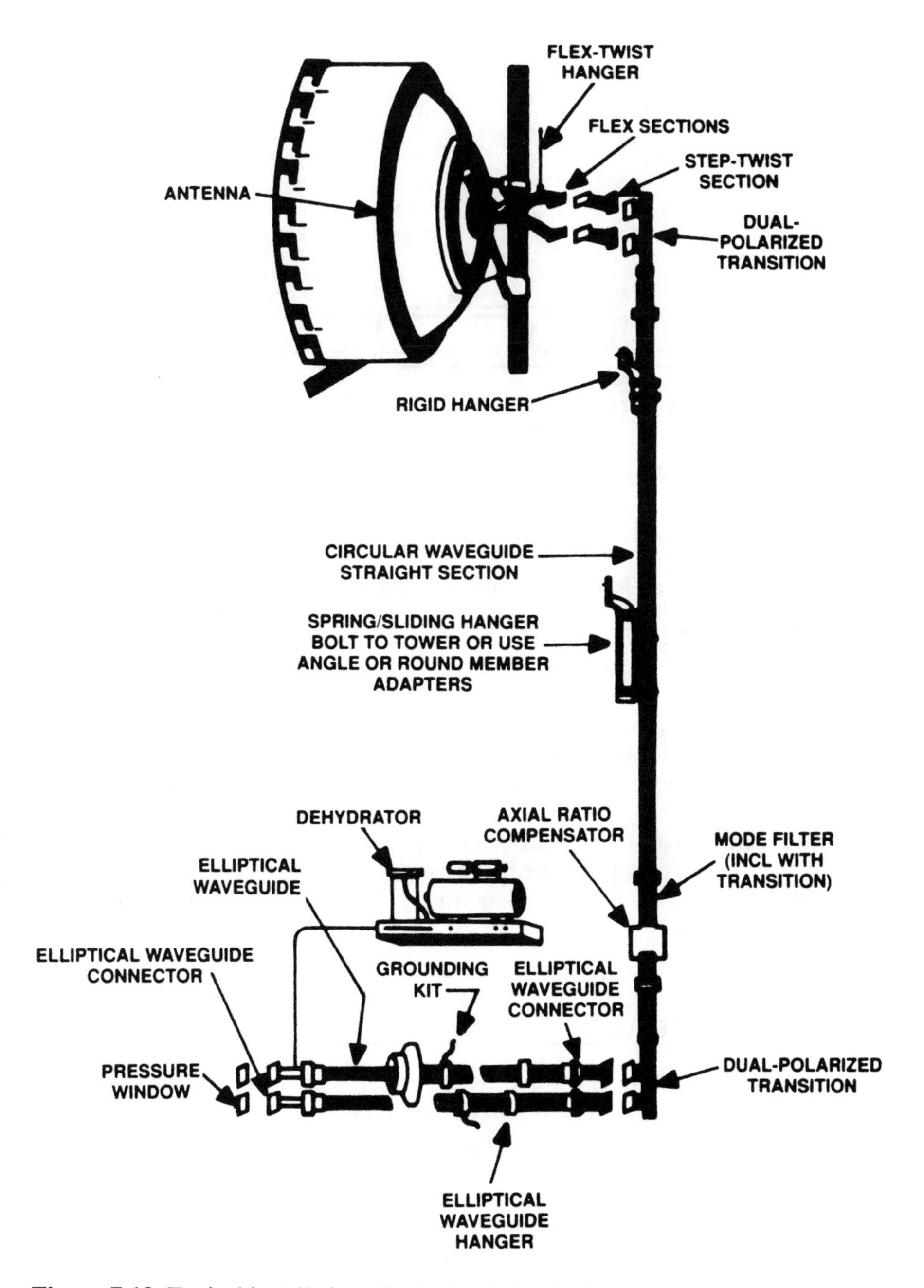

Figure 7.13 Typical installation of a dual-polarized microwave system.

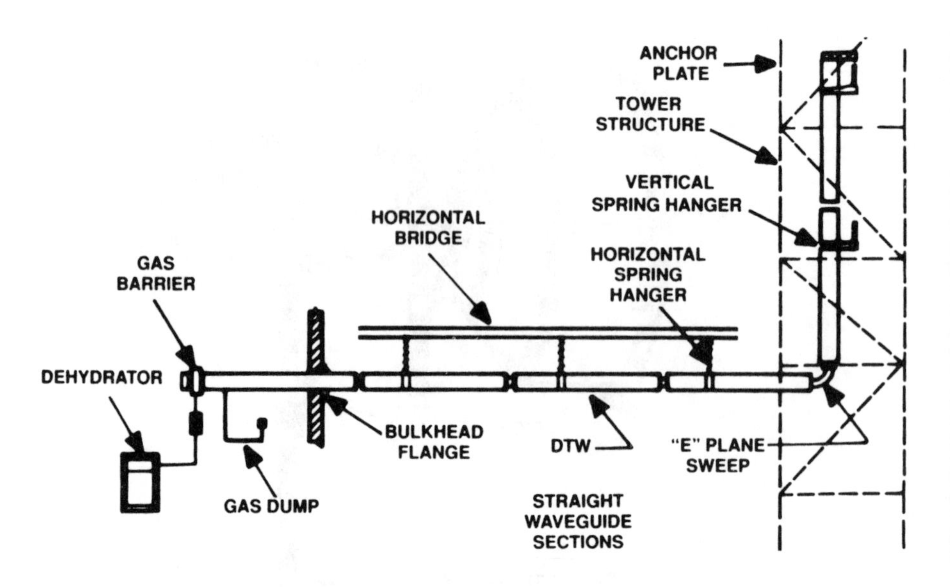

Figure 7.14 Installation hardware for doubly truncated waveguide.

filters, elbows, directional couplers, switches, combiners, and diplexers are currently available. Such hardware permits waveguide to be used from the output of the power-generating device to the antenna. Waveguide-based antennas are also available.

The RF performance of a waveguide component is usually better than the same item in coax. This is especially true in the case of diplexers and filters. Waveguide-based hardware provides lower attenuation and greater power-handling capability for a given physical size.

7.4 Bibliography

Andrew Corporation, "Circular Waveguide: System Planning, Installation and Tuning," Andrew technical bulletin 1061H, Orland Park, IL, 1980.

——— Bulletin 1063H, "Broadcast Transmission Line Systems," Orland Park, IL, 1982.

Ben-Dov, O., and C. Plummer, "Doubly Truncated Waveguide," *Broadcast Engineering* magazine, Intertec Publishing, Overland Park, KS, January 1989.

Cablewave Systems, "Rigid Coaxial Transmission Lines," Cablewave Systems Catalog 700, North Haven, CT, 1989.

——— Technical Bulletin 21A, "The Broadcaster's Guide to Transmission Line Systems," North Haven, CT, 1976.

Fink, D., and D. Christiansen: *Electronics Engineer's Handbook*, 3rd ed., McGraw-Hill, New York, 1989.

Jordan, Edward C.: *Reference Data for Engineers: Radio, Electronics, Computer and Communications*, 7th ed., Howard W. Sams Company, Indianapolis, 1985.

Krohe, Gary L., "Using Circular Waveguide," *Broadcast Engineering* magazine, Intertec Publishing, Overland Park, KS, May 1986.
Perelman, R., and Sullivan, T., "Selecting Flexible Coaxial Cable," *Broadcast Engineering* magazine, Intertec Publishing, Overland Park, KS, May 1988.
Whitaker, Jerry: *Radio Frequency Transmission Systems: Design and Operation*, McGraw-Hill, New York, 1990.

CHAPTER 8

INTERCONNECTING FIBER-OPTIC SYSTEMS

8.1 Introduction

Communicating over a thin strand of glass fiber is literally analogous to communicating by making a flashlight blink. Information transmission using laser technology is accomplished by regulating the flow of quantities of light photons in an optical glass fiber, rather than the amplitude and frequency of electromagnetic waves in free space or the coulombs of electrons in a copper wire.

The electrical signals used to drive the input amplifier interfacing a copper wire circuit or to modulate a radio transmitter power amplifier or oscillator are used instead as a control voltage to vary the light output (intensity) of a light source—an LED (light emitting diode) or ILD (*injection laser diode*)—around a mid-brightness intensity level. The source is never varied to full on or full off, because of the well-known hysteresis (nonlinearity) characteristics common to all transduction processes.

To effect communication transmission, the intensity-varying light is focused onto the end of a minuscule-diameter glass wire (fiber). At the other end of the fiber, the attenuated blinking light is focused onto a photon detector, which translates the light intensity variations back into an electrical signal.

Fiber-optic transmission has many advantages over radio and hard-wired copper transmission. First is its immunity to man made and natural sources of electrical interference—EMI (electromagnetic interference), RFI (radio frequency interference), electrical storms, or differing potentials in electrical ground reference. Immunity to such degradations often makes fiber transmission the only acceptable alternative, regardless of installation and operations costs.

A second, often indispensable, attribute of fiber is its virtual total immunity from transmitted signal jamming and theft. Jamming can be achieved only by physically breaking into the fiber and introducing another light source output. Theft can be accomplished only by breaking into the fiber and tapping off some of the light. Because both processes involve physically tampering with the fiber, they can be immediately detected, and transmissions can be terminated or diverted to other routes.

Fiber transmission has a number of other advantages as well. For example, one fiber can carry multiplexed signals (such as those needed to transmit complex video, audio, and data signals within a plant or around an educational or business campus) that would otherwise require dozens of coax lines and twisted-pair wires. This translates into enormous hardware and labor cost savings—plus time savings and convenience—in both permanent plant and temporary transmission system installation. Compared with copper circuit bandwidth and distance limitations in these applications, fiber transmission has no practical limits.

Common carrier digital transmission circuits currently in use operate at rates of a nominal 200 Mb/s and up. Rates of 565 Mb/s are common throughout the public fiber network. A few carriers are operating 1.2 Gb/s circuits commercially, and circuits operating in excess of 2 Gb/s are currently being tested. The current foreseeable ultimate throughput rate is in the 10 - 12 Gb/s range. The process of inserting customers' digital signal packages into these bit streams is referred to as *time division multiplexing* (TDM). Fiber cables are manufactured in virtually unlimited combinations of fiber quantities and types, some including copper pairs for electrical power distribution. *Strength members* in the cable may be either steel or nonconducting plastic for terminal isolation from different earth ground potentials. Outer sheaths range from materials meeting specific building codes for plenum installation to watertight steel housings for transoceanic cables.

8.1.1 Advantages of Fiber

Fiber-optic technology offers the user a number of benefits:

- Signal-carrying ability. The bandwidth information-carrying capacity of a communications link is directly related to the operating frequency. Light carrier frequencies are several orders of magnitude higher than the highest radio frequencies. Current fiber-optic systems easily surpass the information-carrying capacity of microwave radio and coaxial cable alternatives; and fiber's future carrying capacity has only begun to be used. Fiber provides bandwidths in excess of several gigahertz per kilometer, which allows high-speed transfer of most types of information. Multiplexing techniques allow many signals to be sent over a single fiber.
- Low loss. A fiber circuit provides substantially lower attenuation than copper cables and twisted pairs. It also requires no equalization. In premium cables, attenuation can be below 0.5 dB/km for certain wavelengths.
- Electrical isolation. The fiber and its coating are dielectric material, and the transmitter and receiver in each circuit are electrically isolated from each other. Isolation of separated installations from respective electrical grounds is assured if the strength material (messenger) in the cable is also dielectric. Lightwave transmission is free of spark hazards and creates no EMI. All-dielectric fiber cable may also be installed in hazardous or toxic environments.
- Size and weight. An optical waveguide is less than the diameter of a human hair. A copper cable is many times larger, stiffer, and heavier than a fiber that carries the same quantity of signals. Installation, duct, and handling costs are much lower for a fiber installation than for a similar coaxial system. Fiber cable

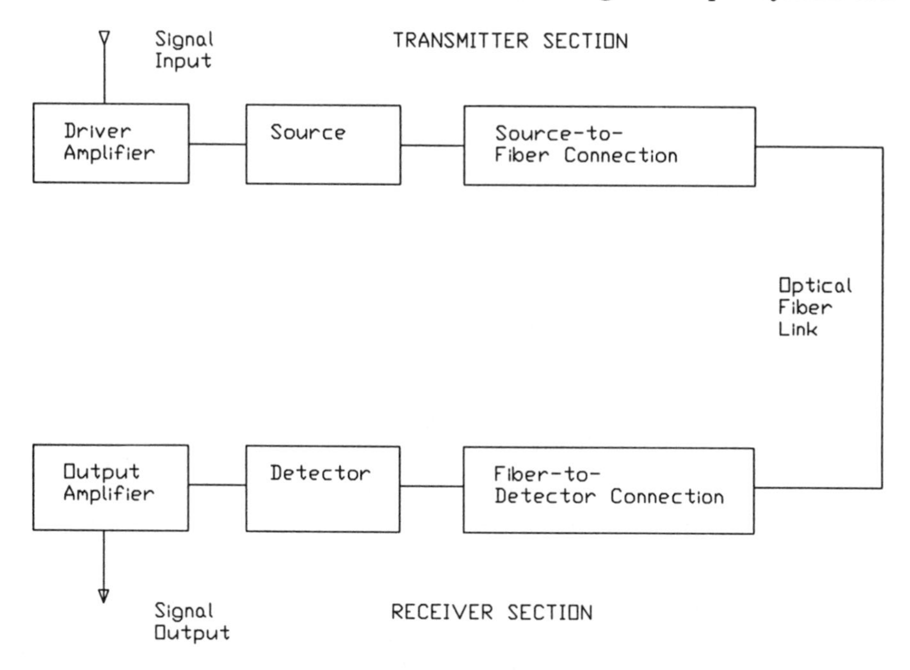

Figure 8.1 A basic fiber-optic transmitter-receiver link.

is the only alternative for circuit capacity expansion when ducts are full of copper.

8.2 Lightwave Transmission Theory[1]

Transmission of information on lightwaves is the most recent dramatic step toward the evolution of communications systems. The heart of this system is an optical fiber (also referred to as an optical waveguide), made of high-purity glass or plastic, through which light from a solid state laser or light-emitting diode is transmitted. Light, which is part of the electromagnetic spectrum, follows the same principles employed in designing microwave transmission systems and television and radio transmitters.

A basic transmission link consists of an optical transmitter terminal, a continuous length of fiber (from a few meters to as long as 30 to 50 km), and an optical receiver terminal (see Figure 8.1). The optical source output intensity of the transmitter (the light power is about 1 mW) is controlled by a driver circuit. This circuit's input is a varying voltage signal that may be analog or digital, with frequencies ranging to 1

1 Portions of this section were adapted from: Brad Dick, "Building Fiber-Optic Transmission Systems," *Broadcast Engineering* magazine, Intertec Publishing, Overland Park, KS, November 1991 and December 1991.

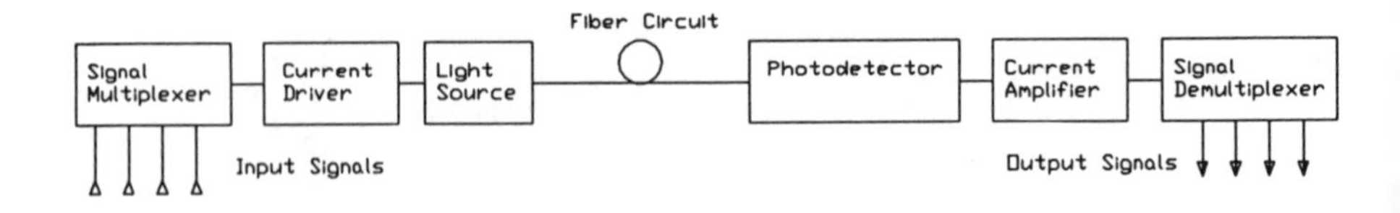

Figure 8.2 Typical end-to-end fiber-optic transmission system.

Gb/s and beyond. The modulating voltage may represent one or many individual signals. The transmitter also contains electronic circuit modules, which condition the electrical signals individually, as required for best end-to-end performance, and a multiplexer, which combines the electrical signals into one, with pre-emphasis characteristics required to maximize optical transmission circuit performance. The receiver contains a light-sensitive solid state detector (photodiode), which converts the received varying intensity light back into an electrical signal, and output circuits, which amplify, demodulate, and reconstruct the signal into its original form. The receiver terminal electrical output modules complement the transmitter electrical input modules, as illustrated in Figure 8.2.

8.2.1 Light and Fiber Characteristics

Light is similar to radio waves, x-rays, and gamma rays in that it is a part of the electromagnetic spectrum. Generally, the frequencies of light used in fiber-optic transmission center around 300 - 400 TeraHertz (10^{14})—several orders of magnitude higher on the electromagnetic spectrum than the highest frequency radio waves. (See Figure 8.3.) Light waves are more generally described in terms of wavelength rather than frequency. The region extending from 800 to 1,600 nm is of greatest interest because current fibers propagate these wavelengths most efficiently.

The speed of light (300,000 km/s) is the velocity of any electromagnetic energy in free space or a vacuum. Light travels slower in all other media, and different wavelengths travel at different speeds in any single medium. When an electromagnetic wave crosses a boundary from one medium to the next, it changes speed. This results in a change of path, called *refraction.* The *index of refraction*, n, is a dimensionless number expressing the ratio of the velocity of light in free space, c, to its velocity in a specific medium. Thus:

$$n = \frac{c}{v}$$

where:
n = the index of refraction
c = the velocity of light in free space
v = the velocity of light in a specific medium

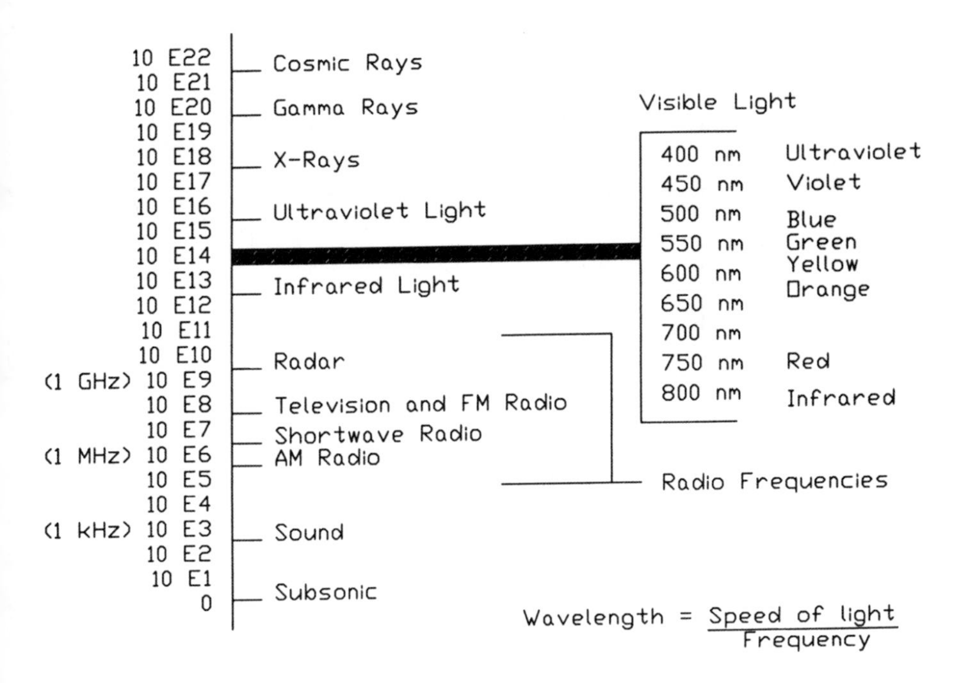

Figure 8.3 The electromagnetic spectrum.

The degree of refraction of a light ray as it passes from one material to the next depends on the refractive index of the material. When discussing refraction, three terms are used:

- The *normal*—the line perpendicular to the interface of the materials.
- The *angle of incidence*—the angle between the incoming ray and normal.
- The *angle of refraction*—the angle between normal and refracted ray.

These terms are illustrated in Figure 8.4.When light passes from a medium with a high index of refraction to a medium with a lower one, the light is refracted toward the normal. As the angle of incidence increases, the angle of refraction approaches 90° with normal. This is called the *critical angle*. If the angle is increased past the critical, the light is totally internally reflected.

8.2.2 Optical Fibers

An optical fiber includes two concentric components:

- The inner core
- The outer cladding

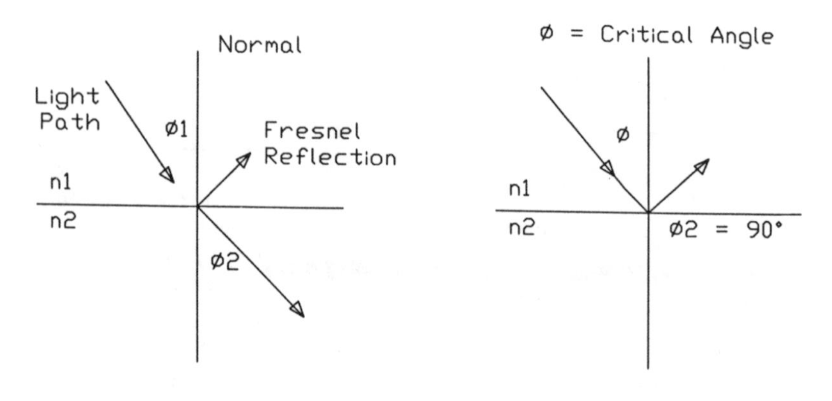

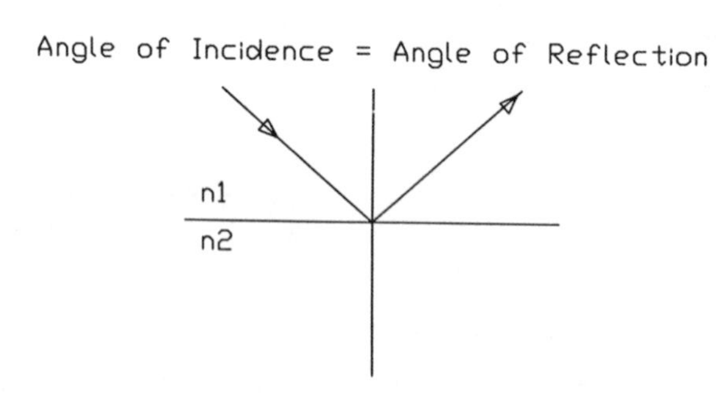

Figure 8.4 Lightwave refraction principles. When the refractive index of medium *n1* (the core) is less than that of medium *n2* (the cladding), light incident on the boundary at an angle less than the critical angle θ propagates through the boundary, but is refracted away from the normal. When the angle of incidence is greater than the critical, the light is totally reflected.

The inner core has a refractive index higher than the outer cladding. A light ray injected into the core at an angle will eventually hit the boundary between the core and the cladding. Rays arriving at an angle greater than the critical angle are reflected back into the core. These rays propagate (bounce their way) down the fiber until they emerge at the receiver terminal. Their time of arrival is—naturally—later than the ray that propagates down the center of the core.

Light striking the interface at less than the critical angle passes into the cladding and is lost. The total of the reflected rays constitutes the *carrier* of the information intensity modulated on it. The exact characteristics of the light ray bundles propagated in a fiber are determined by the fiber's size, construction, and composition. Maxwell's equations show that light does not travel randomly through a fiber; rather, it is channeled into *modes*. A mode is the path of one ray through a fiber.

Optical fibers have at least two, but most often, three distinct regions:

- The core—the central region of the fiber, as illustrated in Figure 8.5. Most of the light travels in the core.

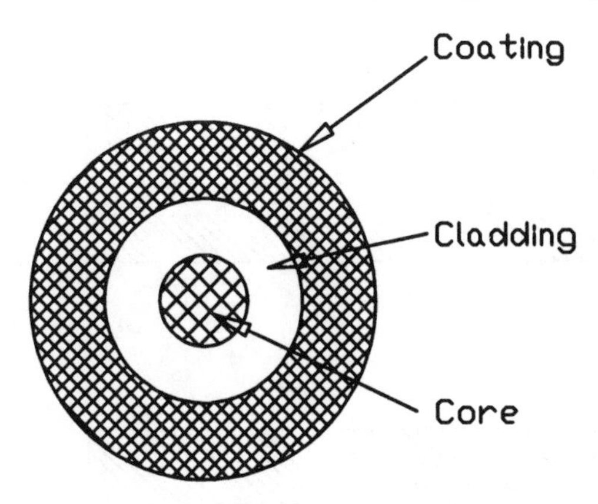

Figure 8.5 The three regions of a fiber-optic cable: the core, the cladding, and the buffer coating.

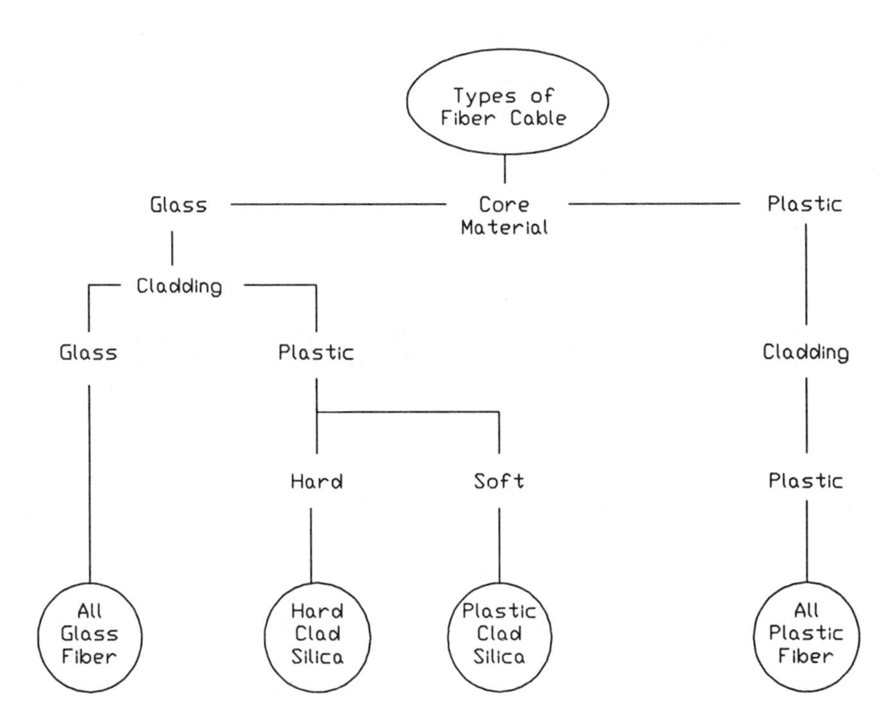

Figure 8.6 Primary types of fiber-optic cables, based on construction elements.

- The cladding (or clad)—the area that surrounds the core, confines the light to the core, and provides additional strength to the fiber through an increase in diameter.

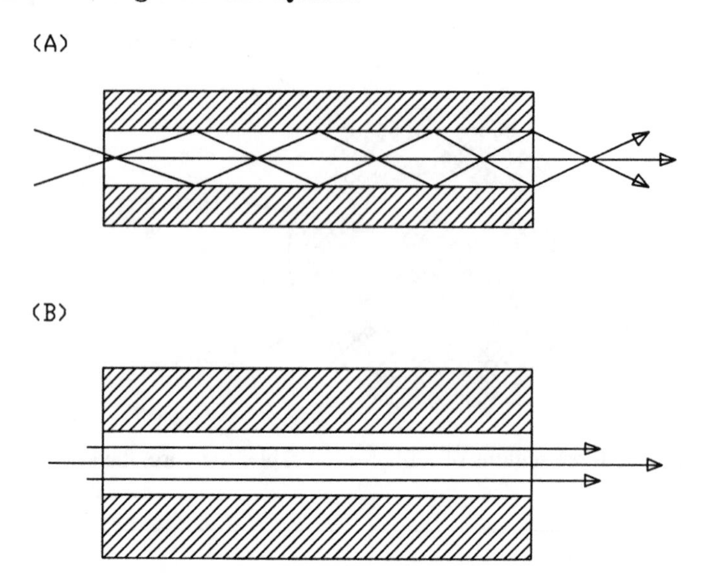

Figure 8.7 Modes of fiber-optic transmission: (a) multi-mode; (b) single mode.

- The buffer coating (or coating)—the material that insulates the core and cladding from the environment. The buffer coating helps the fiber retain its intrinsic strength.

The fiber core, cladding, and buffer coating can be made of various materials. The actual fiber used for communication applications is composed of either glass or plastic materials. The core and cladding also can be made from either glass or plastic. However, the buffer coating is always made of a plastic material. The variety of materials used in the core and cladding results in a range of fibers being produced. The primary types are summarized in Figure 8.6. Most, but not all, fibers used in communications applications have glass cores and glass claddings. The buffer coating may not be required if the cladding is plastic. In this case, the plastic cladding provides additional strength and insulates the fiber from the environment.

Fibers are available in a range of dimensions. Fibers with glass cores and plastic cladding have core diameters of 110 to 200 microns, and cladding diameters of 125 to 380 microns. Fibers with plastic cores and plastic cladding have core diameters of greater than 220 microns and cladding diameters of greater than 250 microns. Buffer coatings for all-glass fibers have standard diameters of 250 or 500 microns. However, other buffer coating diameters are available.[2]

The primary design difference between the two types of fiber—single mode and multi-mode—is the manner in which rays of light travel through the cable. In multi-mode fibers, rays of light travel in paths that are not necessarily parallel to the

2 The term "standard" in this context refers to an industry-accepted standard. It does not mean that the buffer coating diameter meets the requirements of a standards-issuing body.

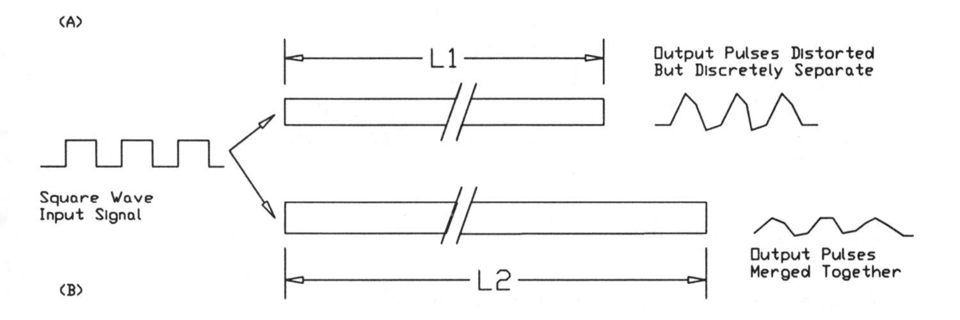

Figure 8.8 The mechanisms of dispersion in a fiber-optic cable: (a) pulse train in a short FO cable; (b) pulse train in a long FO cable.

---

fiber axis. This is shown in Figure 8.7. Some of the light rays move parallel to the axis, and some do not. The result is that some rays of light end up traveling a longer distance than others. Therefore, not all rays arrive at the end point at the same time. In a single-mode fiber, all rays of light travel parallel to the fiber's axis. This is an important difference. The result is that single-mode fiber-optic (FO) cables carry much more information than multi-mode cables. In multi-mode fibers, most of the light energy travels in the core. In single-mode fibers, a significant portion of the light energy travels in the cladding.

Most data communication systems currently use multi-mode fiber. However, single-mode fiber is expected to become the fiber of choice for the majority of communications applications over the next five to 10 years. This is true because of its almost unlimited bandwidth and low attenuation.

8.2.3 Dispersion

Dispersion describes the spreading of a light pulse as it travels down a fiber. A pulse measured at the output will be wider than it was at the input. This limits fiber bandwidth or information-carrying capacity. Pulse rates must be slow enough that dispersion will not cause adjacent pulses to overlap. Consider the example shown in Figure 8.8. Two trains of identical pulses are injected into fibers of differing lengths. In both cases, the pulses are spread by dispersion. For the shorter fiber, the interval between pulses is sufficient to allow each pulse to be distinguished. For the longer fiber, however, the individual pulses tend to merge into an indistinguishable waveform.

Dispersion, which is analogous to pulse rise and fall times in copper cable, is the limiting factor in determining fiber bandwidth. *Modal dispersion* results from differing path lengths of the multiple rays (modes) in a multi-mode fiber. *Chromatic dispersion* limits the bandwidth that can be propagated in a single-mode fiber. This dispersion materializes because no laser light source used in fiber-transmission systems emits light at a single frequency (wavelength). A special class of long-wavelength *distributed feedback* (DFB) lasers with a narrow output spectrum is currently available for propagation of information on single-mode fibers over 50 km. This interrelationship between propagated bandwidth and distance is uniquely determined in each fiber by its core diameter, *step* or *index core grading*, and material purity. The figure of merit for any given fiber is its *bandwidth-distance* rating, specified in megahertz per kilometer (MHz/km). The product of the bandwidth to be transmitted and the length of the optical circuit must be less than the figure-of-merit number.

8.2.4 Types of Fibers

Of the many ways to classify fibers, the most informative is by refractive index profile and number of modes supported. The two main types of index profiles are *step* and *graded*. In a *step index* fiber, the core has a uniform index with a sharp change at the boundary of the cladding. In a *graded index* fiber, the core index is not uniform; it is highest at the center and decreases until it matches the cladding.

8.2.4.A *Step Index Multi-mode Fiber*

A multi-mode step index fiber typically has a core diameter in the 50 to 1,000 micron range. The large core permits many modes of propagation. Because light will reflect differently for different modes, the path of each ray is a different length. The lowest-order mode travels down the center; higher-order modes strike the core-cladding interface at angles near the critical angle. As a result, a narrow pulse of light spreads out as it travels through this type of fiber. This spreading is called *modal dispersion* (see Figure 8.9).

8.2.4.B *Step Index Single (Mono) -Mode Fiber*

Modal dispersion can be reduced by making the fiber core small, typically 5 to 10 microns (1/6 the diameter of a human hair). At this diameter, only one mode propagates efficiently. The small size of the core makes it painstaking to splice. Single mode of propagation permits high-speed, long-distance transmission.

8.2.4.C *Graded Index Multi-mode Fiber*

Like the step index single-mode fiber, a graded index fiber also limits modal dispersion. The core is essentially a series of concentric rings, each with a lower refractive index. Because light travels faster in a lower-index medium, light further from the axis travels faster. Because high-order modes have a faster average velocity than low-order modes, all modes tend to arrive at a given point at nearly the same time. Rays of light are not sharply reflected by the core-cladding interface; they are refracted successively by differing layers in the core.

8.2.5 Characteristics of Attenuation

Attenuation is loss of power. During transit, some of the light in a fiber-optic system is absorbed into the fiber or scattered by impurities. Attenuation for a fiber cable is usually specified in decibels per kilometer (dB/km). For commercially available fibers, attenuation ranges from approximately 0.5 dB/km for premium single-mode fibers to 1,000 dB/km for large-core plastic fibers. Because emitted light represents power, 3 dB represents a doubling or halving of any reference power level.

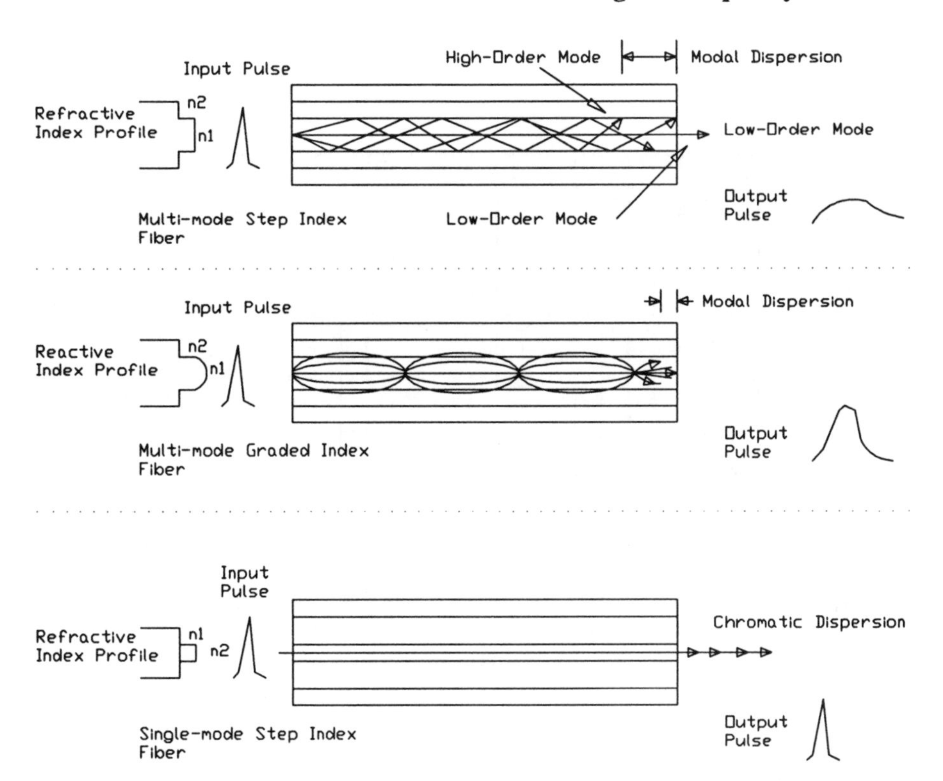

Figure 8.9 Modal dispersion in an FO cable. The core diameter and its refractive index characteristics determine the light propagation path(s) within the fiber core.

Attenuation and light wavelength are also uniquely related in fiber-transmission systems. This is illustrated in Figure 8.10. Most fibers have a medium loss region in the 800 - 900 nm wavelength range (3 - 5 dB/km), a low loss region in the 1,150 - 1,350 nm range (0.6 - 1.5 dB/km), and a very low loss region (less than 0.5 dB/km) in the 1,550 nm range. As a result, optimum performance is achieved by careful balancing of fiber, light source wavelength, and distance requirements.

Light intensity attenuation has no direct effect on the bandwidth of the electrical signals being transported. There is a direct correlation, however, between the S/N of the fiber receiver electronic circuits and the usable recovered optical signal.

8.2.6 Types of Cable

The first step in packaging an optical fiber into a cable is the extrusion of a layer of plastic around the fiber. This layer of plastic is called a *buffer tube*; it should not be confused with the buffer coating. The buffer coating is placed on the fiber by the fiber manufacturer. The buffer tube is placed on the fiber by the cable manufacturer. This extrusion process can produce two different cable designs:

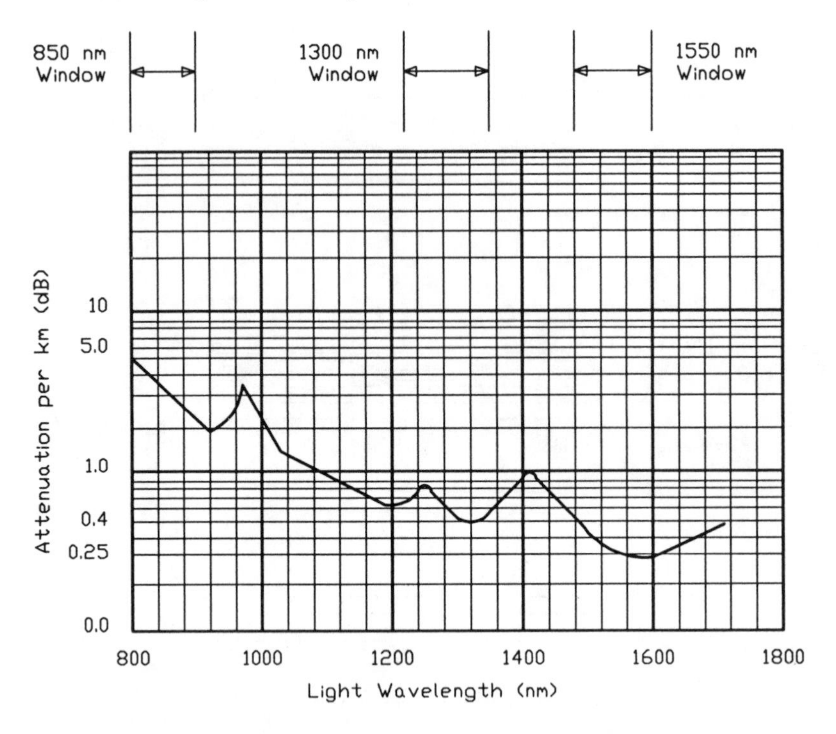

Figure 8.10 Fiber attenuation vs. light wavelength characteristics. Attenuation has been reduced steadily in the last two decades through improved fiber drawing techniques and a reduction in impurities. It has now approached the theoretical limits of silica-based glass at the 1,300 and 1,550 nm wavelengths.

- *Tight tube design*—The inner diameter of the plastic (buffer tube) is the same size as the outer diameter of the fiber, and is in contact with the fiber around its circumference.
- *Loose tube design*—The layer of plastic is significantly larger than the fiber, and, therefore, the plastic is not in contact with the fiber around the circumference of the fiber.

The two types of fiber cable are illustrated in Figure 8.11. Note that the loose tube design is available configured either as a *single-fiber-per-tube* (SFPT) or *multiple-fibers-per-tube* (MFPT) design.

The six-fibers-per-tube MFPT design is often used for data communications, because this is the type specified by Bellcore for use by member telephone companies.

After a fiber (or group of fibers) has been surrounded by a buffer tube, it is called an *element*. The cable manufacturer uses elements to build up the desired type of cable. In building the cable from elements, the manufacturer can create six distinct designs:

- Breakout design
- MFPT, central loose tube design
- MFPT, stranded loose tube design
- SFPT, stranded loose tube design

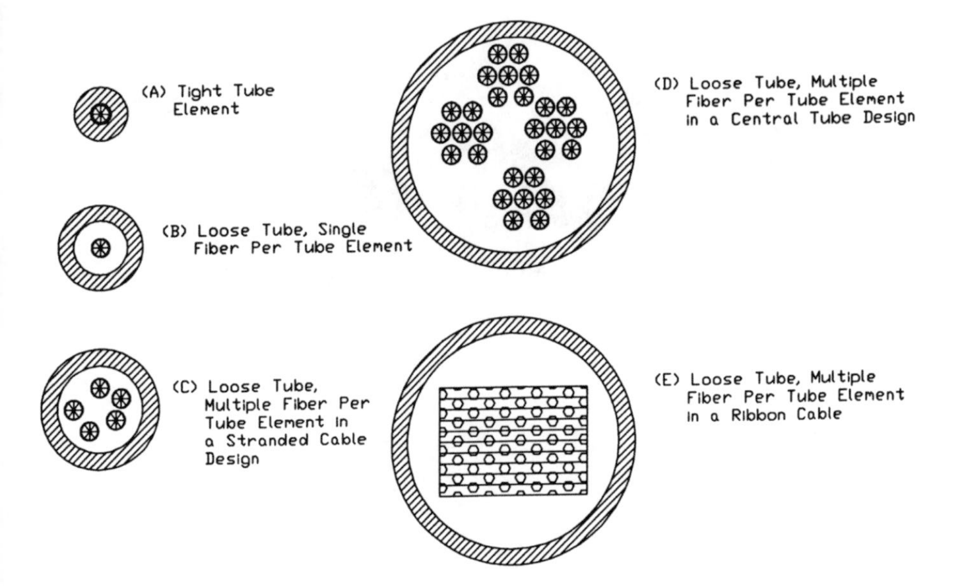

Figure 8.11 Loose-tube cables are available in either single-fiber-per-tube (SFPT) or multiple-fibers-per-tube (MFPT) designs. In both cases, the diameter of the plastic tube surrounding the core is larger than the outside diameter of the core. A tight tube cable is shown in (a); the inner diameter tube is the same as the outer diameter of the fiber.

- *Star*, or *slotted core*, design
- *Tight tube*, or *stuffed*, design

8.2.6.A *Breakout Design*

In the breakout design, shown in Figure 8.12, the element or buffered fiber is surrounded with a flexible-strength member, often Kevlar. The strength member is surrounded by an inner jacket to form a subcable, as shown. Multiple subcables are stranded around a central strength member or filler to form a cable core. This cable core is held together by a binder thread or Mylar wrapping tape. The core is surrounded by an extruded jacket to form the final cable.

Optional steps for this design include additional strength members, jackets, or armor. The additional jackets may be extruded directly on top of one another or separated by additional external strength members.

8.2.6.B *MFPT, Central Loose Tube Design*

Fibers are placed together to form groups. Sometimes, the fibers are laid along a ribbon in groups of 12. These ribbons are then stacked up to 12 high and twisted. This version of the central loose tube design is referred to as a *ribbon design*, and was developed by AT&T. The space between the fibers and the tube can be filled with a water-blocking compound.

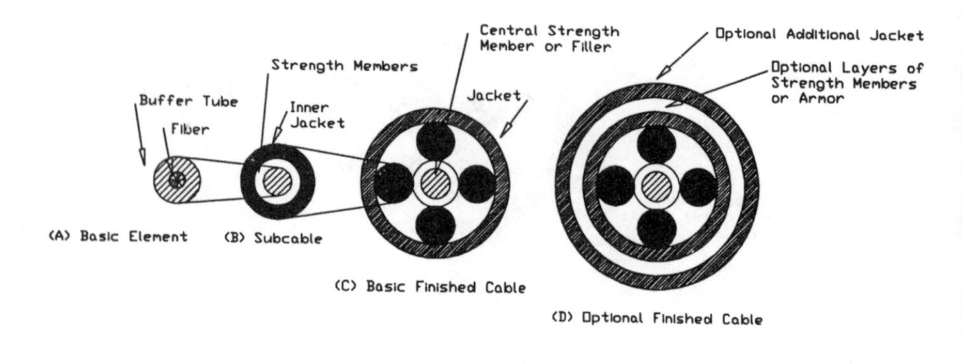

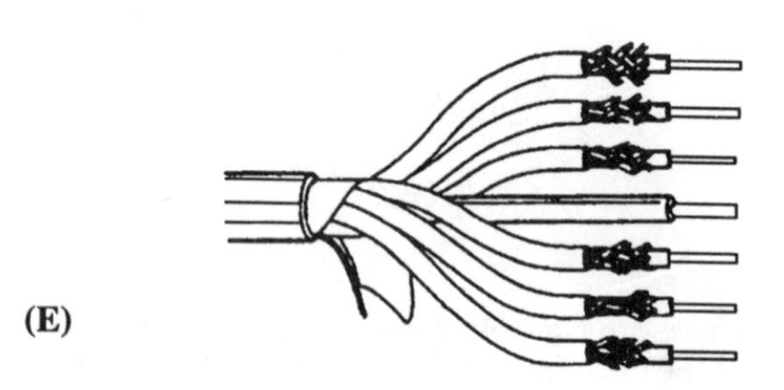

Figure 8.12 In the breakout type of cable, each element is surrounded by a flexible strength member, which is then surrounded by an inner jacket. This forms a subcable, which is incorporated into a larger cable. Optional additional jackets or armor can be applied, as shown in (d). Drawing (e) illustrates the mechanical configuration of a basic FO cable.

8.2.6.C *MFPT, Stranded Loose Tube Design*

Multiple buffer tubes are stranded around a central strength member or filler to form a core, as illustrated in Figure 8.13. This cable core is held together by a binder thread or Mylar wrapping tape. The core is surrounded by an extruded jacket to form a finished cable. Optional jacketing, strength members, or armor can be added.

8.2.6.D *SFPT, Stranded Loose Tube Design*

This type of cable is manufactured similarly to MFPT cable. The primary difference is that the cable has one fiber per tube and smaller-diameter buffer tubes.

8.2.6.E *Star, or Slotted Core, Design*

This design is seldom used in the United States. In this scheme, the buffer tubes (usually MFPT) are laid in helical grooves, which are formed in the filler in the center of the cable. The core is then surrounded by an extruded jacket to form a finished cable. One variation of this design, shown in Figure 8.14, is used by power utility companies. This optical cable provides a conductive ground path from end to end. Instead of a jacket, the cable has helically wrapped wires, some of which are conductors and strength members.

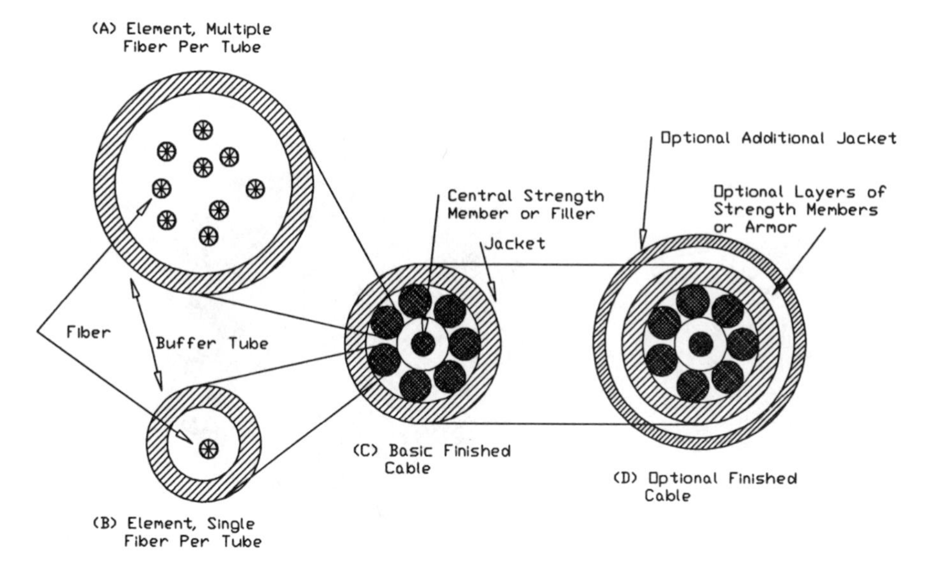

Figure 8.13 The MFPT stranded loose tube design relies on a center strength member to form the cable core. Multiple elements are then added to build up the desired cable capacity.

8.2.6.F *Tight Tube, or Stuffed, Design*

This design is based on the tight tube element. The designs are common in that the core is filled, or stuffed, with flexible strength members, usually Kevlar. The design usually incorporates two or more fibers, as illustrated in Figure 8.15.

8.2.6.G *Application Considerations*

Performance advantages exist for all designs, depending on what parameter is considered. For example, the tight tube design can force the ends of a broken fiber to remain in contact even after the fiber has broken. The result is that transmission may still be possible. When reliability is paramount, this feature may be important.

Loose tube designs have a different performance advantage. They offer a mechanical dead zone, which is not available in tight tube designs. The effect is that stress can be applied to the cable without that stress being transferred to the fiber. This dead zone exists for all mechanical forces, including tensile and crush loads, and bend strains. Tight tube designs do not have this mechanical dead zone. In the tight tube design, any force applied to the cable is also applied to the fibers. Loose tube designs also offer smaller size, lower cost, and smaller bend radii than tight tube designs.

When cable cost alone is considered, loose tube designs have the advantage over tight tube, breakout designs in long-length applications. However, when total installation cost is considered, the loose tube designs may or may not have a cost advantage. This is because loose tube designs have higher connector installation costs. The cost factor is composed of two parts:

- Labor cost
- Equipment cost

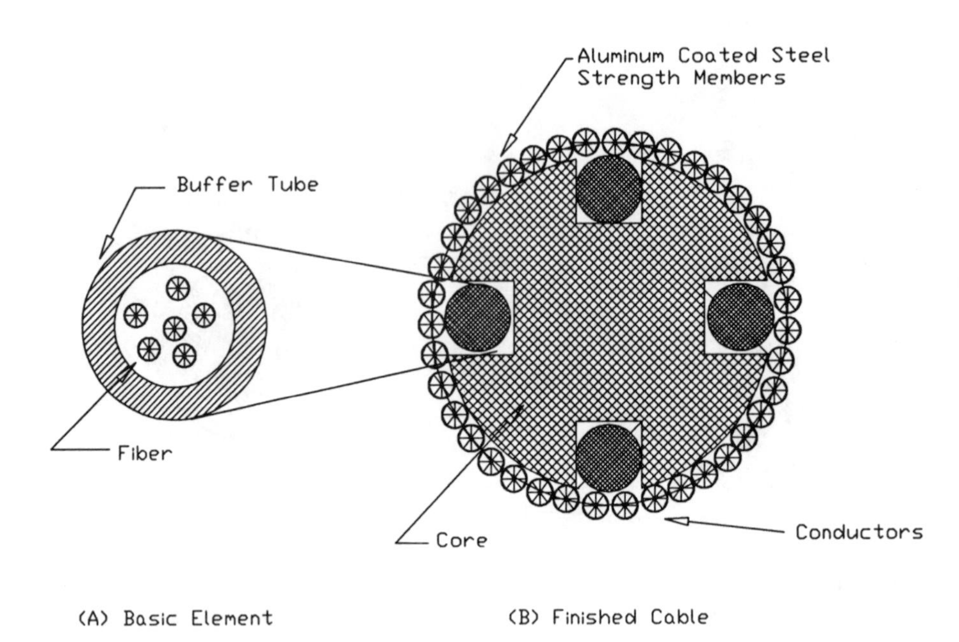

Figure 8.14 Utility companies sometimes use an optical power ground wire type of cable because it incorporates a metallic power ground wire within the design. The cable is based on a slotted-core configuration, but with the addition of helically wrapped wires around the outside for strength and conductivity.

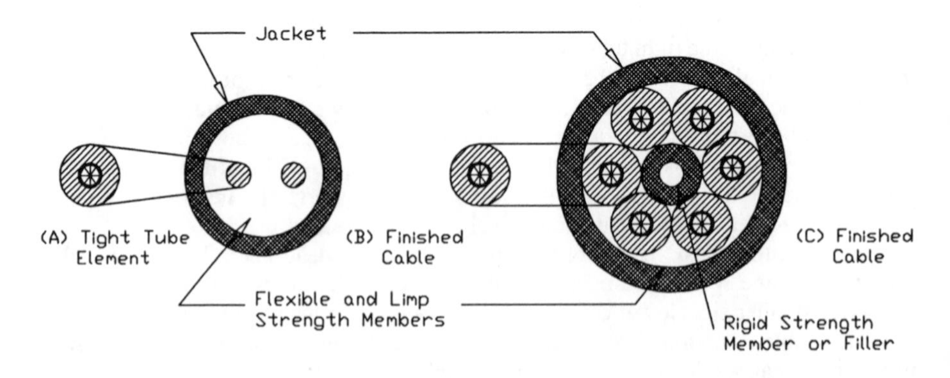

Figure 8.15 The tight tube stuffed design relies on a core filled with flexible strength members, usually Kevlar. Typically, two or more fibers are contained within the cable.

All designs, other than the breakout design, require handling of bare fibers or fibers with tight tubes. During this handling, fibers can be broken, especially where inexperienced personnel are involved.

8.2.7 Specifying Fiber-Optic Cable

In order to completely specify a fiber-optic cable, four primary performance categories must be quantified:

- Installation specifications
- Environmental specifications
- Fiber specifications
- Optical specifications

These criteria are outlined in Table 8.1. Note that not all specifications apply to all situations. The system engineer must review the specific application to determine which of the specifications are applicable. For example, cable installed in conduit or in protected locations will not need to meet a crush load specification.

8.2.7.A *Installation Specifications*

The installation specifications are those that must be met to ensure successful cable installation. There are six:

- Maximum installation load in kilograms-force or pounds-force. This is the maximum tensile load that can be applied to a cable without causing fiber breakage or a permanent change in attenuation. This characteristic must always be specified. Load values for some typical installations are shown in Table 8.2. If the application requires a strength higher than those listed, specify a higher-strength cable. The increased cost of specifying a higher-strength cable is small, typically 5 to 10 percent of the cable cost.
- Minimum installation bend radius in inches or millimeters. This is the minimum radius to which the cable can be bent while loaded at the maximum installation load. This bending can be done without causing a permanent change in attenuation, fiber breakage, or breakage of any portion of the cable structure. The bend radius is usually specified as no less than 20 times the cable diameter. To determine this value, examine the locations where the cable will be installed, and identify the smallest bend the cable will encounter. Conversely, the system engineer can choose the cable and then specify that the conduits or ducts not violate this radius. The radius is actually limited more by the cabling materials than by the bend radius of the fiber.
- Diameter of the cable. Despite the space-effective nature of FO cable, it still must reside in the available space. This is especially true if the cable is to be installed in a partially filled conduit. When faced with limited space, consider first an MFPT design. It requires the least amount of space.
- Diameter of subcables or elements. The diameter of the subcable or the cable elements can become a limiting factor. In the case of a breakout-style cable, the diameter of the subcable must be smaller than the maximum diameter of the connector boot so that the boot will fit on the subcable. Also, the diameter of the element must be less than the maximum diameter acceptable to the

Table 8.1 Fiber cable specification considerations.

Installation specifications:

- Maximum recommended installation load
- Minimum installation bend radius
- Cable diameter
- Diameter of subcables
- Maximum installation temperature range
- Maximum storage temperature range

Environmental specifications:

- Temperature range of operation
- Minimum recommended unloaded bend radius
- Minimum long-term bend radius
- Maximum long-term use load
- Vertical rise
- National Electric Code or local electrical code requirements
- Flame resistance
- UV stability or UV resistance
- Resistance to rodent damage
- Resistance to water damage
- Crushing characteristics
- Resistance to conduction under high-voltage fields
- Toxicity
- High flexibility: static vs. dynamic applications
- Abrasion resistance
- Resistance to solvents, petrochemicals, and other substances
- Hermetically sealed fiber
- Radiation resistance
- Impact resistance
- Gas permeability
- Stability of filling compounds

Fiber specifications:
Dimensional considerations:

- Core diameter
- Clad diameter
- Buffer coating diameter
- Mode field diameter

Optical specifications:
Power considerations:

- Core diameter
- Numerical aperture
- Attenuation rate
- Cut-off wavelength

Table 8.2 Maximum installation loads that fiber cable may be exposed to in various applications.

Typical maximum recommended installation loads	
Application	**Pounds force**
1 fiber in raceway or tray	67 lb
1 fiber in duct or conduit	125 lb
2 fibers in duct or conduit	250 - 500 lb
Multi-fiber (6-12) cables	500 lb
Direct burial cables	600 - 800 lb
Lashed aerial cables	300 lb
Self-supported aerial cables	600 lb

backshell of the connector. Most breakout cables have tight-tube elements, usually with a diameter of 1 mm or less.

- Recommended temperature range for installation (°C). All cables have a temperature range in which they can be installed without damage to either the cable materials or the fibers. Generally, the temperature range is affected more by the the cable materials than the fibers. Not all cable manufacturers include this parameter in their data sheets. If the parameter is not specified, select a conservative temperature range of operation.
- Recommended temperature range for storage (°C). In severe climates, such as deserts and the Arctic, the system engineer must specify a recommended temperature range for storage in °C. This range will strongly influence the materials used in the cable.

8.2.8 Environmental Specifications

Environmental specifications are those that must be met to ensure successful long-term cable operation. Most of the items listed in Table 8.1 are self-explanatory. However, some environmental specifications deserve special attention.

8.2.8.A *Temperature Range of Operation*

This is the temperature range in which the attenuation remains less than the specified value. There are few applications where FO cable cannot be used because of temperature considerations. FO cables composed of plastic materials have maximum and minimum temperature points. If these are exceeded, the materials will not maintain their mechanical properties. After long exposures to high temperatures, plastics deteriorate and become soft. Some materials will begin to crack. After exposures to low temperatures, plastics become brittle and crack when flexed or moved. Under such conditions, the cable coverings will cease to protect the fiber.

Table 8.3 Typical maximum long-term use loads that a cable may be exposed to in various applications.

Typical maximum recommended long-term use loads	
Application	**Pounds force**
1 fiber in raceway or tray	23 - 35 lb
1 fiber in duct or conduit	67 lb
2 fibers in duct or conduit	67 lb
Multi-fiber (6 - 12) cable	33 - 330 lb
Direct burial cable	132 - 180 lb

Another reason for considering the temperature range of operation is the increase in attenuation that occurs when fibers are exposed to temperature extremes. This sensitivity occurs when the fibers are bent. When a cable is subject to extreme temperatures, the plastic materials will expand and contract. The rates at which the expansion and contraction take place are much greater (perhaps 100 times) than the rates of glass fibers. This movement results in the fiber being bent at a microscopic level. The fiber is either forced against the inside of the plastic tube as the plastic contracts, or the fiber is stretched against the inside of the tube as the plastic expands. In either case, the fiber is forced to conform to the microscopically uneven surface of the plastic. On a microscopic level, this is similar to placing the fiber against sandpaper. The bending results in light escaping from the core of the fiber. The result is referred to as a *microbend-induced* increase in attenuation.

8.2.8.B *Minimum Long-Term Bend Radius*

This parameter represents the minimum radius to which the cable can be bent for its entire lifetime. It is usually specified at no less than 10 times the diameter of the cable.

8.2.8.C *Maximum Long-Term Use Load*

Most FO cables are designed for unloaded use. However, substantial physical loading occurs when cables are strung outdoors between poles or hangers, or mounted on a tower. In such cases, the cables are subject to self-loads and to additional loads from the environment, including wind, snow, and ice. The load experienced by the cable depends primarily on the spacing between hangers.

Be sure to specify the correct long-term use load characteristic so that the strain applied to the cable does not exceed a critical value. If this critical value is exceeded, the fiber(s) will probably break spontaneously. The long-term use load depends on the design and construction of the cable. It typically runs 10 to 30 percent of the maximum recommended installation load.

If the cable will experience a significant long-term use load, this specification is more important than the maximum installation load. Self-supporting cables are available from several manufacturers and are used by power utilities for suspensions up to 3,000 ft. In these cases, the maximum span length is specified instead of the long-term use load. Typical long-term use loads are shown in Table 8.3.

Table 8.4 Typical maximum vertical lengths for common types of fiber-optic cable.

Typical maximum vertical rise distances	
Application	**Distance**
1 fiber in raceway or tray	90 ft
2 fibers in duct or conduit	50 - 90 ft
Multi-fiber (6 - 12) cables	50 - 375 ft
Heavy-duty cables	1,000 - 1,640 ft

8.2.8.D *Vertical Rise Distance*

The vertical rise distance is related to the maximum use load. When cables are installed in a riser (within a building) or along a long vertical length, such as up a tower, the self-weight of the cable imposes a load on the cable. This load must be less than the maximum use load. Typical vertical rise distances for a selection of cables are shown in Table 8.4.

8.2.8.E *UV Resistance*

Outdoor installations require that cables be UV-resistant (or UV-stable). Otherwise, the cable jacket will crack and lose flexibility when exposed to sunlight. Most cables for outdoor use rely on a black polyethylene jacketing material, which has built-in UV-absorbing capability and no plasticizers that can evaporate over time.

8.2.8.F *Crush Load*

The crush load is the maximum pressure that can be applied perpendicular to the cable's axis without causing a permanent increase in attenuation or fiber breakage. Crush load specifications include the following:

- Short-term—momentary pressures typically applied during installation.
- Long-term—pressures applied continuously during the life of the cable.

If there is a chance that the cable may be crushed, the system engineer must decide whether the crush load is likely to be a short- or long-term condition. If it is short-term, there are two primary considerations:

- Is the force sufficient to cause the fiber to break?
- Is the force sufficient to cause an unacceptable residual or *hysteresis-type* increase in attenuation?

Table 8.5 lists some typical crush load ratings for FO cable.

8.2.8.G *High-Voltage Resistance*

In some applications, FO cables need to be nonconducting because of their proximity to high voltages. In other applications, FO cables must be resistant to the effects of lightning. In these situations, specify an all-dielectric cable. Such cables are readily available.

Table 8.5 Crush strength ratings for various FO cable types.

Typical cable crush strength ratings		
Characteristic	**Type of cable**	**Force lb/in**
Long-term crush load	6 fibers/cable	57 - 400 lb/in
	1 - 2 fiber cables	314 - 400 lb/in
	Armored cables	450 lb/in
Short-term crush load	6 fibers/cable	343 - 900 lb/in
	1 - 2 fiber cables	300 - 800 lb/in
	Armored cables	600 lb/in

8.2.8.H *Flexibility*

In some applications, cables are subject to repeated bending or flexing. Such uses dictate the need for minimum flexibility requirements for cable materials and fibers. Polyurethane jacketing is commonly used on cables exposed to such stresses. Although the jacketing increases the cost of the cable, it also increases flexibility to 10,000 cycles from the 1,000 cycle level common with the lower-cost PVC and polyethylene jacketing materials.

8.2.9 Optical Specifications

Determining the required optical specifications for a fiber system is the most complex step of the design process. Because system performance depends primarily on these parameters, careful consideration must be given to all of the optical requirements.

The system engineer can often rely on the optoelectronic vendor to recommend specifications for connectors and cables to be used in a system. The vendor should provide the following information:

- Cable core diameter
- Nominal or minimum attenuation rate
- Operating wavelength
- Loss per connector pair

This approach is practical when dealing with simple systems. Ease of use is the obvious advantage. Someone else does the specification work. This is the most conservative approach, but it has the potential disadvantage of excessive system costs. The higher costs result from the vendor's lack of vested interest in recommending the lowest possible cost combination. Or, the vendor may simply be conservative and recommend products with higher performance (and higher costs) than those actually needed.

As an alternative, the system engineer can compile the optical specifications. This approach requires a step-by-step determination of the basic optical power or loss requirements for each component. When determining the power budget, the designer considers separately the optical power losses from fiber, connectors, and any other passive components.

8.2.9.A *Evaluating Competitive Products*

After the basic design steps have been completed, the system engineer must select the FO cable and hardware manufacturers. Competitive products should be examined within the following parameters:

- *Uniformity of attenuation.* Although this parameter is not universally included in cable specifications, there are some situations in which it can become important. If jumper cables are to be cut from a longer length of cable and used in a system with a total optical power budget close to the limit of the optoelectronics, nonuniformity in the longer length can result in high-loss jumpers. A typical uniformity specification has no local losses greater than 0.1 dB.
- *Microbend sensitivity.* Microbend sensitivity can be tested easily. Connect a section of cable to a power meter and stabilized light source. Now bend the cable. If properly designed and manufactured, the cable will show no change in power transmission until the cable is bent to below its minimum recommended bend radius.
- *Consistency of stripability.* This is an important characteristic for tight tube designs. It is one that can be checked easily. Using the proper buffer tube stripper, repeatedly remove the buffer tube from the fiber. The best cables (in terms of stripability) will be those that consistently allow stripping without fiber damage on the first attempt. Poorly made cables are those that frequently break during the stripping process.
- *Ease of jacket removal.* The ease with which the jacket can be removed from the cable varies among manufacturers. Experience has shown that cables with rip cords take much less time to prepare for connector installation than those without. Cables of the same design, but from different manufacturers, usually have subtle differences. These differences can affect the overall installation time and cost.

8.2.9.B *Spare Lines*

Most users of fiber-optic technology discover that their needs increase with time. Therefore, it should be standard practice to install more fibers than initially required. Two techniques are commonly used to determine the number of spare fibers to install. The first is to estimate the growth needs of the facility. This is a difficult process to perform accurately. The second technique is to use an industry rule of thumb, which was developed by telephone companies. It requires the installation of two to three times the number of fibers required. The current industry ratio is approximately three times.

8.3 Fiber Transmission System Components

The optical source in a fiber transmission system is either a light-emitting diode (LED) or an injection laser diode (ILD). The LED is an incoherent source (with broad emitted spectral bandwidth) compared with an ILD. It is characterized by output power levels well below 1 mW, on the order of -10 to -30 dBm. It operates at slower speeds, but it also costs considerably less than ILDs. The LED is suitable for applications requiring transmission over less than 10 km and modulation bandwidths less than 100 Mb/s. LEDs operate in the 840 and 1,300 nm windows.

A laser light source requires more electronics to operate, but is more powerful, with outputs of approximately 0 dBm and higher. The practical upper power limit for a given application is a function of many considerations, including device cost and permissible radiation level. ILDs cost substantially more than LEDs. However, they operate in the 1,300 nm and low-attenuation 1,550 nm windows. The combination of low attenuation and high modulation frequency limits makes ILDs effective as sources in systems that must operate over extended distances, or that must transport large numbers of signals, or both.

Selecting an optical source for a fiber transmission system requires careful evaluation to ensure that its modulation frequency limit is greater than the bandwidth to be transmitted, and that it provides enough optical power at a wavelength that will satisfy the distance requirement. Another important consideration is that optical devices do not turn on with the linear characteristics of other electrical devices. Figure 8.16 shows the electrical current to optical power transfer characteristics of a typical LED. It is evident from the chart that modulation techniques that require extreme linearity in the electro-optical transducer are not suitable for fiber-optic applications.

8.3.1 Light Detector

A detector performs a complementary function to the source: It converts incident optical energy back to electrical energy. Detectors in fiber transmission systems are PIN (*positive intrinsic negative*) or APD (*avalanche photodiode*) semiconductors. The limiting sensitivity (threshold) in detecting weak incoming signals determines link performance.

In an ideal PIN diode, each incident photon creates an electron-hole pair in the semiconductor lattice. This, in turn, sets one electron flowing in the external circuit. If the received light is weak, the generated current may not be strong enough to overcome noise that is inherent in the diode and receiver circuit. In such cases, it is best to increase the detector output before amplification by the receiver. Such gain is generated in an avalanche photodiode. A reverse bias adds several electron energy volts to each liberated electron to provide amplification of the received signal, which is not linear with received levels.

8.3.2 Modulation Techniques

Information can be modulated onto an optical carrier in three ways:

- *Intensity modulation*, where light amplitude carries the information.
- *Frequency modulation*, where pulse speeds directly correspond to input signal levels.
- *Pulse code modulation* (PCM), where a digital code is used to convey input information

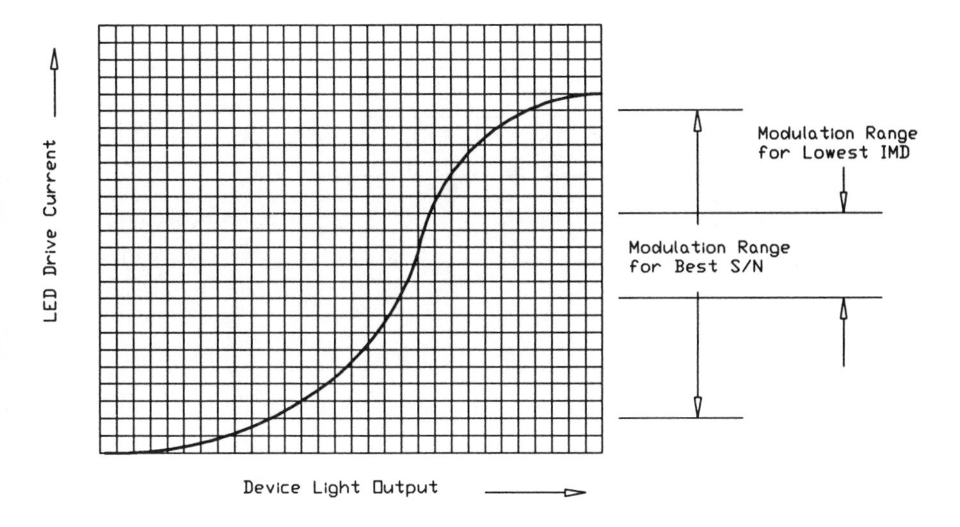

Figure 8.16 LED drive current/light output transfer characteristic. A restricted modulation index is necessary in analog signal modulation to limit harmonic and intermodulation distortion products.

8.3.2.A *Intensity Modulation*

Intensity modulation (referred to as IM or AM) is the simplest form of modulation. The instantaneous amplitude of the input signal directly controls the output intensity of the light source. Unfortunately, as shown in Figure 8.16, optical devices do not turn on in a linear fashion. As a result, *intermodulation distortion* (IMD) may be introduced into the signal when such modulation characteristics are implemented. Also, the *signal-to-noise ratio* (S/N) of the signal is directly correlated to the optical power. As a result, attenuation in a system lowers S/N directly. Further, because most optical devices do not completely turn off (particularly lasers), there is always an optical noise floor present.

8.3.2.B *Frequency Modulation*

Frequency modulation overcomes the S/N and IMD limitations of intensity modulation. FM is costly and requires more electronics, but it removes the problems of signal quality from the optical domain to the electrical domain. Required S/N and IMD performance specifications are achieved in the design of the frequency modulator. The highest performance output is a square pulse train, where the input signal information is carried in the time domain as pulse spacing.

8.3.2.C *PCM*

A fiber transmission system is most transparent to the signals being transported when it is operated in a pulsed mode. It is, therefore, particularly complementary to signals that are presented to it as fixed amplitude FM or a digital bit stream. In telephony applications, the relatively narrow bandwidth of a single voice channel can be efficiently digitized and multiplexed with up to 10,000 or more other similar channels

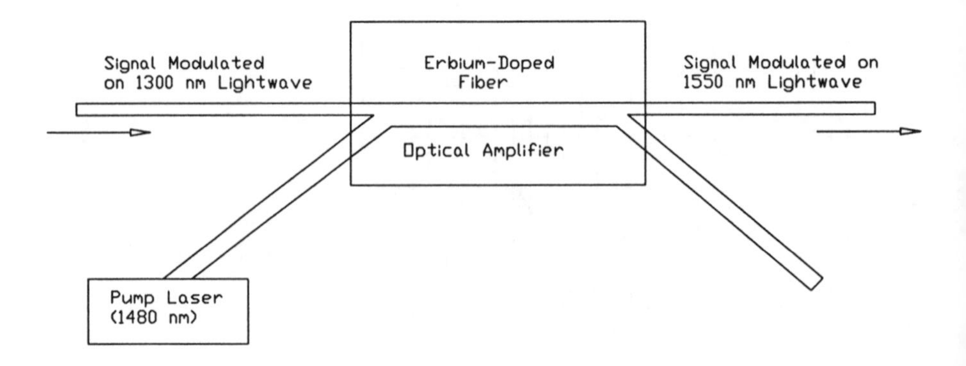

Figure 8.17 Block diagram of an optical amplifier (repeater).

on a single fiber. The received quality of a digitally transmitted voice channel is noticeably better than analog, because of the complete absence of impulse noise and distortion endemic to analog transmission.

8.3.3 Repeater

A fiber transmission circuit maximum acceptable length is a function of the following elements:

- Signal bandwidth to be transported
- Transmitter *launch power*
- Circuit fiber and interconnection component attenuation
- Fiber modal and chromatic dispersion characteristics
- Receiver threshold sensitivity
- End-to-end transmission circuit performance requirements set by the user

Analog circuits that are 50 km long are currently possible. Digital circuits equipped with bit error detection and correction overhead can operate for perhaps double this distance.

Analog and digital repeaters have been developed to extend transmission distances. First-generation repeaters operate at electrical baseband. The repeater terminal is literally a receiver whose demodulator is hardwired to the transmitter modulator. This design is crudely equivalent to creating a long-distance telephone circuit by holding the receiver of one old-fashioned two-handed telephone to the mouthpiece of another.

Analog-domain repeatering in this fashion has a limit of four or five tandem sections, and a transmission maximum of 150 to 200 km. Digital-domain repeaters (regenerators) have no limit, as long as transmission circuit bit errors are detected and corrected at each regenerator before new pulse train creation.

Second-generation repeaters increase the optical signal power in the light domain. Optical amplifiers (in prototype testing as of this writing) that accomplish this feat have no direct equivalent in electrical signal transmission. As illustrated in Figure

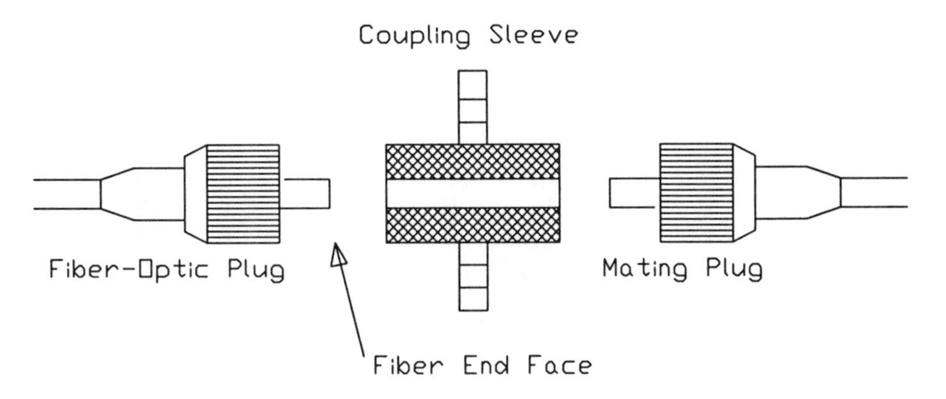

Figure 8.18 The mechanical arrangement of a simple fiber-optic connector.

8.17, they are constructed of a nominal 8-meter length of Erbium-doped fiber inserted in the optical transmission path. The 10 to 50 mW output of a pump laser operating at 980 or 1,490 nm is coupled into the fiber via a passive coupler. Its interaction with the optical carrier results in production of a new carrier modulated with the original information, but operating at a nominal 1,550 nm wavelength.

This presents a problem, because most of the fiber in the long-haul public network is optimized for 1,300 nm carrier transmission. The circuit performance payoff—optical light power increases of about 20 to 30 dB—is spurring researchers to find solutions to this problem.

8.3.4 Switching

First-generation fiber transmission circuit switchers operate at electrical baseband. Efforts to develop optical switchers are underway in telecommunications and computer industry research labs. Scientific conference reports indicate that second-generation optical-fiber switches will reach the market later this decade.

8.3.5 Connectors

The purpose of a fiber-optic connector is to efficiently convey the optical signal from one link or element to the next. Most connectors share a design similar to the assembly shown in Figure 8.18. Typically, connectors are plugs (male) and are mated to precision couplers or sleeves (female). While the specific mechanical design of each connector type varies from one manufacturer to the next, the basic concept is the same: provide precise alignment of the optical fiber cores through a ferrule in the coupler. Some connectors are designed to keep the fiber ends separated, while other designs permit the fiber ends to touch in order to reduce reflections resulting from the glass-to-air-to-glass transition.

The fiber is prepared and attached to the connector, usually with an adhesive or epoxy cement, and polished flush with the connector tip. Factory-installed connectors

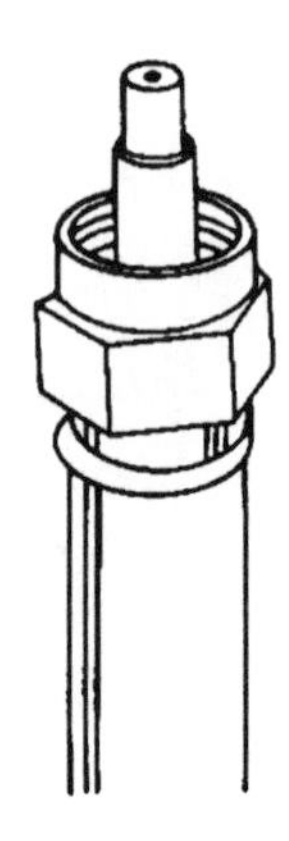
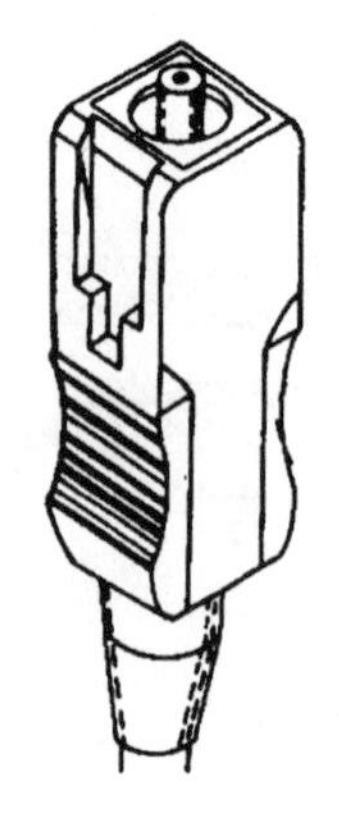
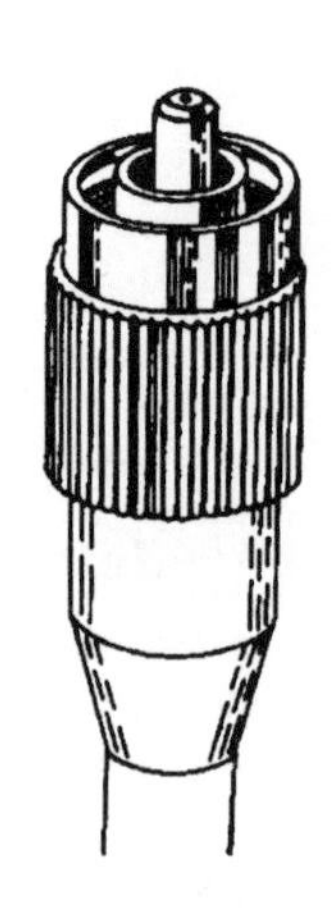

Figure 8.19 Common types of fiber-optic connectors.

typically use heat-cured epoxy and hand or machine polishing. Field installable connectors include epoxy-and-polish types, and crimp-on types. The crimp-on connector simplifies field assembly considerably.

The ferrule is a critical element of the connector. The ferrule functions to hold the fiber in place for optimum transmission of light energy. Ferrule materials include ceramics, stainless steel alloys, and glass.

8.3.5.A *Connector Properties*

There are many types of fiber-optic connectors. Each design has evolved to fill a specific application, or class of applications. Figure 8.19 shows three common fiber optic connectors.

The selection of a connector should take into consideration the following:

- Insertion loss
- Allowable loss budget for the fiber system
- Consistent loss characteristics over a minimum number of connect-disconnect cycles
- Sufficiently high return loss for proper system operation
- Ruggedness
- Compatibility with fiber connectors of the same type
- High tensile strength
- Stable thermal characteristics

As with any system that transports energy, the fewer number of connectors and/or splices, the better. Pigtail leads are often required between a fiber termination panel and the transmission/reception system; however, keep such links to a minimum. Figure 8.20 shows how the fiber-optic light source may be terminated as a panel-mounted connector in order to minimize the number of pigtail links.

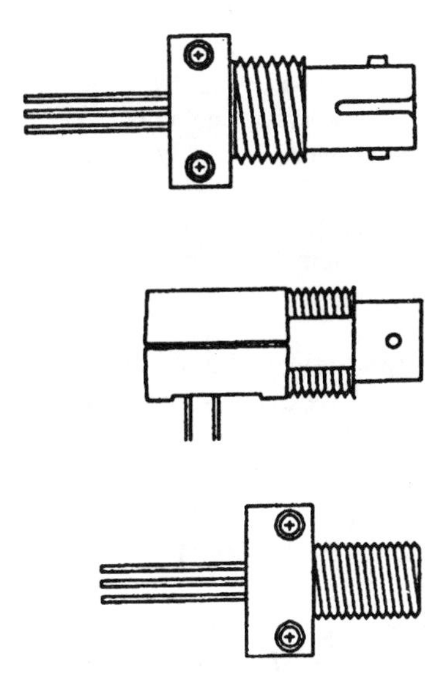

Figure 8.20 Circuit board mounted LED optical sources with connector terminations.

8.3.5.B *Performance Considerations*

The optical loss in a fiber optic connector is the primary measure of device quality. Connector loss specifications are derived by measuring the optical power through a length of fiber. Next, the fiber is cut in the center of its length and the connectors are attached and mated with a coupler. The power is then measured again at the end of the fiber. The additional loss in the link represents the loss in the connector.

Return loss is another important measurement of connector quality. Return loss is the optical power that is reflected toward the source by a connector. Connector return loss in a single-mode link, for example, can diffuse back into the laser cavity, degrading its stability. In a multi-mode link, return loss can cause extraneous signals, reducing overall performance.

8.4 Transmission System Options

Fiber-based data networks have evolved into six major operational classes:

- *Local area network* (LAN)—designed for inside plant or office communications.
- *Campus area network* (CAN)—designed for communications within an industrial or educational campus.
- *Metropolitan area network* (MAN)—designed for communications between different facilities within a given metropolitan area. MANs generally operate over common-carrier-owned switched networks that are installed in and over public rights of way.

- *Regional area network* (RAN)—interconnecting MANs within a unified geographic area, generally installed and owned by interexchange carriers (IECs).
- *Wide area network* (WAN)—communications systems operating over large geographic areas. Common-carrier networks interconnect MANs and RANs within a contiguous land mass, generally within a country's political boundaries.
- *Global area network* (GAN)—networks interconnecting WANs, both across countries' terrestrial borders and across ocean floors, including between continents.

LAN and CAN circuits carry a variety of multiplexed analog and/or digital signal transmissions on a single fiber. MAN and WAN circuits are also used to transmit either analog and digital signals, or both.

8.4.1 System Design Alternatives

The signal form at the input and/or output interface of a fiber system may be either analog or digital, and the number of independent electrical signals that can be transmitted on a fiber may be one or many.

Independent electrical signals may be combined into one signal for optical transmission by virtually unlimited combinations of electrical analog frequency division (analog AM and/or FM carriers) multiplexing (integration into one complex electrical signal) and digital bit stream multiplexing. Three primary multiplexing schemes are used for fiber transmission:

- *Frequency division multiplexing* (FDM)
- *Time division multiplexing* (TDM)
- *Wave(length) division multiplexing* (WDM)

8.4.1.A *Frequency Division Multiplexing*

The FDM technique of summing multiple AM or FM carriers is widely used in coaxial cable distribution. Unfortunately, the nonlinearity of optical devices operated in the intensity-modulation mode results in substantial, and often unacceptable, noise and intermodulation distortion in the delivered signal channel. Wide and selective spacing of carriers ameliorates this problem to some degree.

8.4.1.B *Time Division Multiplexing*

TDM involves sampling the input signals at a high rate, converting the samples to high-speed digital codes, and interleaving the codes into pre-determined time slots.

The principles of digital TDM are straightforward. Specific-length bit groups in a high-speed digital bit stream are repetitively allocated to carry the digital representations of individual analog signals and/or the outputs of separate digital devices.

8.4.1.C *Wave(length) Division Multiplexing*

This multiplexing technique, illustrated in Figure 8.21, reduces the number of optical fibers required to meet a specific transmission requirement. Two or more complete and independent fiber transmission systems, operating at different optical wavelengths, can be transported over a single fiber by combining them in a passive optical multiplexer. This is an assembly in which pigtails from multiple optical transmitters

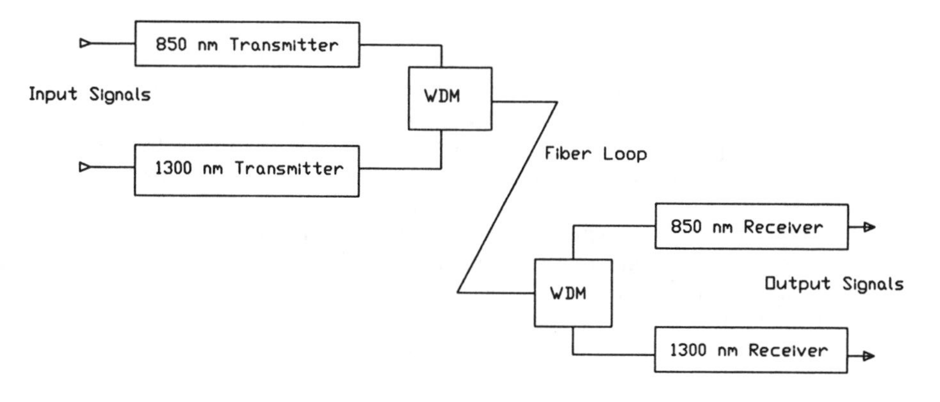

Figure 8.21 Block diagram of a wave division multiplexer—a passive optical assembly created by fusing optical fiber pigtails.

---

are fused and spliced into the transporting fiber. Demultiplexing these optical signals at the receiver is accomplished in an opposite-oriented passive optical multiplexer. The pigtails are coupled into photodetectors through wavelength-selective optical filters.

The ground rules for coupling light sources at different wavelengths are quite loose. A fiber might have two lightwave circuits of different wavelengths traveling in opposite directions. Three circuits can be transported without receiver optical filtering by using light sources from the 840, 1,300, and 1,550 nm wavelength windows. On the other hand, three discrete wavelengths from the 1,300 nm window have been used in some experiments. The light sources were selected at nominal 1,280, 1,300, and 1,320 nm wavelengths, with sharply tuned optical filters that provide receive-end separation.

Several considerations are important in designing such systems. First, adequate separation (measured in dB) of each optical channel is required. Optical crosstalk between channels results in lowering the effective S/N of the link. Second, passive multiplexers and optical filters result in some optical attenuation when used (1 to 5 dB per device). As a result, effective transmission distance is reduced significantly in WDM multiplexing. Finally, optical light sources change output wavelength over temperature. The tuned bandwidth of the WDM components must be wide enough to accommodate any expected drift.

Constantly dropping fiber costs currently make WDM effective when it is used to add capacity to existing installations. Further, because satisfactory optical isolation is costly and technically difficult to achieve, WDM techniques are only appropriate for FM and digital systems whose carrier-to-noise characteristics can accommodate the power losses created in the WDM system.

8.4.2 Operating Modes

Multi-mode fibers are used in transmission applications for which business and industrial customers now use coaxial cable and point-to-point microwave. Distances can range from a few hundred feet to many miles without a repeater. Single-mode fibers are now almost exclusively used by common carriers to provide transmission services over rights of way and in cables that they have installed and own. Optical repeaters extend the transmission range to 200 km.

Multi-mode transmitters typically use LEDs with a *launch* power of -15 dBm and lower, at wavelengths of either 850 or 1,300 nm. Receiver sensitivity must be 67 dBr for short-haul purposes. Typical receiver thresholds are -25 to -36 dBm. The *loss budget* available for establishing a circuit is the difference between these two values. Attenuation of multi-mode fiber is typically 3 to 4 dB/km at the 850 nm (short) wavelength, and 1 to 2 dB at the 1,300 nm (long) wavelength.

Single-mode transmitters are equipped with either LED or ILD light sources, operating at 1,300 and 1,550 nm wavelengths. ILD launch power ranges from a nominal 0 dBm (typically a -3 dBm consistently attainable level) to -13 dBm. Receivers for these wavelengths have nominal -25 to -36 dBm thresholds, providing system loss budgets from 12 to 36 dB. Attenuation of single-mode fiber is typically 0.5 dB/km at 1,300 nm, and as low as 0.25 dB/km at 1,550 nm.

8.4.3 Loss Budget Calculations

Designing a fiber transmission system begins with establishing the end-to-end performance requirements. After the overall system performance of the link is defined, the *optical loss budget* must be calculated. Items in the loss budget begin with the transmission circuit length, multiplied by the attenuation per kilometer of the selected fiber at the operating transmission wavelength. The loss budget must also include allowances for the following:

- Power losses through each connector pair and fusion splice in the circuit (normally well under 1 dB per junction).
- An allowance for light source power output decrease and fiber attenuation increase with age.
- An allowance for additional splices required by physical damage to the cable at some point in its useful life.

This allowance should represent at least a 3 to 6 dB deduction from the calculated loss budget, made before other maximum distance calculations begin. These calculations define the output power from the optical transmitter, and the receiver sensitivity needed to meet or exceed end-to-end signal transmission specifications.

An optical power meter can be used to measure the end-to-end attenuation of new construction after the optical transmitter is on line. Connecting the meter to the light source through an optical patch cord accounts for the first connector interface loss and establishes the actual power to be coupled into the fiber circuit. Connecting the meter to the receive end of the fiber verifies that the loss budget has been calculated accurately. A measured loss exceeding calculation by 1 dB or more should initiate further tests to verify that the fiber circuit installation meets expectations.

It is important to note that a fiber transmission circuit does not fail if its loss budget is exceeded. But for every decibel of arriving signal power level below the operating threshold of the receiver, subtract 3 dB from the system S/N for short-haul operation.

8.4.3.A *Evaluating Existing Systems*

Two testing techniques can be used to evaluate the physical condition and losses in an existing fiber circuit. An *optical time domain reflectometer* (OTDR) is used to measure attenuation and identify the location of breaks and bad splices in a fiber circuit. The OTDR operates on the principle of transmitting pulses of light and measuring the return time of all *back-scattered* (reflected) light from each pulse. This instrument provides a graphical representation of the fiber's condition as a function of distance. It can literally pinpoint a break in a fiber to approximately 1 m over tens of kilometers. By selecting a test light wavelength not present in the fiber, the measurement can be performed while the circuit is in use. This testing is important in the evaluation of circuits that are candidates for WDM capacity expansion. It is also useful before specifying new transmission systems using existing fiber. The testing will uncover routing, splices, and damage that are not documented.

8.4.4 Transmission System Parameters

In a fiber-optic transmission system, optimum performance specifications are achieved over wide analog bandwidths or at high digital transmission rates by designing the transmitter/modulator, fiber, and receiver detector/current amplifier as a matched set of components, to carry one precisely conditioned signal. The fiber has incredible bandwidth-distance characteristics. The bandwidth characteristics can be controlled in the manufacturing process. The limiting transmission distance is determined by a combination of the core diameter of the fiber and the spectral bandwidth of the light source. But the signal quality received is rigorously limited by nonlinearities in the electron-to-photon and photon-to-electron energy conversion (*transduction*) processes.

A broadband coaxial cable transmission system is conceptually quite different. The cable has a distance (resistive) attenuation determined by the copper conductors, and a frequency vs. attenuation characteristic determined by conductor-to-shield spacing and the dielectric material. These characteristics do not change with changes in the quantity of input channels or the spectral bandwidth of the channels. If each transmitted channel output is confined to its assigned spectral bandwidth, and is detected and demodulated in an equally well-designed receiver, that channel output quality will not be degraded by the presence of other channels on the cable. If, however, the entire spectrum of individual channels must be processed through broadband amplifiers (repeaters) before reaching their common destination, the result is no better and may often be worse than mixing all of the signals initially into one light source driver.

8.5 Bibliography

Ajemian, Ronald G., "Fiber Optic Connector Considerations for Professional Audio," *Journal of the Audio Engineering Society*, Audio Engineering Society, New York, June 1992.

Crutchfield, E. B.: *NAB Engineering Handbook*, 8th ed., National Association of Broadcasters, Washington, D.C., 1992.

Pearson, Eric: *How to Specify and Choose Fiber-Optic Cables,* Pearson Technologies, Acworth, GA, 1991.

CHAPTER 9

GROUNDING PRACTICES

9.1 Introduction

The primary purpose of grounding electronic hardware is to prevent electrical shock. The National Electrical Code (NEC) and local building codes are designed to provide for the safety of the workplace. Local codes must always be followed. Occasionally, code sections are open for some interpretation. If in doubt, consult a field inspector.

Codes constantly are being changed or expanded, because new situations arise that were not anticipated when the codes were written. Sometimes, an interpretation will depend on whether the governing safety standard applies to building wiring or to a factory-assembled product to be installed in a building. Underwriters Laboratories (UL) and other qualified testing organizations examine products at the request and expense of manufacturers or purchasers. They "list" products if the examination reveals that the device or system presents no significant safety hazard when installed and used properly.

Municipal and county safety inspectors generally accept UL and other qualified testing laboratory certification listings as evidence that a product is safe to install. Without a listing, the end user may not be able to obtain the necessary wiring permits and inspection sign-off. On-site wiring must conform to local wiring codes. Most codes are based on the NEC. Electrical codes specify wiring materials, wiring devices, circuit protection, and wiring methods.

9.1.1 Planning the Ground System

The attention given to the design and installation of a facility ground system is a key element in the day-to-day reliability of the plant. A well-designed and -installed ground network is invisible to the engineering staff. A marginal ground system, however, can cause problems on a regular basis. Grounding schemes range from simple to complex, but any system serves three primary purposes:

- It provides for operator safety.
- It protects electronic equipment from damage caused by transient disturbances.
- It diverts stray RF energy from sensitive audio, video, control, and computer equipment.

Most engineers view grounding mainly as a method to protect equipment from damage or malfunction. However, the most important element is operator safety. The 120 or 208 Vac line current that powers most equipment can be dangerous—even deadly—if handled improperly. Grounding of equipment and structures provides protection against wiring errors or faults that could endanger human life.

Proper grounding is basic to protection against ac line disturbances. This applies whether the source of the disturbance is lightning, power-system switching activities, or faults in the distribution network. Proper grounding is also a key element in preventing RF interference in transmission or computer equipment. A facility with a poor ground system may experience RFI problems on a regular basis.

Implementing an effective ground network is not an easy task. It requires planning, quality components, and skilled installers. It is not inexpensive. However, proper grounding is an investment that will pay off in facility reliability.

A ground system consists of two key elements:

- The earth-to-grounding electrode interface outside the facility.
- The ac power and signal-wiring systems inside the facility.

9.2 Establishing an Earth Ground

The grounding electrode is the primary element of any ground system. The electrode can take many forms. In all cases, its purpose is to interface the electrode (a conductor) with the earth (a semiconductor). Grounding principles have been refined to a science. Still, however, many misconceptions exist. An understanding of proper grounding procedures begins with the basic earth-interface mechanism.

9.2.1 Grounding Interface

The grounding electrode (or ground rod) interacts with the earth to create a hemisphere-shaped volume. This is illustrated in Figure 9.1. The size of this volume is related to the size of the grounding electrode. The length of the electrode has a much greater effect than the diameter. Studies have demonstrated that the earth-to-electrode resistance from a driven ground rod increases exponentially with the distance from that rod. At a given point, the change becomes insignificant. It has been found that for maximum effectiveness of the earth-to-electrode interface, each ground rod must have a hemisphere-shaped volume with a diameter that is approximately 2.2 times the rod's length.

The constraints of economics and available real estate place practical limitations on the installation of a ground system. It is important, however, to keep the 2.2 rule in mind, because it allows the facility design engineer to take advantage of the available resources. Figure 9.2 illustrates the effects of locating ground rods too close (less than 2.2 times the rod length). An overlap area is created that effectively wastes

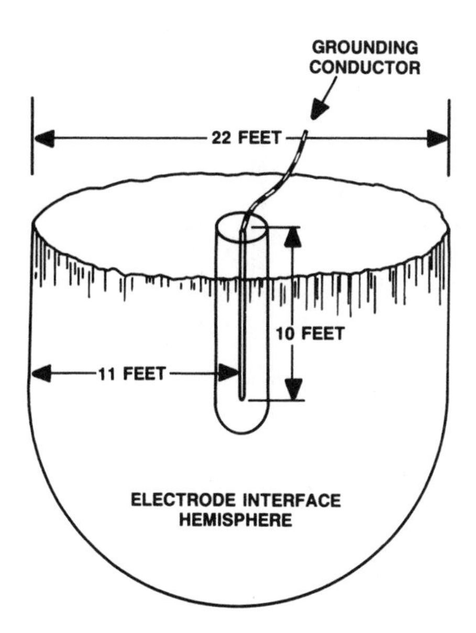

Figure 9.1 The effective earth-interface hemisphere resulting from a single driven ground rod. The 90 percent effective area of the rod extends to a radius of approximately 1.1 times the rod's length.

---

some of the earth-to-electrode capabilities of the two ground rods. Research has shown, for example, that two 10 ft ground rods driven only 1 ft apart provide about the same resistivity as a single 10 ft rod.

There are two schools of thought with regard to ground-rod length. The first is that extending ground-rod length beyond about 10 ft is of little value for most types of soil. The reason is presented in Figure 9.3, where ground resistance is plotted as a function of ground-rod length. Beyond 10 ft, a point of diminishing returns is reached. The second school of thought is that optimum earth-to-electrode interface is achieved with long (40 ft or greater) rods, driven to penetrate the local water table. With this type of installation, consider the difficulty in attempting to drive long ground rods. This discussion assumes that the composition of the soil around the grounding electrode is reasonably uniform. Depending on the location, however, this may not be the case.

Horizontal grounding electrodes provide essentially the same resistivity as an equivalent-length vertical electrode. As Figure 9.4 demonstrates, the difference between a 10 ft vertical and a 10 ft horizontal ground rod is negligible (275 Ω vs. 250 Ω). This comparison includes the effects of the vertical connection element from the surface of the ground to the horizontal rod. By itself, the horizontal ground rod provides an earth-interface resistivity of approximately 308 Ω when buried at a depth of 36-in.

Ground rods come in many sizes and lengths. The more popular sizes are 1/2-, 5/8-, 3/4-, and 1-in. The 1/2-in size is available in steel, with stainless-clad, galvanized, or copper-clad rods. All-stainless-steel rods also are available. Ground rods can be purchased in unthreaded or threaded (sectional) lengths. The sectional sizes are typically 9/16- or 1/2-in rolled threads. Couplers are made from the same materials as

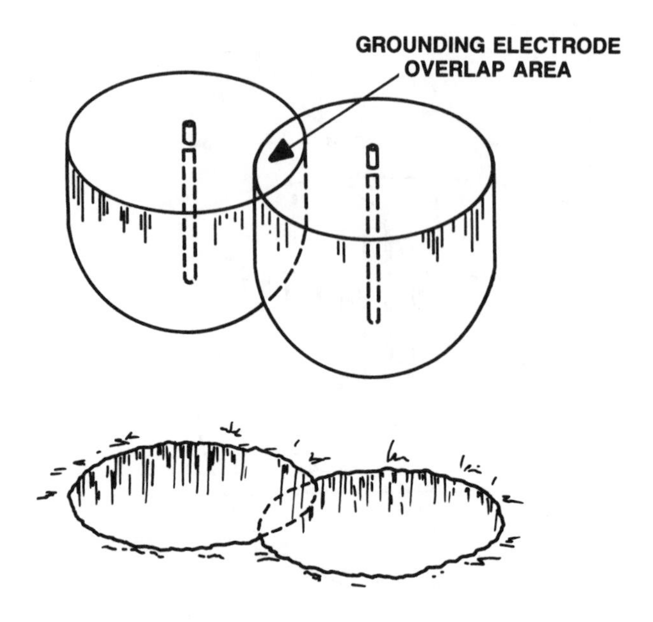

Figure 9.2 The effect of overlapping earth-interface hemispheres by placing two ground rods at a spacing less than 2.2 times the length of either rod. The overlap area represents wasted earth-to-grounding electrode interface capability.

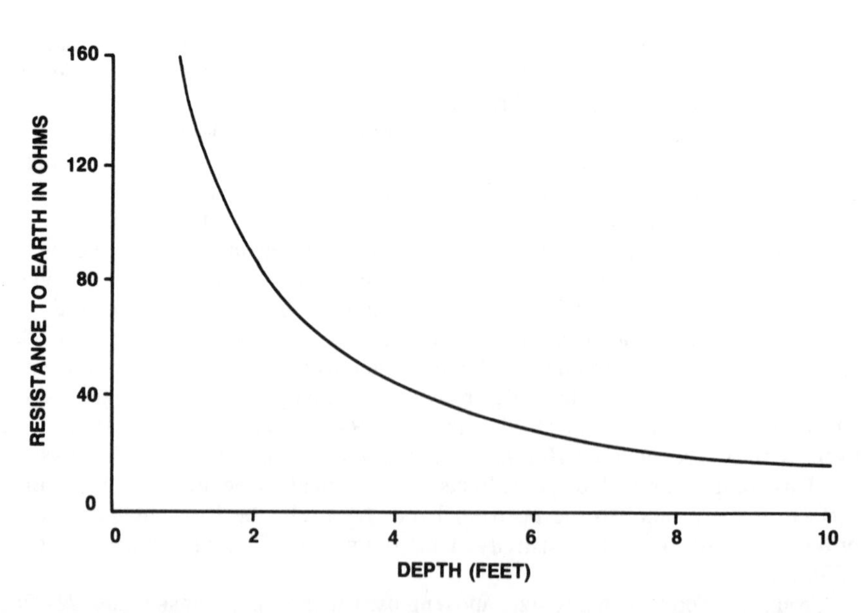

Figure 9.3 Charted grounding resistance as a function of ground-rod length. Ground-rod length in excess of 10 ft produces diminishing returns. The chart applies to a 1-in-diameter rod.

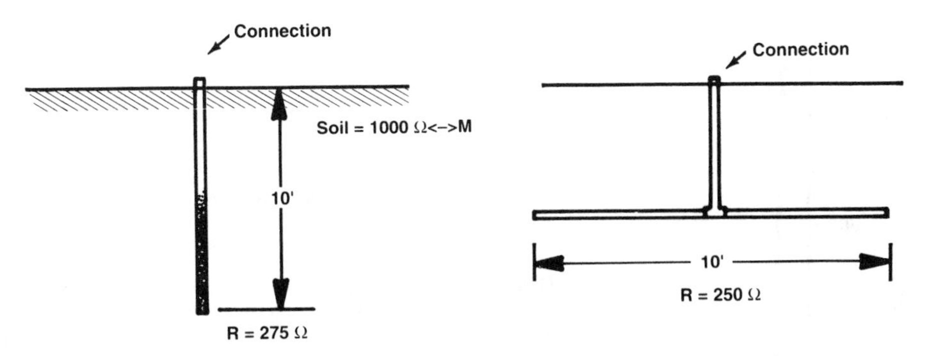

Figure 9.4 The effectiveness of vertical ground rods compared with horizontal ground rods.

the rods. These couplers can be used to join 8 or 10 ft rods together. A 40 ft ground rod is driven one 10 ft section at a time.

The type and size of ground rod used is determined by how many sections are to be connected and how hard or rocky the soil is. Copper-clad 5/8-in x 10 ft rods are probably the most popular. Rod diameter has minimal effect on final ground impedance. Copper cladding is designed to prevent rust, not for better conductivity. Although the copper certainly provides a better conductor interface to earth, the steel that it covers is also an excellent conductor when compared with ground conductivity. The thickness of the cladding is important only as far as rust protection is concerned.

9.2.1.A *Soil Resistivity*

Wide variations in soil resistivity can be found within a given geographic area, as documented in Table 9.1. The wide range of values results from differences in moisture content and mineral content, and from temperature.

9.2.2 Chemical Ground Rods

A chemically activated ground system is a common alternative to the conventional ground rod. The idea behind it is to increase the earth-to-electrode interface by

Table 9.1 Typical resistivity of common soil types

Type of soil	Resistivity in Ω/cm		
	Average	Minimum	Maximum
Filled land, ashes, salt marsh	2,400	600	7,000
Top soils, loam	4,100	340	16,000
Hybrid soils	16,000	1,000	135,000
Sand and gravel	90,000	60,000	460,000

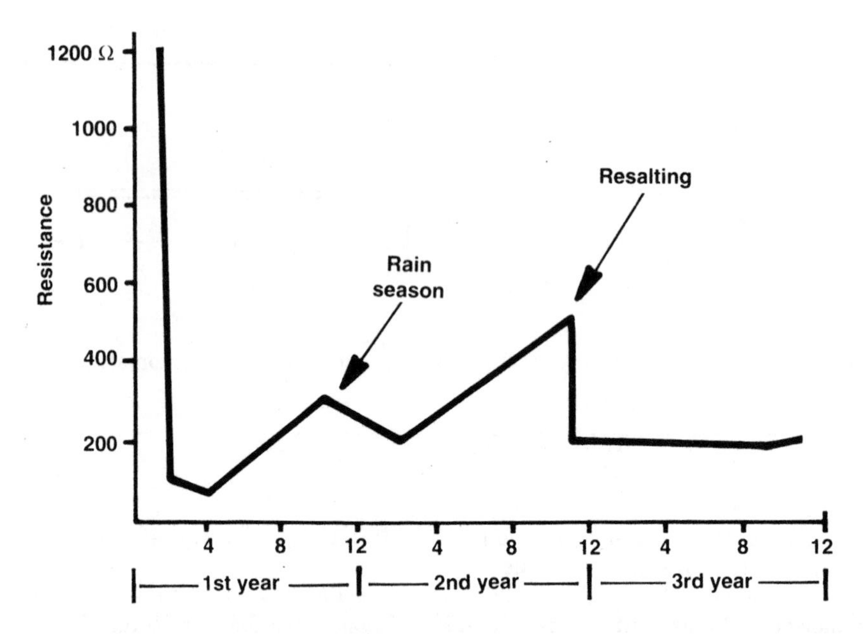

Figure 9.5 The effect of soil salting on ground-rod resistance with time. The expected resalting period, shown here as two years, varies depending on the local soil conditions and the amount of moisture present.

---

conditioning the soil surrounding the rod. Experts have known for many years that the addition of ordinary table salt (NaCl) to soil will reduce the resistivity of the earth-to-ground electrode interface. With the proper soil moisture level (4 to 12 percent), *salting* can reduce soil resistivity from 10,000 Ω/m to less than 100 Ω/m. Salting the area surrounding a ground rod (or group of rods) follows a predictable life-cycle pattern, which is illustrated in Figure 9.5. Subsequent salt applications are rarely as effective as the initial salting.

Various approaches have been tried over the years to solve this problem. One product is shown in Figure 9.6. This chemically activated grounding electrode consists of a 2 1/2-in-diameter copper pipe filled with rock salt. Breathing holes are provided on the top of the assembly, and seepage holes are located at the bottom. The theory of operation is simple: Moisture is absorbed from the air (when available) and is then absorbed by the salt. This creates a solution that seeps out of the base of the device and conditions the soil in the immediate vicinity of the rod.

Another approach is shown in Figure 9.7. This device incorporates a number of ports (holes) in the assembly. Moisture from the soil (and rain) is absorbed through the ports. The metallic salts subsequently absorb the moisture, forming a saturated solution that seeps out of the ports and into the earth-to-electrode hemisphere. Tests have shown that if the moisture content is within the required range, earth resistivity can be reduced by as much as 100:1. Figure 9.8 shows the measured performance of a typical chemical ground rod in three types of soil.

Implementations of chemical ground-rod systems vary depending on the application. Figure 9.9 illustrates a counterpoise ground consisting of multiple leaching

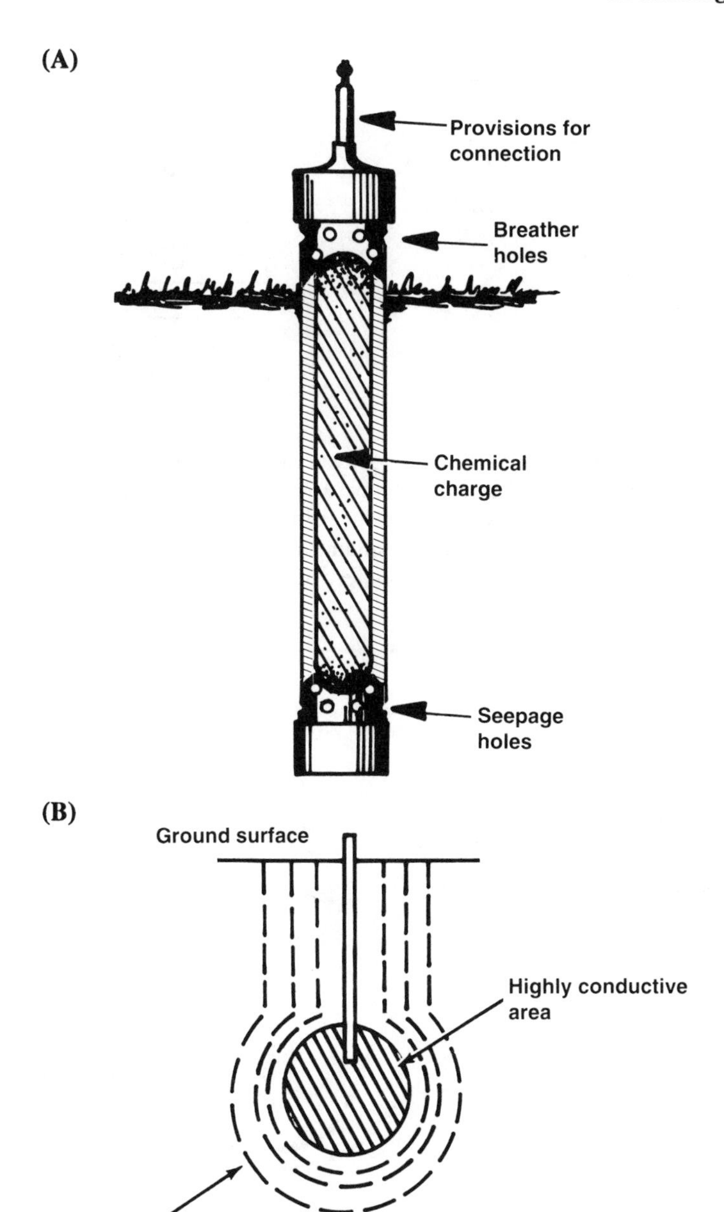

Figure 9.6 An air-breathing chemically activated ground rod: (a) breather holes at the top of the device permit moisture penetration into the chemical charge section of the rod; (b) a salt solution seeps out of the bottom of the unit to form a conductive shell. (Adapted from: Roy Carpenter, "Improved Grounding Methods for Broadcasters," *Proceedings of the SBE National Convention*, Society of Broadcast Engineers, Indianapolis, 1987.)

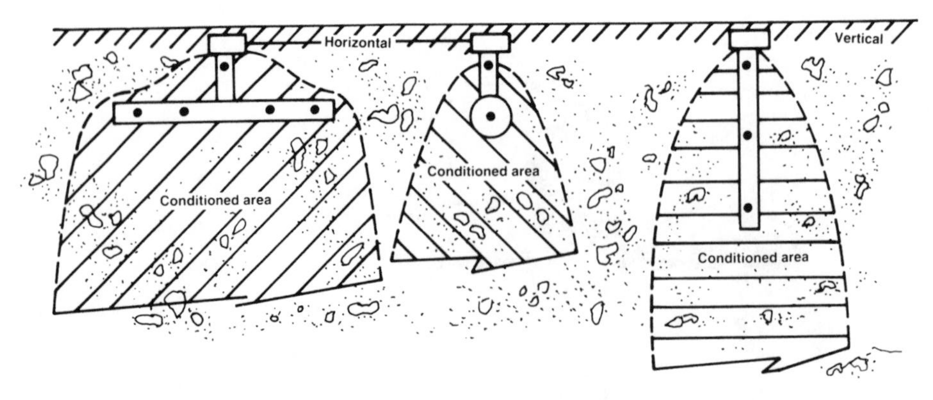

Figure 9.7 An alternative approach to the chemically activated ground rod. Multiple holes are provided on the ground-rod assembly to increase the effective earth-to-electrode interface. Note that chemical rods can be produced in a variety of configurations. (Adapted from: Roy Carpenter, "Improved Grounding Methods for Broadcasters," *Proceedings of the SBE National Convention*, Society of Broadcast Engineers, Indianapolis, 1987.)

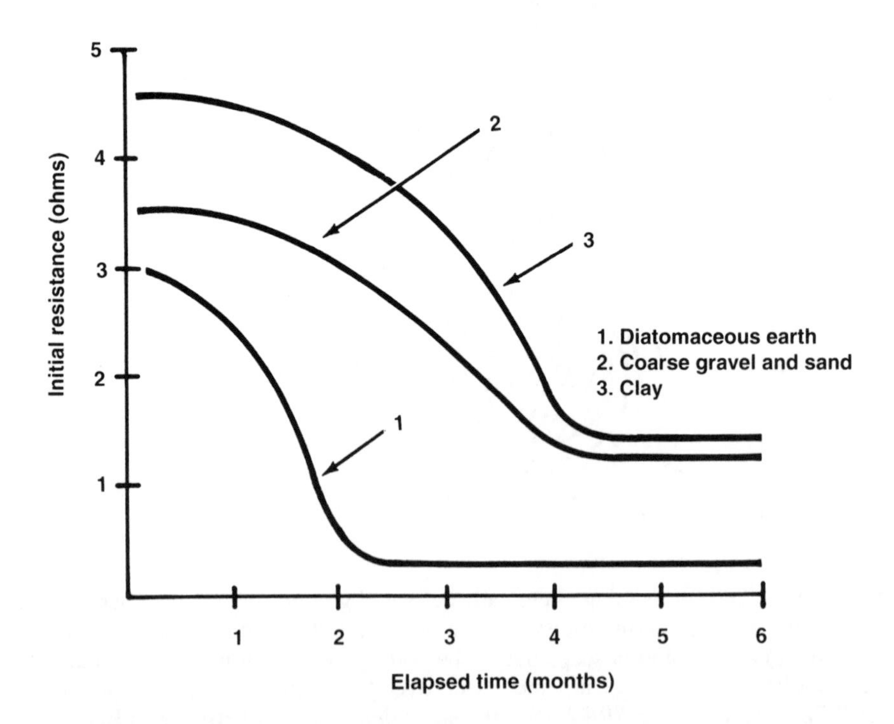

Figure 9.8 Measured performance of a chemical ground rod. (Adapted from: Roy Carpenter, "Improved Grounding Methods for Broadcasters," *Proceedings of the SBE National Convention*, Society of Broadcast Engineers, Indianapolis, 1987.)

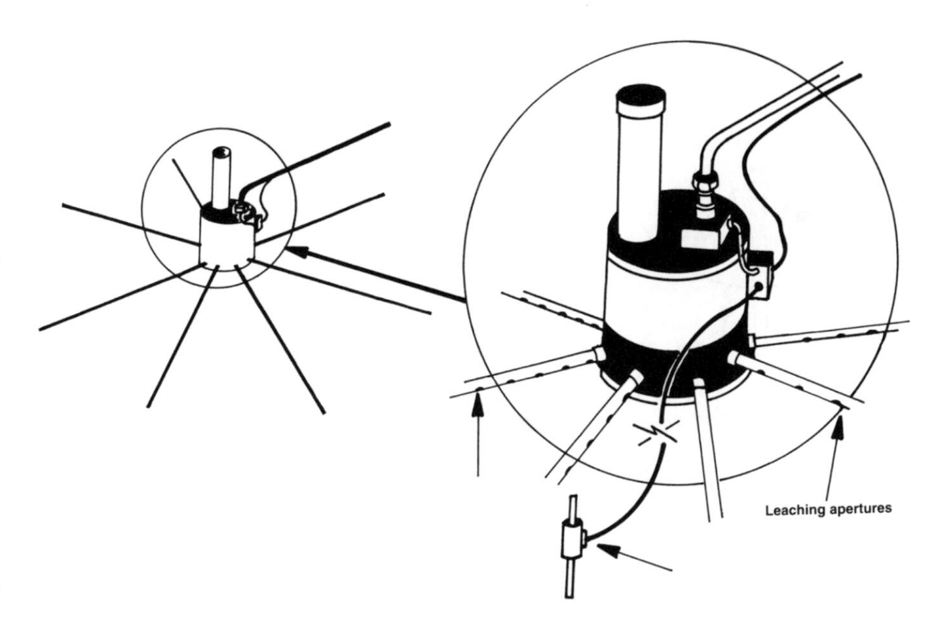

Figure 9.9 Hub and spoke counterpoise ground system.

apertures connected to a central hub in a spoke fashion. The system is serviceable because additional salt compound can be added to the hub at required intervals to maintain the effectiveness of the ground. Figure 9.10 shows a counterpoise system consisting of individual chemical ground rods interconnected with radial wires that are buried below the surface.

9.2.3 Ufer Ground System

Driving ground rods is not the only method of achieving a good earth-to-electrode interface. The concept of the *Ufer ground* has gained interest because of its simplicity and effectiveness. The Ufer approach (named for its developer), however, must be designed into a new structure. It cannot be added later. The Ufer ground takes advantage of the natural chemical- and water-retention properties of concrete to provide an earth ground. Concrete retains moisture for 15 to 30 days after a rain. The material has a ready supply of ions to conduct current because of its moisture-retention properties, mineral content, and inherent pH. The large mass of any concrete foundation provides a good interface to ground.

A Ufer system, in its simplest form, is made by routing a solid-copper wire (no. 4 gauge or larger) within the foundation footing forms before concrete is poured. Figure 9.11 shows one such installation. The length of the conductor run within the concrete is important. Typically, a 20 ft run (10 ft in each direction) provides a 5 Ω ground in 1,000 Ω/m soil.

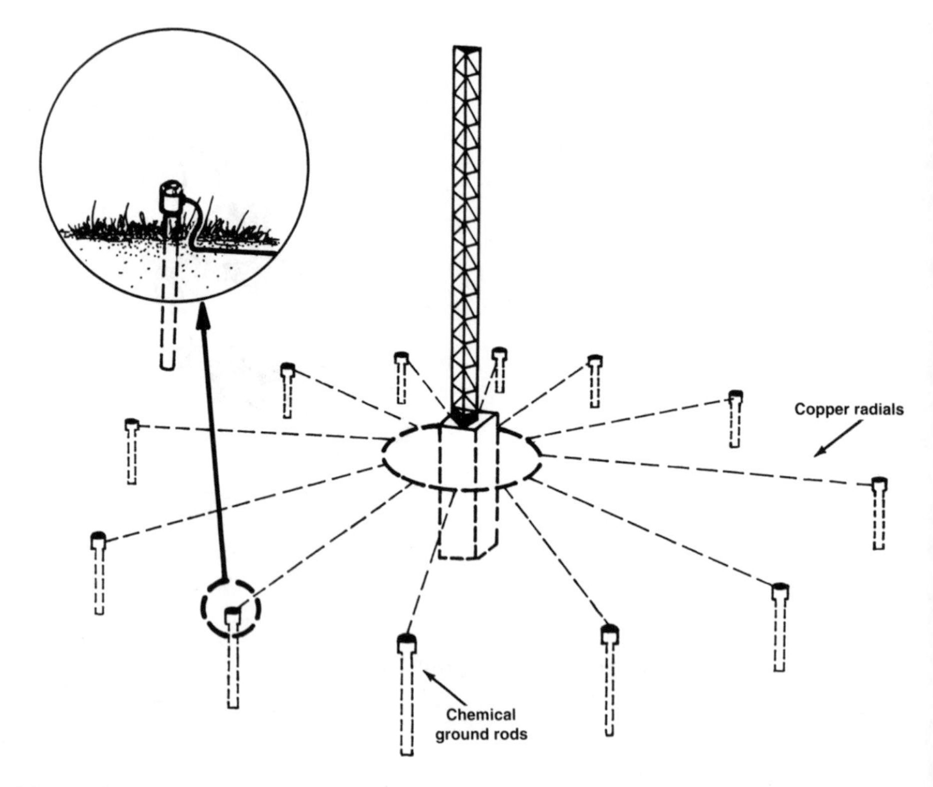

Figure 9.10 Tower grounding scheme using buried copper radials and chemical ground rods.

As an alternative, steel reinforcement bars (rebar) can be welded together to provide a rigid, conductive structure. A ground lug is provided to tie equipment to the ground system in the foundation. The rebar must be welded, not tied, together. If it is only tied, the resulting poor connections between rods can result in arcing during a current surge. This can lead to deterioration of the concrete in the affected areas.

The design of a Ufer ground is not to be taken lightly. Improper installation can result in a ground system that is subject to problems. The grounding electrodes must be kept a minimum of 3-in from the bottom and sides of the concrete to avoid the possibility of foundation damage during a large lightning strike. If an electrode is placed too near the edge of the concrete, a surge could turn the water inside the concrete to steam and break the foundation apart.

The Ufer approach also can be applied to guy-anchor points or the tower base, as illustrated in Figure 9.12. Welded rebar or ground rods sledged in place after the rebar cage is in position may be used. By protruding below the bottom concrete surface, the ground rods add to the overall electrode length to help avoid thermal effects that may crack the concrete. The maximum length necessary to avoid breaking the concrete under a lightning discharge is determined by the following:

- Type of concrete (density, resistivity, and other factors)

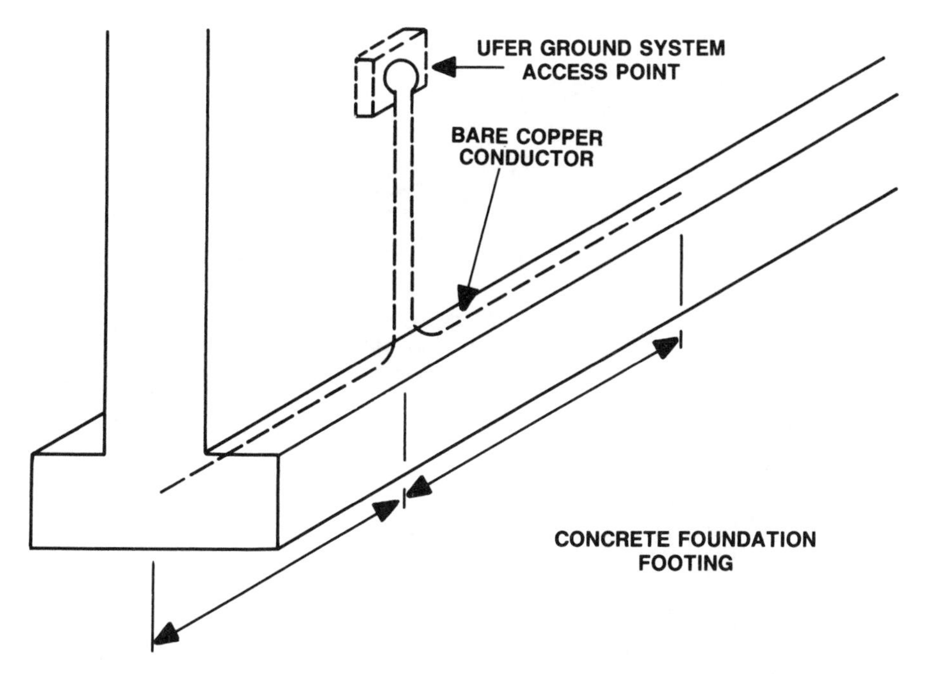

Figure 9.11 The basic concept of a Ufer ground system, which relies on the moisture-retentive properties of concrete to provide a large earth-to-electrode interface. Design of such a system is critical. Do not attempt to build a Ufer ground without the assistance of an experienced contractor.

- Water content of the concrete
- How much of the buried concrete surface area is in contact with the ground
- Ground resistivity
- Ground water content
- Size and length of the ground rod
- Size of lightning flash

The last variable is a gamble. The 50 percent mean occurrence of lightning strikes is 18 A, but super strikes can occur that approach 100 to 200 kA.

Before implementing a Ufer ground system, consult a qualified contractor. Because the Ufer ground system will be the primary grounding element for the facility, it must be installed correctly.

9.3 Bonding Ground-System Elements

A ground system is only as good as the methods used to interconnect the component parts. Do not use soldered-only connections outside the equipment building. Crimped/brazed and *exothermic* (*Cadwelded*) connections are preferred. To make a proper bond, all metal surfaces must be cleaned, any finish removed to bare metal,

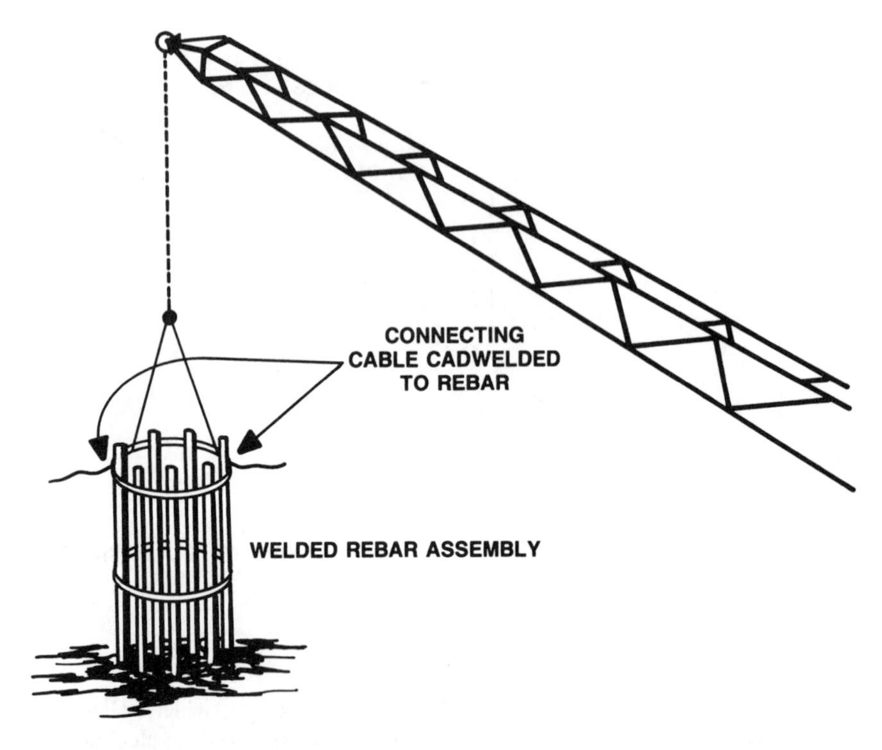

Figure 9.12 The Ufer ground system as applied to a transmission-tower base or guy-wire anchor point. When using this type of ground system, bond all rebar securely to prevent arcing in the presence of large surge currents.

---

and surface preparation compound applied. Protect all connections from moisture by appropriate means—usually sealing compound and heat-shrink tubing.

It is not uncommon for an untrained installer to use soft solder to connect the elements of a ground system. Such a system is doomed from the start. Soft-soldered connections cannot stand up to the acid and mechanical stress imposed by the soil. The most common method of connecting the components is silver-soldering. The process requires the use of brazing equipment, which may be unfamiliar to many system engineers. The process uses a high-temperature/high-conductivity solder to complete the bonding process. For most grounding systems, however, the best bonding approach is the Cadwelding process. (Cadweld is a registered trademark of Erico Corporation.)

9.3.1 Cadwelding

Cadwelding is the preferred method of connecting the elements of a ground system. Molten copper is used to melt connections together, forming a permanent bond. This process is particularly useful in joining dissimilar metals. In fact, if copper and galvanized cable must be joined, Cadwelding is the only acceptable means. The

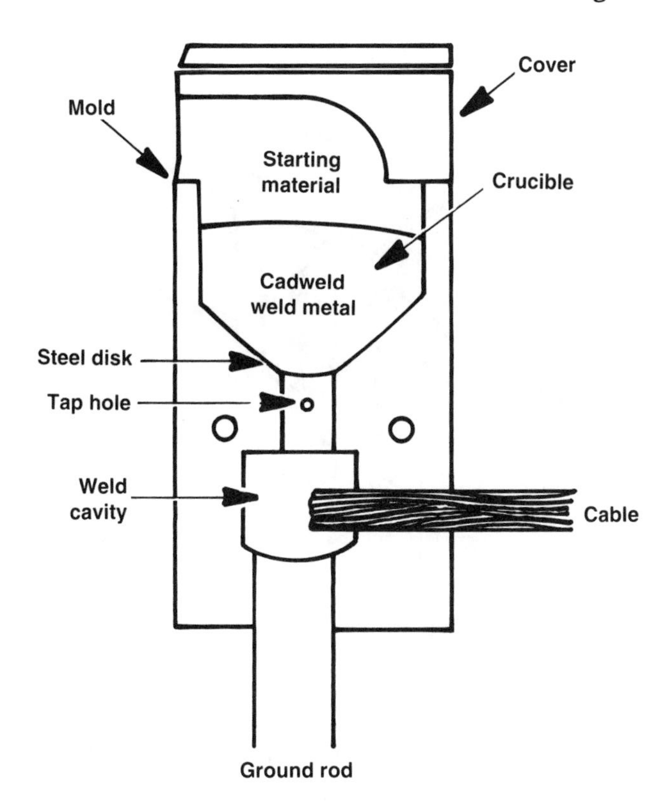

Figure 9.13 Typical Cadweld mold for connecting a cable to a ground rod.

completed connection will not loosen or corrode and will carry as much current as the cable connected to it.

Cadwelding is accomplished by dumping powdered metals (copper oxide and aluminum) from a container into a graphite crucible and igniting the material with a flint lighter. Reduction of the copper oxide by the aluminum produces molten copper and aluminum oxide slag. The molten copper flows over the conductors, bonding them together.

Figure 9.13 shows a typical Cadweld mold. A variety of special-purpose molds are available to join different-size cables and copper strap. Figure 9.14 shows a bonding form for a copper-strap-to-ground-rod interface.

9.3.2 Ground-System Inductance

Conductors interconnecting sections or components of an earth ground system must be kept as short as possible to be effective. The inductance of a conductor is a major factor in its characteristic impedance to surge energy. For example, consider a no. 6 AWG copper wire 10 m long. The wire has a dc resistance of 0.013 Ω and an inductance of 10 μH. For a 1,000 A lightning surge with a 1 μs rise time, the resistive

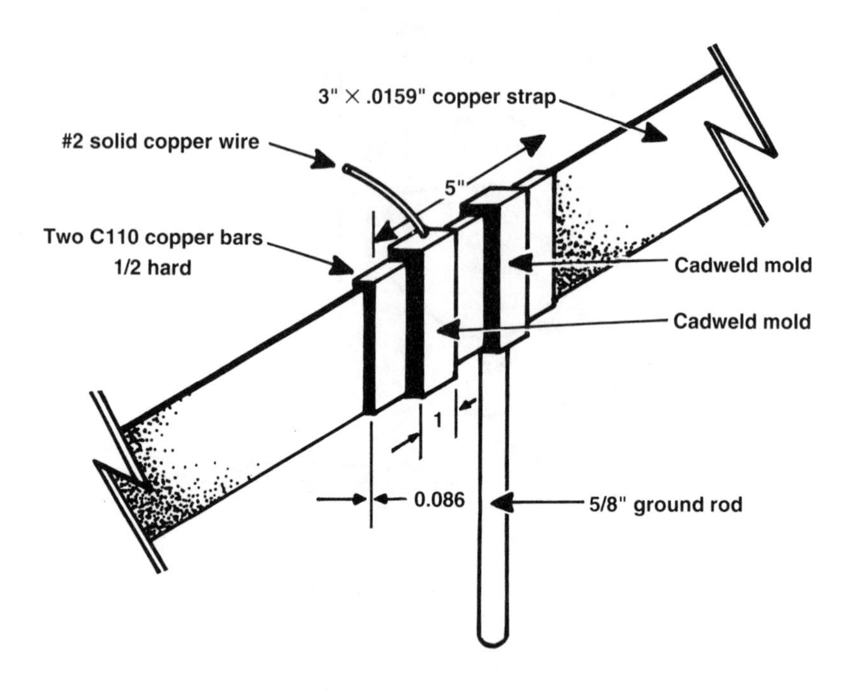

Figure 9.14 Cadweld mold for connecting a copper strap to a ground rod.

voltage drop will be 13 V, but the reactive voltage drop will be 10 kV. Further, any bends in the conductor will increase its inductance and further decrease the effectiveness of the wire. Bends in ground conductors should be gradual. A 90° bend is electrically equivalent to a 1/4-turn coil. The sharper the bend, the greater the inductance.

Because of the fast rise time of most lightning discharges and power-line transients, the *skin effect* plays an important role in ground-conductor selection. When planning a facility ground system, view the project from an RF standpoint.

9.4 Designing a Building Ground System

After determining the required grounding elements, they must be connected together in a unified system. Many different approaches may be taken, but the goal is the same: Establish a low-resistance, low-inductance path to surge energy. Figure 9.15 shows a building ground system using a combination of ground rods and buried bare-copper radial wires. This design is appropriate when the building is large or when it is located in an urban area. This approach also may be used when the facility is located in a high-rise building that requires a separate ground system. Most newer office buildings have ground systems designed into them. If a comprehensive building ground system is provided, use it. For older structures (constructed of wood or brick), a separate ground system is required.

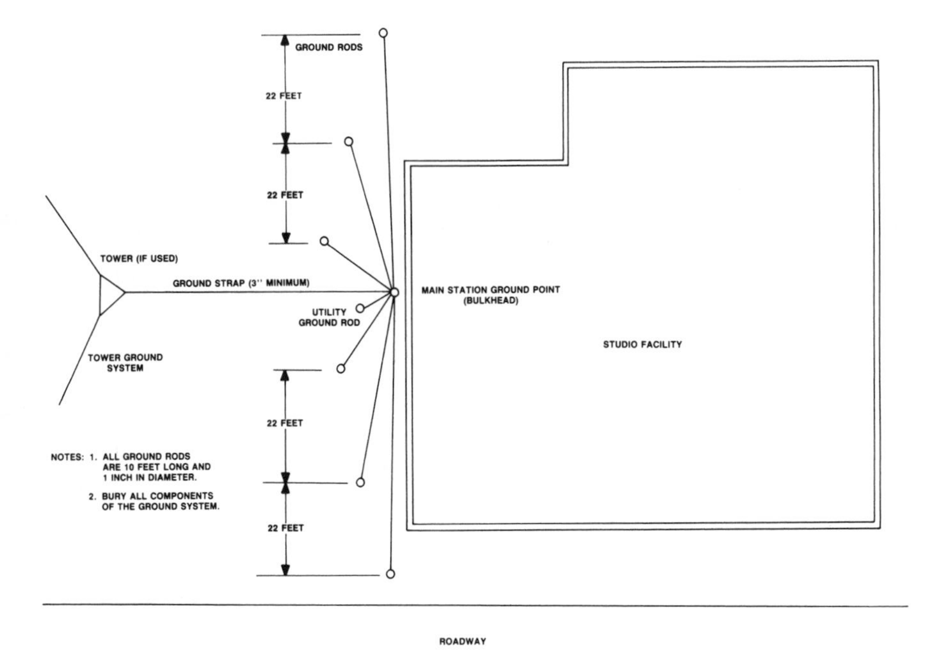

Figure 9.15 A facility ground system using the hub-and-spoke approach. The available real estate at the site will dictate the exact configuration of the ground system. If a tower is located at the site, the tower ground system is connected to the building ground, as shown.

Figure 9.16 shows another approach in which a perimeter ground strap is buried around the building and ground rods are driven into the earth at regular intervals (2.2 times the rod length). The ground ring consists of a one-piece copper conductor that is bonded to each ground rod.

If a transmission or microwave tower is located at the site, connect the tower ground system to the main ground point via a copper strap. The width of the strap must be at least 1 percent of the length and, in any event, not less than 3-in wide. The building ground system is not a substitute for a tower ground system, no matter the size of the tower. The two systems are treated as independent elements, except for the point at which they interconnect.

Tie the utility company power system ground rod to the main facility ground point as required by the local electrical code. Do not consider the building ground system to be a substitute for the utility company ground rod. The utility rod is important for safety reasons and must not be disconnected or moved. Do not remove any existing earth ground connections to the power line neutral connection. Doing so may violate local electrical code.

Bury all elements of the ground system to reduce the inductance of the overall network. Do not make sharp turns or bends in the interconnecting wires. Straight, direct wiring practices reduce the overall inductance of the system and increase its

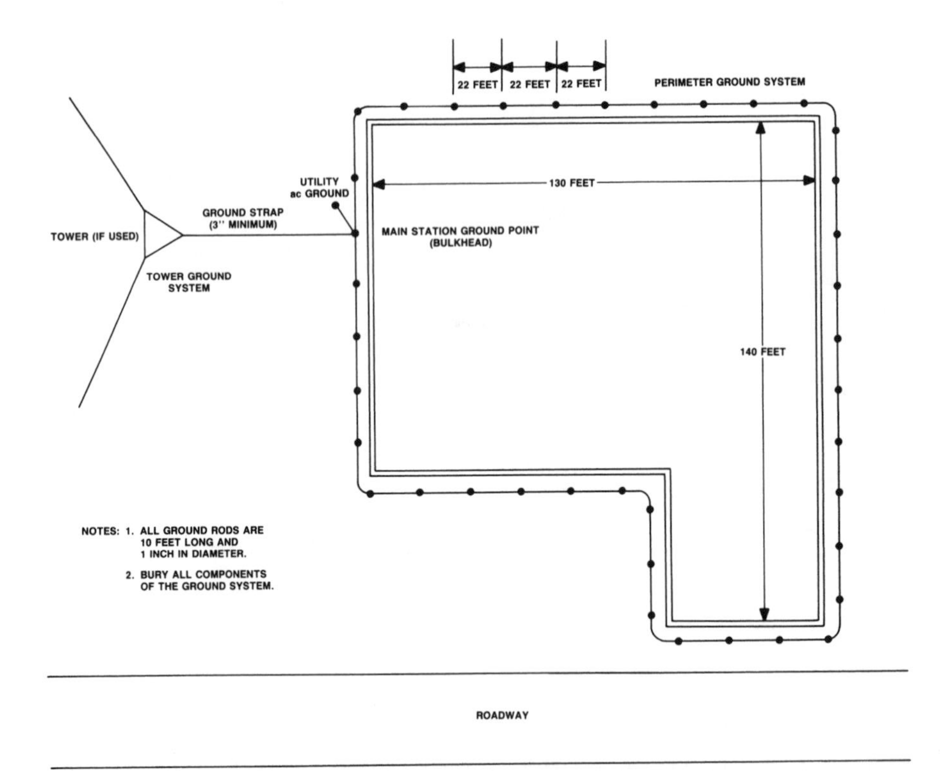

Figure 9.16 Facility ground using a perimeter ground-rod system. This approach works well for buildings with limited available real estate.

effectiveness in shunting fast rise-time surges to earth. Figure 9.17 illustrates the interconnection of a tower and building ground system. In most areas, soil conductivity is high enough to permit rods to be connected with no. 6 or larger bare-copper wire. In areas of sandy soil, use copper strap. A wire buried in low-conductivity, sandy soil tends to be inductive and less effective in dealing with fast-rise-time current surges. Again, make the width of the ground strap at least 1 percent of its overall length. Connect buried elements of the system, as shown in Figure 9.18.

9.4.1 Bulkhead Panel

The bulkhead panel is the cornerstone of an effective facility grounding system. The concept of the bulkhead is simple: Establish one reference point to which all cables entering and leaving the equipment building are grounded and to which all transient-suppression devices are mounted. Figure 9.19 shows a typical bulkhead installation for a communications facility. The panel size depends on the spacing, number, and dimensions of the coaxial lines, power cables, and other conduit entering or leaving the building.

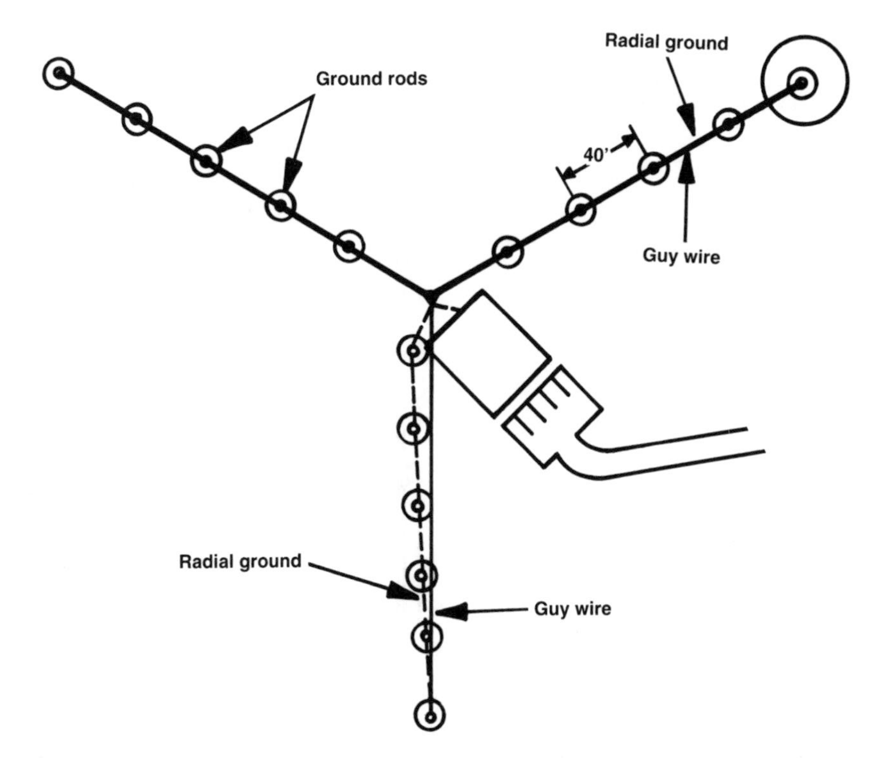

Figure 9.17 A typical guy-anchor and tower-radial grounding scheme. The radial ground is no. 6 copper wire. The ground rods are 5/8 in x 10 ft.

To provide a weatherproof point for mounting transient-suppression devices, the bulkhead can be modified to accept a subpanel, as shown in Figure 9.20. The subpanel is attached so that it protrudes through an opening in the wall and creates a secondary plate on which transient suppressors are mounted and grounded. A typical cable/suppressor-mounting arrangement for a communications site is shown in Figure 9.21. To handle the currents that may be experienced during a lightning strike or large transient on the utility company ac line, the bottom-most subpanel flange (which joins the subpanel to the main bulkhead) must have a total surface-contact area of at least 0.75-in^2 per transient suppressor.

Because the bulkhead panel carries significant current during a lightning strike or ac line disturbance, it must be constructed of heavy material. The recommended material is 1/8-in C110 (solid copper) 1/2 hard. Use 18-8 stainless-steel mounting hardware to secure the subpanel to the bulkhead.

Because the bulkhead panel establishes the central grounding point for all equipment within the building, it must be tied to a low-resistance (and low-inductance) perimeter ground system. The bulkhead establishes the *main facility ground point*, from which all grounds inside the building are referenced. A typical bulkhead installation for a small communications site is shown in Figure 9.22.

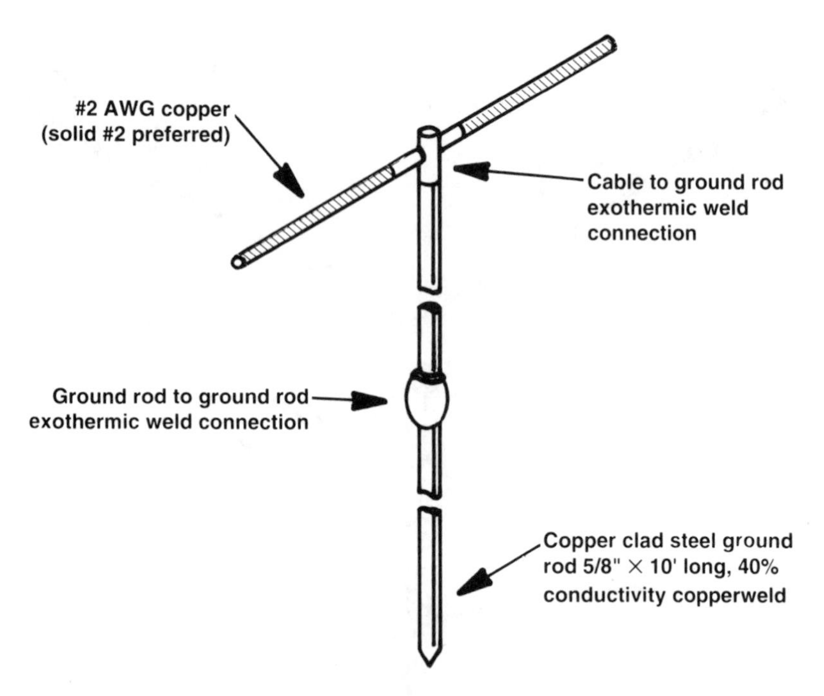

Figure 9.18 Preferred bonding method for below-grade elements of the ground system.

9.4.2 Bulkhead Grounding

A properly installed bulkhead panel will exhibit lower impedance and resistance to ground than any other equipment or cable-grounding point at the facility. Because the bulkhead panel will be used as the central grounding point for all of the equipment inside the building, the lower the inductance to the perimeter ground system, the better. The best arrangement is to simply extend the bulkhead panel down the outside of the building, below grade, to the perimeter ground system. This approach is illustrated in Figure 9.23.

If cables are used to ground the bulkhead panel, secure the interconnection to the outside ground system along the bottom section of the panel. Use multiple no. 1/0 or larger copper wire or several solid-copper straps. If using strap, attach with stainless-steel hardware, and apply joint compound for aluminum bulkhead panels. Clamp and Cadweld, or silver-solder, for copper/brass panels. If no. 1/0 or larger wire is used, employ crimp lug and stainless-steel hardware. Measure the dc resistance. It should be no greater than 0.01 Ω between the ground system and the panel. Repeat this measurement on an annual basis.

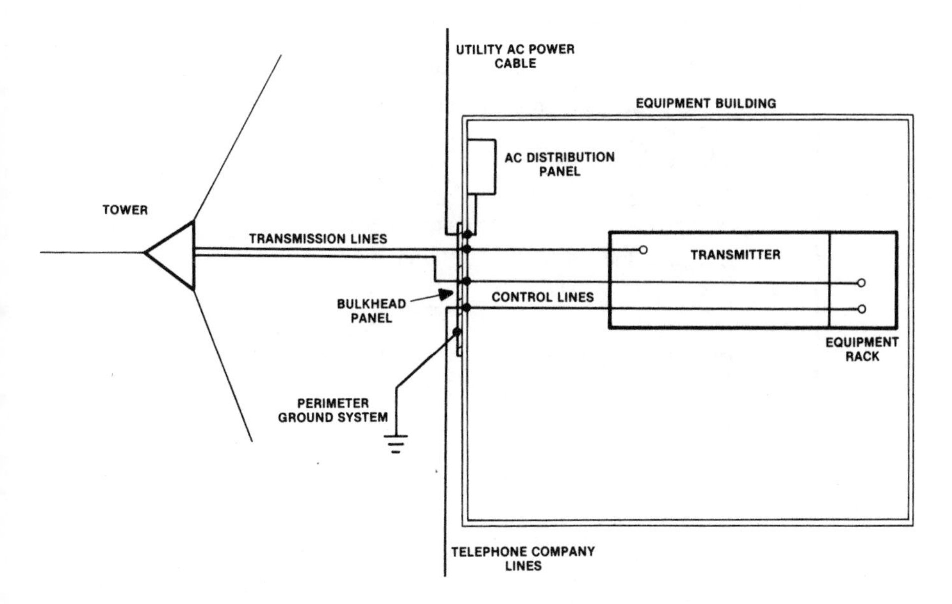

Figure 9.19 The basic design of a bulkhead panel for a facility. The bulkhead establishes the grounding reference point for the plant.

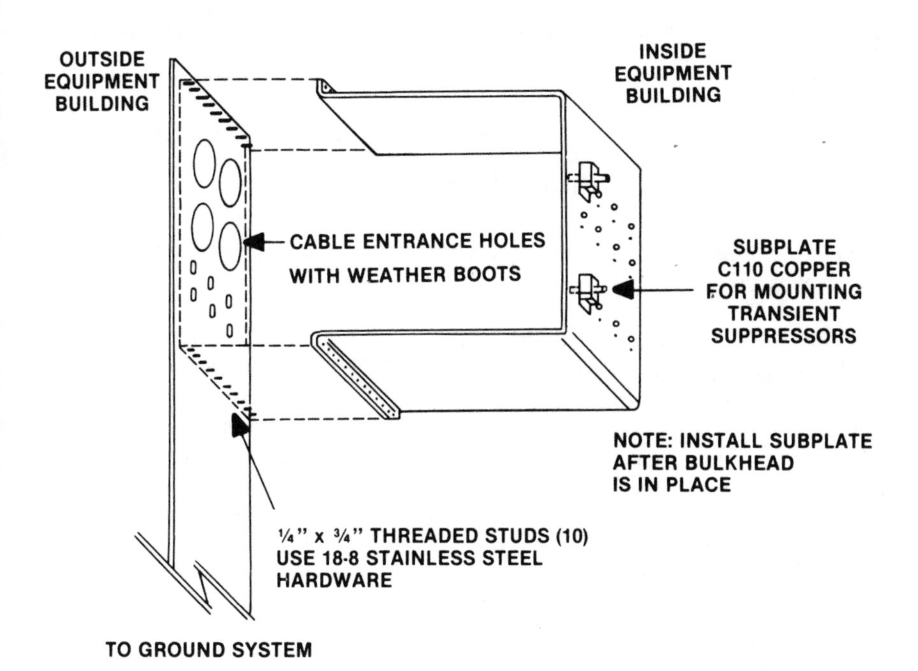

Figure 9.20 The addition of a subpanel to a bulkhead as a means of providing a mounting surface for transient-suppression components. To ensure that the bulkhead is capable of handling high surge currents, use the hardware shown.

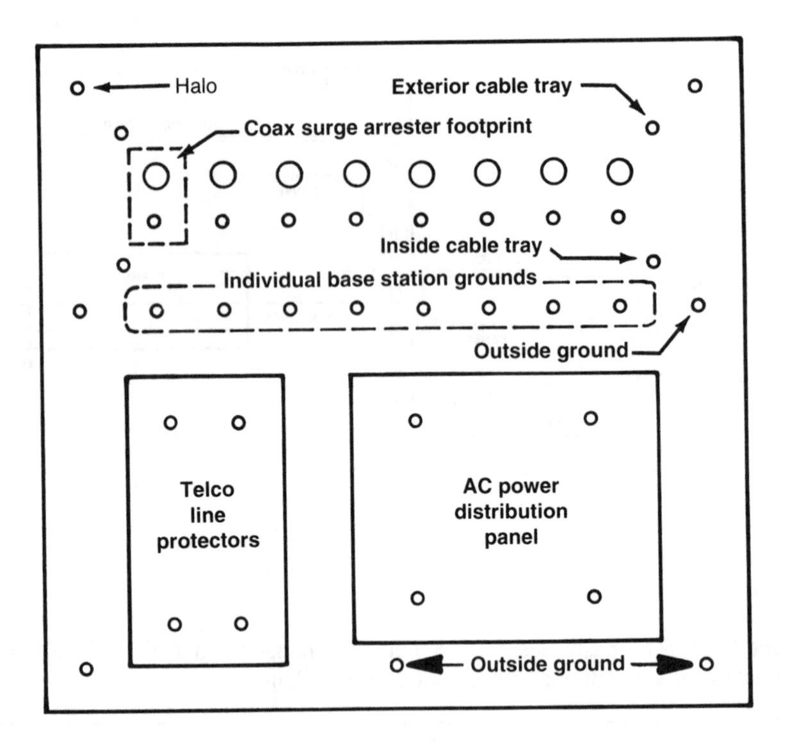

Figure 9.21 Mounting-hole layout for a communications site bulkhead subpanel.

9.4.3 Checklist for Proper Grounding

A methodical approach is necessary in the design of a facility ground system. Consider the following points:

1. Install a bulkhead panel to provide mechanical support, electric grounding, and lightning protection for coaxial cables, power feeds, and telephone lines entering the equipment building.
2. Install an internal ground bus using no. 2 or larger solid-copper wire. (At transmission facilities, use copper strap that is at least 3-in wide.) Form a *star* grounding system. At larger installations, form a *star-of-stars* configuration. Do not allow ground loops to exist in the internal ground bus. Connect the following items to the building internal ground system:
   - Chassis racks and cabinets of all hardware
   - All auxiliary equipment
   - Battery charger
   - Switchboard
   - Conduit
   - Metal raceway and cable tray
3. Connect outside metal structures to the earth ground array (towers, metal fences, metal buildings, and guy-anchor points).
4. Connect the power-line ground to the array. Strictly follow local electrical code.

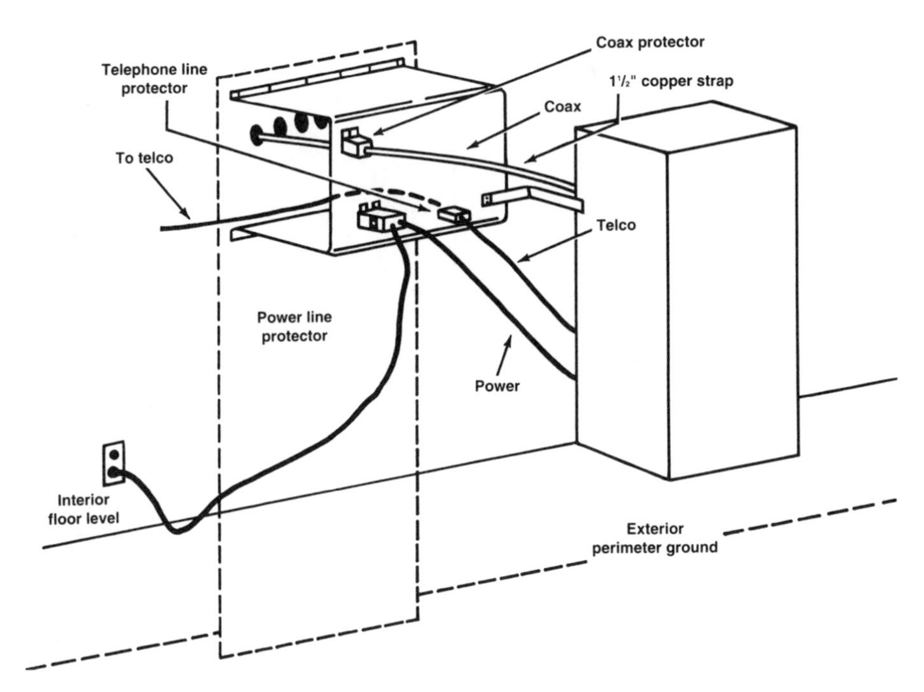

Figure 9.22 Bulkhead installation at a small communications site.

5. Connect the bulkhead to the ground array through a low-inductance, low-resistance bond.
6. Do not use soldered-only connections outside the equipment building. Crimped, brazed, and exothermic (Cadwelded) connections are preferable. For a proper bond, all metal surfaces must be cleaned, finishes removed to bare metal, and surface preparation compound applied (where necessary). Protect all connections from moisture by appropriate means (sealing compound and heat sink tubing).

9.5 Bibliography

Block, Roger: "How to Ground Guy Anchors and Install Bulkhead Panels," *Mobile Radio Technology* magazine, Intertec Publishing, Overland Park, KS, February 1986.

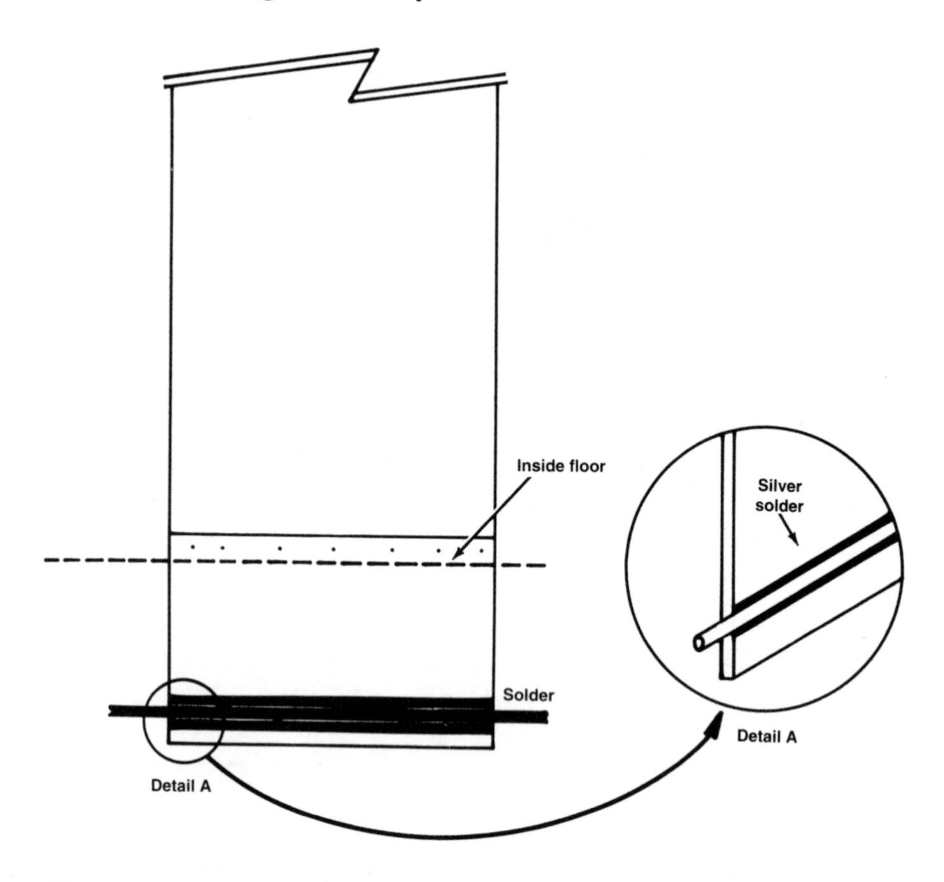

Figure 9.23 The proper way to ground a bulkhead panel and provide a low-inductance path for surge currents stripped from cables entering and leaving the facility. The panel extends along the building exterior to below grade. It is silver-soldered to a no. 2/0 copper wire that interconnects with the outside ground system.

Block, Roger: "The Grounds for Lightning and EMP Protection," PolyPhaser Corporation, Gardnerville, NV, 1987.

Carpenter, R.B.: "Improved Grounding Methods for Broadcasters," *Proceedings of the SBE National Convention*, Society of Broadcast Engineers, Indianapolis, 1987.

Defense Civil Preparedness Agency, "EMP Protection for AM Radio Stations," Washington, D.C., TR-61-C, May 1972.

Hill, Mark: "Computer Power Protection," *Broadcast Engineering* magazine, Intertec Publishing, Overland Park, KS, April 1987.

Little, Richard: "Surge Tolerance: How Does Your Site Rate?," *Mobile Radio Technology* magazine, Intertec Publishing, Overland Park, KS, June 1988.

Schneider, John: "Surge Protection and Grounding Methods for AM Broadcast Transmitter Sites," *Proceedings of the SBE National Convention*, Society of Broadcast Engineers, Indianapolis, 1987.

Technical Reports LEA-9-1, LEA-0-10 and LEA-1-8, Lightning Elimination Associates, Santa Fe Springs, CA.
Whitaker, Jerry: *AC Power Systems Handbook*, CRC Press, Boca Raton, FL, 1991.

CHAPTER 10

SAFETY CONSIDERATIONS

10.1 Introduction

Safety is critically important to engineering personnel who work around powered hardware, especially if they work under time pressures. Safety shouldn't be taken lightly. *Life safety* systems are those designed to protect life and property. Such systems include emergency lighting, fire alarms, smoke exhaust and ventilating fans, and site security.

10.1.1 Facility Safety Equipment

Personnel safety is the responsibility of the facility manager. Proper life safety procedures and equipment must be installed. Safety-related hardware includes the following:

- *Emergency power off (EPO) button.* EPO push buttons are required by safety codes for data processing (DP) centers. One must be located at each principal exit from the DP room. Other EPO buttons may be located near operator workstations. The EPO system, intended only for emergencies, disconnects all power to the room, except for lighting.
- *Smoke detector.* Two basic types of smoke detectors are common. The first compares the transmission of light through air in the room with light through a sealed optical path into which smoke cannot penetrate. Smoke causes a differential or *backscattering* effect that, when detected, triggers an alarm after a preset threshold has been exceeded. The second type senses the ionization of combustion products, rather than visible smoke. A mildly radioactive source, usually nickel, ionizes the air passing through a screened chamber. A charged probe captures ions and detects the small current that is proportional to the rate of capture. When combustion products or material other than air

molecules enter the probe area, the rate of ion production changes abruptly, generating a signal that triggers the alarm.

- *Flame detector*. The flame sensor responds not to heated surfaces or objects, but to infrared, when it flickers with the unique characteristics of a fire. Such detectors, for example, will respond to a lighted match, but not to a cigarette. The ultraviolet light from a flame also is used to distinguish between hot, glowing objects and open flame.
- *Halon*. The Halon fire-extinguishing agent is a low-toxicity, compressed gas that is contained in pressurized vessels. Discharge nozzles in DP and other types of equipment rooms are arranged to dispense the entire contents of a central container or of multiple smaller containers of Halon when actuated by a command from the fire control system. The discharge is sufficient to extinguish flame and stop combustion of most flammable substances. Halon is one of the more common fire-extinguishing agents used for DP applications. Halon systems are usually not practical, however, in large, open-space computer or communications centers.
- *Water sprinkler*. Although water is an effective agent against a fire, activation of a sprinkler system will cause damage to the equipment it is meant to protect. Interlock systems must drop all power (except for emergency lighting) before the water system is discharged. Most water systems use a two-stage alarm. Two or more fire sensors, often of different design, must signal an alarm condition before water is discharged into the protected area. Where sprinklers are used, floor drains and EPO controls must be provided.
- *Fire damper*. When a fire is detected, dampers are used to block ventilating passages in strategic parts of the system. This prevents fire from spreading through the passages and keeps fresh air from fanning the flames. A fire damper system, combined with the shutdown of cooling and ventilating air, enables Halon to be retained in the protected space until the fire is extinguished.

Many life safety system functions can be automated. The decision of what to automate and what to operate manually requires considerable thought. If the life safety control panels are accessible to a large number of site employees, most functions should be automatic. Alarm-silencing controls should be kept locked away. A mimic board can be used to identify problem areas readily. Figure 10.1 illustrates a well-organized life safety control system. Note that fire, HVAC (heating, ventilation, and air conditioning), security, and EPO controls all are readily accessible. Note also that operating instructions are posted for life safety equipment, and an evacuation route is shown. Important telephone numbers are posted, and a direct-line telephone (not via the building switchboard) is provided. All equipment is located adjacent to a lighted emergency exit door.

Life safety equipment must be maintained just as diligently as the electronic hardware that it protects. Conduct regular tests and drills. It is, obviously, not necessary or advisable to discharge Halon or water during a drill.

Configure the life safety control system to monitor the premises for dangerous conditions and also to monitor the equipment designed to protect the facility. Important monitoring points include HVAC machine parameters, water and/or Halon pressure, emergency battery-supply status, and other elements of the system that could compromise the ability of life safety equipment to function properly. Basic guidelines for life safety systems include the following:

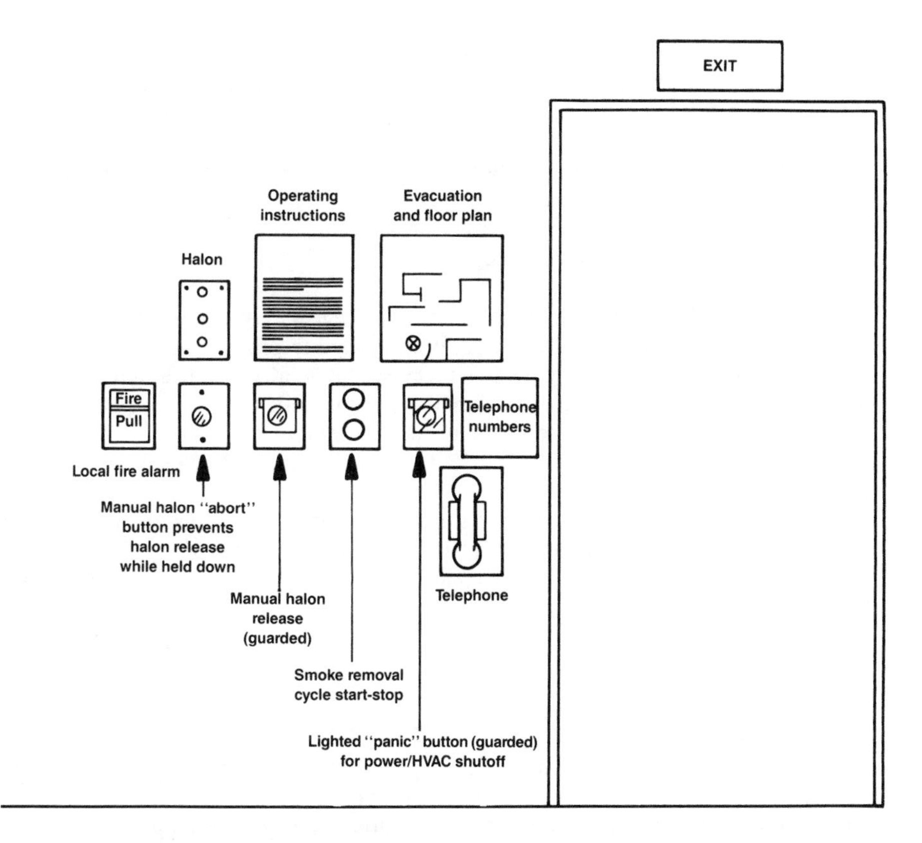

Figure 10.1 A well-organized life safety control station. (Adapted from: Federal Information Processing Standards Publication No. 94, *Guideline on Electrical Power for ADP Installations*, U.S. Department of Commerce, National Bureau of Standards, Washington, DC, 1983.)

- Carefully analyze the primary threats to life and property within the facility. Develop contingency plans to meet each threat.
- Prepare a life safety manual, and distribute it to all employees at the facility. Require them to read it.
- Conduct drills for employees at random times without notice. Require acceptable performance from employees.
- Prepare simple, step-by-step instructions on what to do in an emergency. Post the instructions in a conspicuous place.
- Assign after-hours responsibility for emergency situations. Prepare a list of supervisors that operators should contact if problems arise. Post the list with phone numbers. Keep the list accurate and up-to-date. Always provide the names of three individuals who may be contacted in an emergency.
- Work with a life safety consultant to develop a coordinated control and monitoring system for the facility. Such hardware will be expensive, but it

must be provided. The facility may be able to secure a reduction in insurance rates if comprehensive safety efforts can be demonstrated.

- Interface the life safety system with automatic data-logging equipment so that documentation can be assembled on any event.
- Insist on complete, up-to-date schematic diagrams for all facility hardware. Insist that the diagrams include any changes made during installation or subsequent modification.
- Provide sufficient emergency lighting.
- Provide easy-access emergency exits.

The importance of providing standby power for sensitive loads at commercial and industrial facilities is obvious. It is equally important to provide standby power for life safety systems. A lack of ac power must not render the life safety system inoperative. Sensors and alarm control units should include their own backup battery supplies. In a properly designed system, all life safety equipment will be fully operational despite the loss of all ac power to the facility, including backup power for sensitive loads.

Place cables linking the life safety control system with remote sensors and actuators in a separate conduit containing only life safety conductors. Study the National Electrical Code and all applicable local and federal codes relating to safety. Follow them strictly.

10.2 Electrical Shock

It takes surprisingly little current to injure a person. Studies at Underwriters Laboratories (UL) show that the electrical resistance of the human body varies with the amount of moisture on the skin, the muscular structure of the body, and the applied voltage. Typical hand-to-hand resistance ranges from 500 Ω to 600 kΩ, depending on the conditions. Higher voltages have the ability to break down the outer layers of the skin, which can reduce the overall resistance value. UL uses the lower value, 500 Ω, as the standard resistance between major extremities, as from the hand to the foot. This value generally is considered the minimum that would be encountered. In fact, this may not be unusual, because wet conditions or a cut or other break in the skin significantly reduce human body resistance.

10.2.1 Effects on the Human Body

Table 10.1 lists some effects that typically result when a person is connected across a current source with a hand-to-hand resistance of 2.4 kΩ. The table shows that a current of 50 mA will flow between the hands, if one hand is in contact with a 120 Vac source and the other hand is grounded. The table indicates that even the relatively small current of 50 mA can produce *ventricular fibrillation* of the heart and perhaps death. Medical literature describes ventricular fibrillation as very rapid, uncoordinated contractions of the ventricles of the heart, which result in loss of synchronization between heartbeat and pulse beat. The electrocardiograms shown in Figure 10.2 compare a healthy heart rhythm with one in ventricular fibrillation. Unfortunately, once ventricular fibrillation occurs, it will continue. Barring resuscitation techniques, death will ensue within a few minutes.

Table 10.1 The effects of current on the human body.

Current	Effect
1 mA or less	No sensation, not felt
More than 3 mA	Painful shock
More than 10 mA	Local muscle contractions, sufficient to cause "freezing" to the circuit for 2.5 percent of the population
More than 15 mA	Local muscle contractions, sufficient to cause "freezing" to the circuit for 50 percent of the population
More than 30 mA	Breathing is difficult, can cause unconsciousness
50 mA to 100 mA	Possible ventricular fibrillation of the heart
100 mA to 200 mA	Certain ventricular fibrillation of the heart
More than 200 mA	Severe burns and muscular contractions; heart more apt to stop than to go into fibrillation
More than a few amperes	Irreparable damage to body tissues

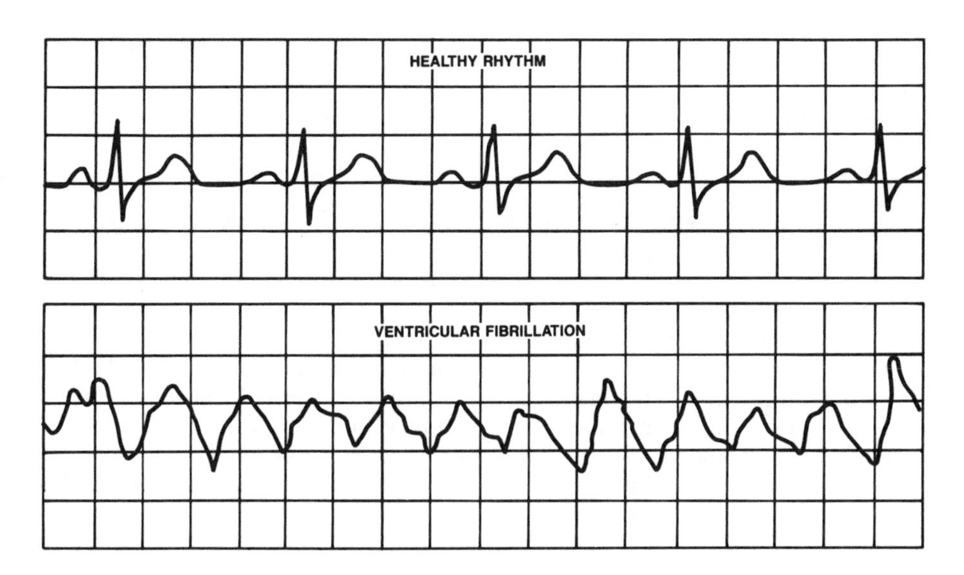

Figure 10.2 Electrocardiograms showing the healthy rhythm of a heart (*top*), and ventricular fibrillation of the heart (*bottom*).

The route taken by the current through the body greatly affects the degree of injury. Even a small current, passing from one extremity through the heart to another extremity, is dangerous and capable of causing severe injury or electrocution. There

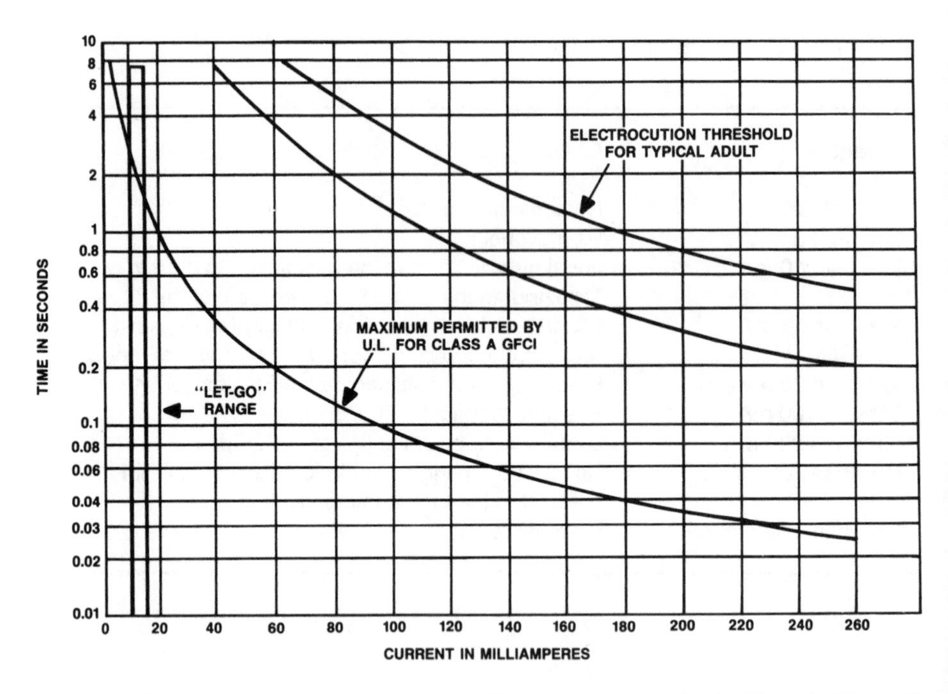

Figure 10.3 Effects of electric current and time on the human body. Note the "let-go" range.

---

are cases in which a person has contacted extremely high current levels and lived. However, when this happens, it is usually because the current passes only through a single limb and not through the entire body. In these instances, the limb is often lost, but the person survives.

Current is not the only factor in electrocution. Figure 10.3 summarizes the relationship between current and time on the human body. The graph shows that 100 mA flowing through an adult human body for 2 s will cause death by electrocution. An important factor in electrocution, the *let-go range*, also is shown on the graph. This point marks the amount of current that causes *freezing*, or the inability to let go of a conductor. At 10 mA, 2.5 percent of the population would be unable to let go of a live conductor; at 15 mA, 50 percent of the population would be unable to let go of an energized conductor. It is apparent from looking at the graph that even a small amount of current can freeze someone to a conductor. The objective for those who must work around electric equipment is to protect themselves from electrical shock. Table 10.2 lists required precautions for maintenance personnel working near high voltages.

10.2.2 Circuit-Protection Hardware

A common primary panel or equipment circuit breaker or fuse will not protect an individual from electrocution. However, the *ground-fault interrupter* (GFI), used

Table 10.2 Required safety practices for engineers working around high-voltage equipment.

- Remove all ac power from the equipment. Do not rely on internal contactors or SCRs to remove dangerous ac.
- Trip the appropriate power-distribution circuit breakers at the main breaker panel.
- Place signs as needed to indicate that the circuit is being serviced.
- Switch the equipment being serviced to the *local control* mode, as provided.
- Discharge all capacitors using the discharge stick provided by the equipment manufacturer.
- Do not remove, short-circuit, or tamper with interlock switches on access covers, doors, enclosures, gates, panels, or shields.
- Keep away from live circuits.
- Allow any component to cool completely before attempting to replace it.
- If a leak or bulge is found on the case of an oil-filled or electrolytic capacitor, do not attempt to service the part until it has cooled completely.
- Know which parts in the system contain PCBs. Handle them appropriately.
- Minimize exposure to RF radiation.
- Avoid contact with hot surfaces within the system.

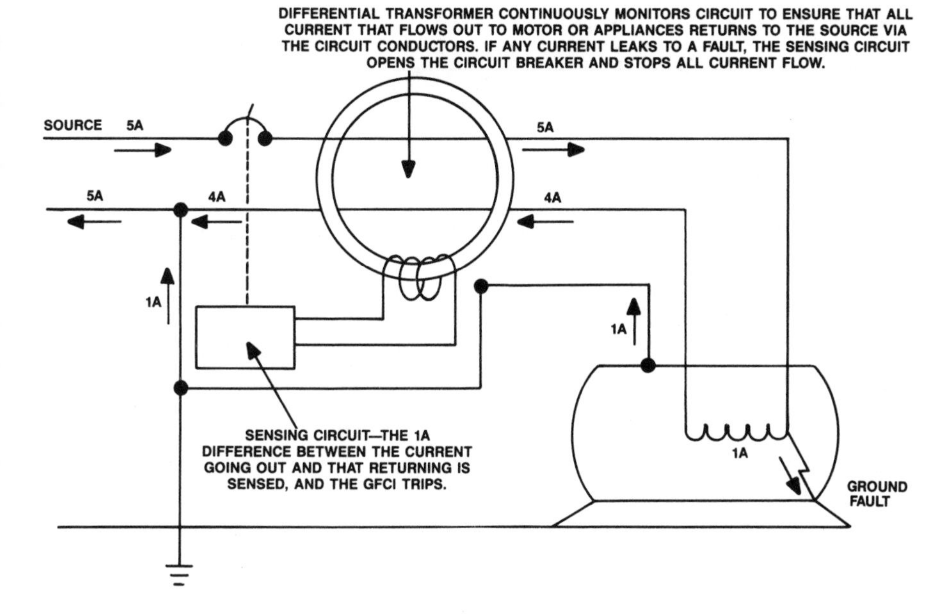

Figure 10.4 Basic design of a ground-fault interrupter (GFI).

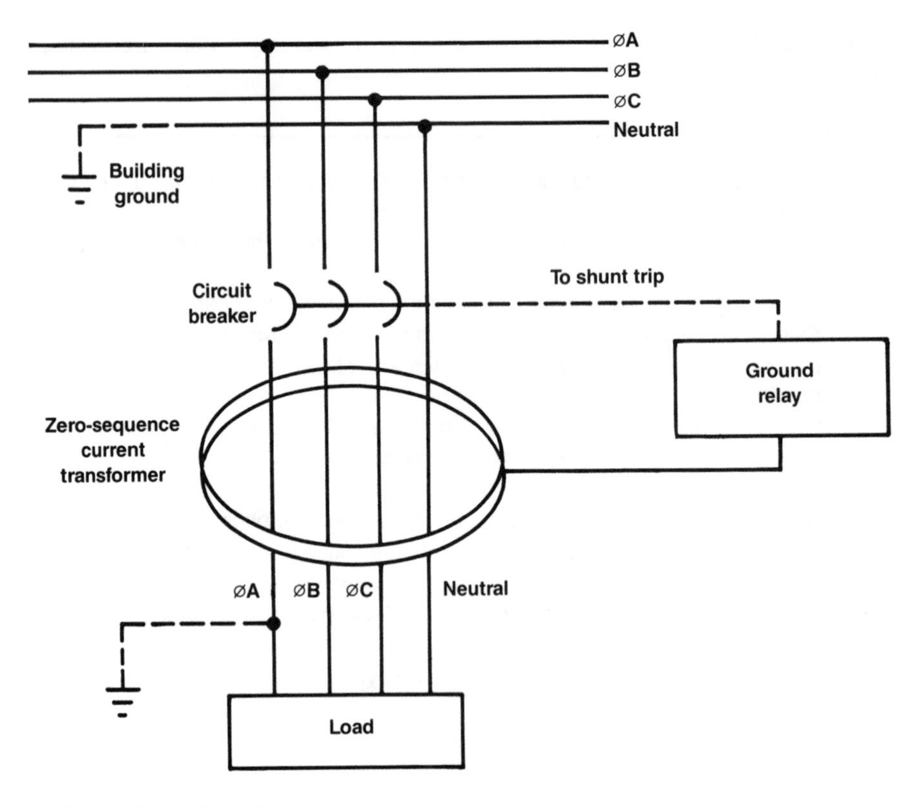

Figure 10.5 Ground-fault detection in a three-phase ac system.

properly, can help prevent electrocution. Shown in Figure 10.4, the GFI works by monitoring the current being applied to the load. It uses a differential transformer that senses an imbalance in load current. If a current (typically 5 mA, ±1 mA on a low-current 120 Vac line) begins flowing between the neutral and ground or between the hot and ground leads, the differential transformer detects the leakage and opens the primary circuit (usually within approximately 2.5 ms).

OSHA (Occupational Safety and Health Administration) rules specify that temporary receptacles (those not permanently wired) and receptacles used on construction sites be equipped with GFI protection. Receptacles on two-wire, single-phase portable and vehicle-mounted generators of not more than 5 kW, where the generator circuit conductors are insulated from the generator frame and all other grounded surfaces, need not be equipped with GFI outlets.

GFIs will not protect a person from every type of electrocution. If the victim becomes connected to both the neutral and the hot wire, the GFI will not detect an imbalance, and will not open the primary circuit.

10.2.2.A *Three-Phase Systems*

For large, three-phase loads, detecting ground currents and interrupting the circuit before injury or damage can occur is a more complicated proposition. The classic

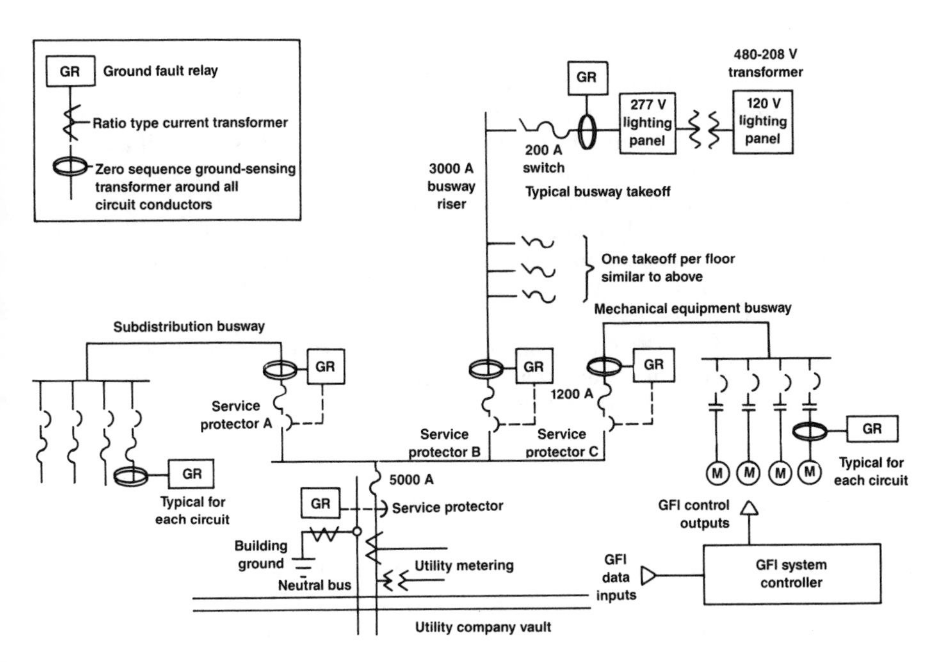

Figure 10.6 Ground-fault protection system for a large, multistory building.

method of protection involves the use of a zero-sequence *current transformer* (CT). Such devices are basically an extension of the single-phase GFI circuit, shown in Figure 10.4. Three-phase CTs have been developed to fit over bus ducts, switchboard buses, and circuit-breaker studs. Rectangular core-balanced CTs are able to detect leakage currents as small as several milliamperes when the system carries as much as 4 kA. "Doughnut-type" toroidal zero-sequence CTs also are available in varying diameters.

The zero-sequence current transformer is designed to detect the magnetic field surrounding a group of conductors. As shown in Figure 10.5, in a properly operating three-phase system, the current flowing through the conductors of the system—including the neutral—travels out and returns along those same conductors. The net magnetic flux detected by the CT is zero. No signal is generated in the transformer winding, regardless of current magnitudes—symmetrical or asymmetrical. If one phase conductor is faulted to ground, however, the current balance will be upset. The ground-fault detection circuit then will trip the breaker and open the line.

For optimum protection in a large facility, GFI units are placed at natural branch points of the ac power system. Obviously, it is preferable to lose only a small portion of a facility in the event of a ground fault than it is to have the entire plant dropped. Figure 10.6 illustrates such a distributed system. Sensors are placed at major branch points to isolate any ground fault from the remainder of the distribution network. In this way, the individual GFI units can be set for higher sensitivity and shorter time delays than would be practical with a large, distributed load. The technology of GFI devices has improved significantly in the past few years. New integrated circuit devices and improved CT designs have provided improved protection components at a lower cost.

Sophisticated GFI monitoring *systems* are available that analyze ground-fault currents and isolate the faulty branch circuit. This feature prevents needless tripping of GFI units up the line toward the utility service entrance. For example, if a ground fault is sensed in a fourth-level branch circuit, the GFI system controller automatically locks out first-, second-, and third-level devices from operating to clear the fault. The problem, therefore, is safely confined to the fourth-level branch. The GFI control system is designed to operate in a fail-safe mode. In the event of a control-system shutdown, the individual GFI trip relays would operate independently to clear whatever fault currents may exist.

The system engineer should hire an experienced electrical contractor to conduct a full ground-fault protection study. Direct the contractor to identify possible failure points, and to recommend corrective actions.

10.2.3 Working with High Voltage

Rubber gloves are a common safety measure used by engineers working on high-voltage equipment. These gloves are designed to provide protection from hazardous voltages when the wearer is working on "hot" circuits. Although the gloves may provide some protection from these hazards, placing too much reliance on them poses the potential for disaster. There are several reasons why gloves should be used only with a lot of caution and respect. A common mistake made by engineers is to assume that the gloves always provide complete protection. The gloves found in many facilities may be old and untested. Some may even have been "repaired" by users, perhaps with electrical tape. Few tools could be more hazardous than such a pair of gloves.

Know the voltage rating of the gloves. Gloves are rated differently for ac and dc voltages. For instance, a *class 0* glove has a minimum dc breakdown voltage of 35 kV; the minimum ac breakdown voltage, however, is only 6 kV. Further, high-voltage rubber gloves are not tested at RF frequencies, and RF can burn a hole in the best of them. Working on live circuits involves much more than simply wearing a pair of gloves. It involves a frame of mind—an awareness of everything in the area, especially ground points.

Gloves alone may not be enough to protect an individual in certain situations. Recall the axiom of keeping one hand in a pocket while working on a device with current flowing? The axiom actually is based on simple electricity. The hot connection is not what causes the problem; it is the ground connection that permits current flow. A study in California showed that more than 90 percent of electrical equipment fatalities occurred when the grounded person contacted a live conductor. Line-to-line electrocution accounted for less than 10 percent of the deaths.

When working around high voltages, always look for grounded surfaces—and keep away from them. Even concrete can act as a ground if the voltage is high enough. If work must be conducted in live cabinets, consider using—in addition to rubber gloves—a rubber floor mat, rubber vest, and rubber sleeves. Although this may seem to be a lot of trouble, consider the consequences of making a mistake. Of course, the best troubleshooting advice is never work on any circuit unless no hazardous voltages are present. Also, any circuits or contactors that normally contain hazardous voltages should be grounded firmly before work begins.

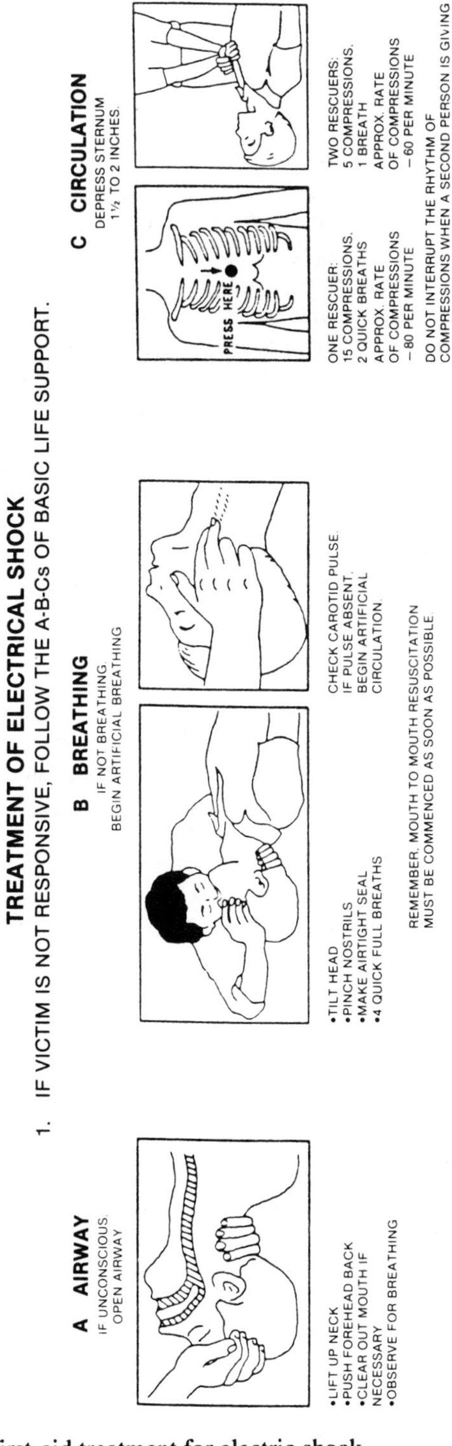

Figure 10.7 Basic first-aid treatment for electric shock.

Table 10.3 Basic first-aid procedures.

For extensively burned and broken skin:

- Cover affected area with a clean sheet or cloth.
- Do not break blisters, remove tissue, remove adhered particles of clothing, or apply any salve or ointment.
- Treat victim for shock, as required.
- Arrange for transportation to a hospital as quickly as possible.
- If victim's arms or legs are affected, keep them elevated.
- If medical help will not be available within an hour and the victim is conscious and not vomiting, prepare a weak solution of salt and soda. Mix 1 teaspoon of salt and 1/2 teaspoon of baking soda to each quart of tepid water. Allow the victim to sip slowly about 4 oz (half a glass) over a period of 15 min. Discontinue fluid intake if vomiting occurs. (Do not allow alcohol consumption.)

For less severe burns (first- and second-degree):

- Apply cool (not ice-cold) compresses, using the cleanest available cloth article.
- Do not break blisters, remove tissue, remove adhered particles of clothing, or apply salve or ointment.
- Apply clean, dry dressing, if necessary.
- Treat victim for shock, as required.
- Arrange for transportation to a hospital as quickly as possible.
- If victim's arms or legs are affected, keep them elevated.

Another important safety rule is to never work alone. Even if a trained assistant is not available when maintenance is performed, someone should be on hand to help in an emergency.

10.2.4 First-Aid Procedures

Be familiar with first-aid treatment for electrical shock and burns. Always keep a first-aid kit on hand at the facility. Figure 10.7 illustrates the basic treatment for electrical shock victims. Copy the information, and post it in a prominent location. Better yet, obtain more detailed information from the local heart association or Red Cross chapter. Personalized instruction on first aid usually is available locally. Table 10.3 lists basic first-aid procedures for burns.

10.3 OSHA Safety Requirements

The federal government has taken a number of steps to help improve safety within the workplace. OSHA is probably the most visible. It helps industries monitor and correct safety practices. The agency's records show that electrical standards are among the

Table 10.4 Sixteen common OSHA violations *(National Electrical Code, NFPA no. 70).*

Fact sheet number	Subject	NEC reference
1	Guarding of live parts	110-17
2	Identification	110-22
3	Uses allowed for flexible cord	400-7
4	Prohibited uses of flexible cord	400-8
5	Pull at joints and terminals must be prevented	400-10
6-1	Effective grounding, Part 1	250-51
6-2	Effective grounding, Part 2	250-51
7	Grounding of fixed equipment, general	250-42
8	Grounding of fixed equipment, specific	250-43
9	Grounding cord and plug-connected equipment	250-45
10	Grounding cord and plug-connected equipment	250-59
11	AC circuits and systems to be grounded	250-5
12	Location of overcurrent devices	240-24
13	Splices in flexible cords	400-9
14	Electrical connections	110-14
15	Marking equipment	110-21
16	Working clearances about electric equipment	110-16

most frequently violated of all safety standards. Table 10.4 lists 16 of the most common electrical violations, which include these areas:

- Protective covers
- Identification and marking
- Extension cords
- Grounding

10.3.1 Protective Covers

Exposure of live conductors is a common safety violation. All potentially dangerous electric conductors should be covered with protective panels. The danger is that someone may come into contact with the exposed, current-carrying conductors. It also is possible for metallic objects, such as ladders, cable, or tools to come in contact with a hazardous voltage, creating a life-threatening condition. Open panels also present a fire hazard.

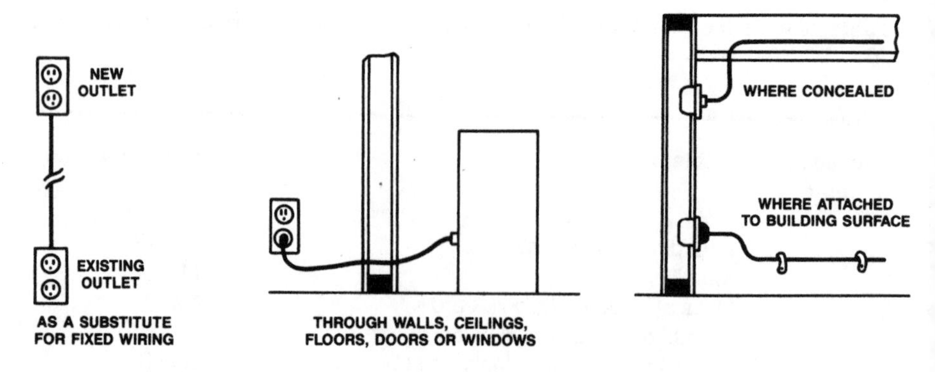

Figure 10.8 Flexible cord uses prohibited under NEC rules.

10.3.2 Identification and Marking

Properly identify and label all circuit breakers and switch panels. The labels for breakers and equipment switches may be years old, and may no longer describe the equipment that is actually in use. This confusion poses a safety hazard. Improper labeling of the circuit panel may lead to unnecessary damage—or worse, casualties—if the only person who understands the system is unavailable in an emergency. If a number of devices are connected to a single disconnect switch or breaker, provide a diagram or drawing for clarification. Label with brief phrases, and use clear, permanent, legible markings.

Equipment marking is a closely related area of concern. This is not the same as equipment identification. Marking equipment means labeling the equipment breaker panels and ac disconnect switches according to device rating. Breaker boxes should contain a nameplate showing the manufacturer name, rating, and other pertinent electrical data. The intent of this rule is to prevent devices from being subjected to excessive loads or voltages.

10.3.3 Extension Cords

Extension (flexible) cords often are misused. Although it may be easy to connect a new piece of equipment with a flexible cord, be careful. The National Electrical Code lists only eight approved uses for flexible cords.

The use of a flexible cord where the cable passes through a hole in the wall, ceiling, or floor is a common violation. Running the cord through doorways, windows, or similar openings also is prohibited. A flexible cord should not be attached to building surfaces or concealed behind building walls or ceilings. These common violations are illustrated in Figure 10.8.

Failure to provide adequate strain relief on connectors is another common problem. Whenever possible, use manufactured cable connections.

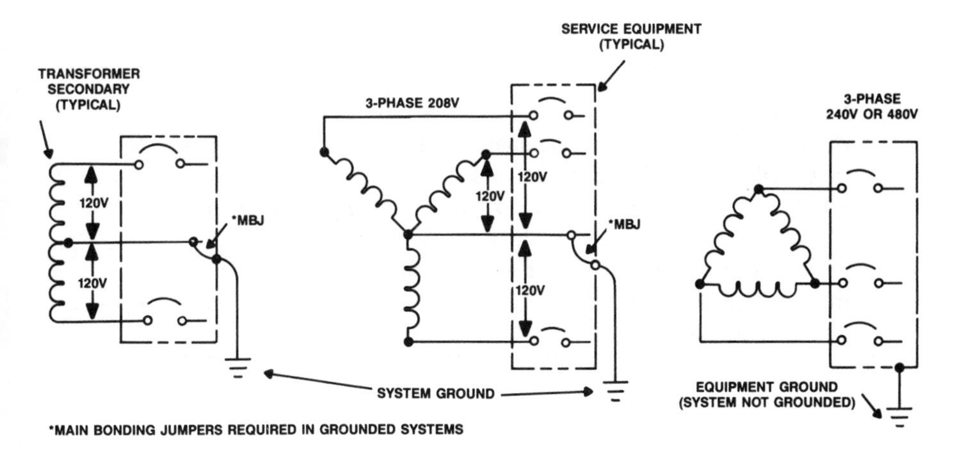

Figure 10.9 Primary electrical service grounding. Although regulations have been in place for many years, OSHA inspections still uncover violations in the grounding of primary electrical service systems.

10.3.4 Grounding

OSHA regulations describe two types of grounding: *system grounding* and *equipment grounding*. System grounding refers to the connection of one of the current-carrying conductors (such as the terminals of a supply transformer) to ground. (See Figure 10.9.) Equipment grounding refers to the connection of all noncurrent-carrying metal surfaces together and to ground. From a grounding standpoint, the only difference between a grounded and ungrounded electrical system is that the *main-bonding jumper* from the service equipment ground to a current-carrying conductor is omitted in the ungrounded system.

The system ground performs two tasks:

- It provides the final connection from equipment-grounding conductors to the grounded-circuit conductor, thus completing the ground-fault loop.
- It solidly ties the electrical system and its enclosures to their surroundings (usually earth, structural steel, and plumbing). This prevents voltages at any source from rising to harmfully high voltage-to-ground levels.

Note that equipment grounding—bonding all electric equipment to ground—is required whether or not the system is grounded. System grounding should be handled by the electrical contractor installing the power feeds.

Equipment grounding serves two important functions:

- It bonds all surfaces together so that there can be no harmful voltage differences among them.

Table 10.5 Major points to consider when developing a facility safety program.

- Management assumes the leadership role regarding safety policies.
- Responsibility for safety- and health-related activities is clearly assigned.
- Hazards are identified, and steps are taken to eliminate them.
- Employees at all levels are trained in proper safety procedures.
- Thorough accident/injury records are maintained.
- Medical attention and first aid is readily available.
- Employee awareness and participation is fostered through incentives and an ongoing, high-profile approach to workplace safety.

- It provides a ground-fault current path from a fault location back to the electrical source, so that if a fault current develops, it will rise to a level sufficient to operate the breaker or fuse.

The National Electrical Code is complex, and it contains numerous requirements concerning electrical safety. If the facility electric wiring system has gone through many changes over the years, have the system inspected by a qualified consultant. The fact sheets listed in Table 10.4 provide a good starting point for a self-evaluation. The fact sheets are available from any local OSHA office.

10.3.5 Management Responsibility

The key to operating a safe facility is diligent management. A carefully thought-out plan ensures a coordinated approach to protecting staff members from injury, and the facility from potential litigation. Although the details and overall organization may vary from workplace to workplace, some general guidelines can be stated. Common practices are summarized in Table 10.5.

If managers are concerned about safety, employees probably also will be. Display safety pamphlets, and recruit employee help in identifying hazards. Reward workers for good safety performance. Often, an incentive program will help to encourage safe work practices. Eliminate any hazards that have been identified, and obtain OSHA forms and any first-aid supplies that would be needed in an emergency. The OSHA "Handbook for Small Business" outlines the legal requirements imposed by the Occupational Safety and Health Act of 1970. The handbook, which is available from OSHA, also suggests ways in which a company can develop an effective safety program.

Free on-site consultations also are available from OSHA. A consultant will tour the facility and offer practical advice about safety. These consultants do not issue citations, propose penalties, or routinely provide information about workplace conditions to the federal inspection staff. Contact the nearest OSHA office for additional information. Table 10.6 provides a basic checklist of safety points for consideration.

Table 10.6 Sample checklist of important safety items.

Refer regularly to this checklist to maintain a safe facility. For each category shown, be sure that:

Electrical safety:

- Fuses of the proper size have been installed.
- All ac switches are mounted in clean, tightly closed metal boxes.
- Each electrical switch is marked to show its purpose.
- Motors are clean and free of excessive grease and oil.
- Motors are maintained properly and provided with adequate overcurrent protection.
- Bearings are in good condition.
- Portable lights are equipped with proper guards.
- All portable equipment is double-insulated or properly grounded.
- The facility electrical system is checked periodically by a contractor competent in the NEC.
- The equipment-grounding conductor or separate ground wire has been carried all the way back to the supply conductor.
- All extension cords are in good condition, and the grounding pin is not missing or bent.
- Ground-fault interrupters are installed as required.

Exits and access:

- All exits are visible and unobstructed.
- All exits are marked with a readily visible, properly illuminated sign.
- There are sufficient exits to ensure prompt escape in the event of an emergency.

Fire protection:

- Portable fire extinguishers of the appropriate type are provided in adequate numbers.
- All remote vehicles have proper fire extinguishers.
- Fire extinguishers are inspected monthly for general condition and operability, which is noted on the inspection tag.
- Fire extinguishers are mounted in readily accessible locations.
- The fire alarm system is tested annually.

10.4 Bibliography

Code of Federal Regulations, 40, Part 761, Washington DC.

"Current Intelligence Bulletin #45," National Institute for Occupational Safety and Health Division of Standards Development and Technology Transfer, Feb. 24, 1986.

"Electrical Standards Reference Manual," U.S. Department of Labor, Washington, DC.

Federal Information Processing Standards Publication No. 94, *Guideline on Electrical Power for ADP Installations*, U.S. Department of Commerce, National Bureau of Standards, Washington, DC, 1983.

Hammar, Willie: *Occupational Safety Management and Engineering*, Prentice-Hall, New York.

Lawrie, Robert: *Electrical Systems for Computer Installations*, McGraw-Hill, New York, 1988.

"Occupational Injuries and Illnesses in the United States by Industry," OSHA Bulletin 2278, U.S. Department of Labor, Washington, DC, 1985.

OSHA "Handbook for Small Business," U.S. Department of Labor, Washington, DC.

——— "Electrical Hazard Fact Sheets," U.S. Department of Labor, Washington, DC, January 1987.

Whitaker, Jerry: *AC Power Systems*, CRC Press, Boca Raton, FL, 1991.

CHAPTER 11

REFERENCE DATA

11.1 Standard Units

Name	Symbol	Quantity
ampere	A	electric current
ampere per meter	A/m	magnetic field strength
ampere per square meter	A/m^2	current density
becquerel	Bg	activity (of a radionuclide)
candela	cd	luminous intensity
coulomb	C	electric charge
coulomb per kilogram	C/kg	exposure (x and gamma rays)
coulomb per sq. meter	C/m^2	electric flux density
cubic meter	m^3	volume
cubic meter per kilogram	m^3/kg	specific volume
degree Celsius	˚C	Celsius temperature
farad	F	capacitance
farad per meter	F/m	permittivity
henry	H	inductance
henry per meter	H/m	permeability
hertz	Hz	frequency
joule	J	energy, work, quantity of heat
joule per cubic meter	J/m^3	energy density
joule per kelvin	J/K	heat capacity
joule per kilogram K	J/(kg•K)	specific heat capacity
joule per mole	J/mol	molar energy
kelvin	K	thermodynamic temperature
kilogram	kg	mass
kilogram per cubic meter	kg/m^3	density, mass density
lumen	lm	luminous flux
lux	lx	luminance

Standard Units, Continued

Name	Symbol	Quantity
meter	m	length
meter per second	m/s	speed, velocity
meter per second sq.	m/s^2	acceleration
mole	mol	amount of substance
newton	N	force
newton per meter	N/m	surface tension
ohm	Ω	electrical resistance
pascal	Pa	pressure, stress
pascal second	Pa•s	dynamic viscosity
radian	rad	plane angle
radian per second	rad/s	angular velocity
radian per second squared	rad/s^2	angular acceleration
second	s	time
siemens	S	electrical conductance
square meter	m^2	area
steradian	sr	solid angle
tesla	T	magnetic flux density
volt	V	electrical potential
volt per meter	V/m	electric field strength
watt	W	power, radiant flux
watt per meter kelvin	W/(m•K)	thermal conductivity
watt per square meter	W/m^2	heat (power) flux density
weber	Wb	magnetic flux

11.2 Standard Prefixes

Multiple	Prefix	Symbol
10^{18}	exa	E
10^{15}	peta	P
10^{12}	tera	T
10^{9}	giga	G
10^{6}	mega	M
10^{3}	kilo	k
10^{2}	hecto	h
10	deka	da
10^{-1}	deci	d
10^{-2}	centi	c
10^{-3}	milli	m
10^{-6}	micro	μ
10^{-9}	nano	n
10^{-12}	pico	p
10^{-15}	femto	f
10^{-18}	atto	a

11.3 Common Standard Units

Unit	Symbol
centimeter	cm
cubic centimeter	cm^3
cubic meter per second	m^3/s
gigahertz	GHz
gram	g
kilohertz	kHz
kilohm	kΩ
kilojoule	kJ
kilometer	km
kilovolt	kV
kilovoltampere	kVA
kilowatt	kW
megahertz	MHz
megavolt	MV
megawatt	MW
megohm	MΩ
microampere	μA
microfarad	μF
microgram	μg
microhenry	μH
microsecond	μs
microwatt	μW
milliampere	mA
milligram	mg
millihenry	mH
millimeter	mm
millisecond	ms
millivolt	mV
milliwatt	mW
nanoampere	nA
nanofarad	nF
nanometer	nm
nanosecond	ns
nanowatt	nW
picoampere	pA
picofarad	pF
picosecond	ps
picowatt	pW

11.4 Conversion Reference Data

A

To Convert	Into	Multiply By
abcoulomb	statcoulombs	2.998×10^{10}
acre	sq. chain (Gunters)	10
acre	rods	160
acre	square links (Gunters)	1×10^{5}
acre	Hectare or sq. hectometer	0.4047
acre-feet	cubic feet	43,560.0
acre-feet	gallons	3.259×10^{5}
acres	sq. feet	43,560.0
acres	sq. meters	4,047
acres	sq. miles	1.562×10^{-3}
acres	sq. yards	4,840
ampere-hours	coulombs	3,600.0
ampere-hours	faradays	0.03731
amperes/sq. cm	amps/sq. in	6.452
amperes/sq. cm	amps/sq. meter	10^{4}
amperes/sq. in	amps/sq. cm	0.1550
amperes/sq. in	amps/sq. meter	1,550.0
amperes/sq. meter	amps/sq. cm	10^{-4}
amperes/sq. meter	amps/sq. in	6.452×10^{-4}
ampere-turns	gilberts	1.257
ampere-turns/cm	amp-turns/in	2.540
ampere-turns/cm	amp-turns/meter	100.0
ampere-turns/cm	gilberts/cm	1.257
ampere-turns/in	amp-turns/cm	0.3937
ampere-turns/in	amp-turns/m	39.37
ampere-turns/in	gilberts/cm	0.4950
ampere-turns/meter	amp-turns/cm	0.01
ampere-turns/meter	amp-turns/in	0.0254
ampere-turns/meter	gilberts/cm	0.01257
Angstrom unit	inch	3937×10^{-9}
Angstrom unit	meter	1×10^{-10}
Angstrom unit	micron or (Mu)	1×10^{-4}
are	acre (U.S.)	0.02471
ares	sq. yards	119.60
ares	acres	0.02471
ares	sq. meters	100.0
astronomical unit	kilometers	1.495×10^{8}
atmospheres	ton/sq. in	0.007348
atmospheres	cm of mercury	76.0
atmospheres	ft of water (at 4°C)	33.90
atmospheres	in of mercury (at 0°C)	29.92
atmospheres	kg/sq. cm	1.0333
atmospheres	kg/sq. m	10,332

atmospheres	pounds/sq. in	14.70
atmospheres	tons/sq. ft	1.058

B

To Convert	Into	Multiply By
barrels (U.S., dry)	cubic inches	7056
barrels (U.S., dry)	quarts (dry)	105.0
barrels (U.S., liquid)	gallons	31.5
barrels (oil)	gallons (oil)	42.0
bars	atmospheres	0.9869
bars	dynes/sq. cm	10^4
bars	kg/sq. m	1.020×10^4
bars	pounds/sq. ft	2,089
bars	pounds/sq. in	14.50
Baryl	Dyne/sq. cm	1.000
bolt (U.S. cloth)	meters	36.576
Btu	liter-atmosphere	10.409
Btu	ergs	1.0550×10^{10}
Btu	foot-lb	778.3
Btu	gram-calories	252.0
Btu	horsepower-hr	3.931×10^{-4}
Btu	joules	1,054.8
Btu	kilogram-calories	0.2520
Btu	kilogram-meters	107.5
Btu	kilowatt-hr	2.928×10^{-4}
Btu/hr	foot-pounds/s	0.2162
Btu/hr	gram-calories/s	0.0700
Btu/hr	horsepower-hr	3.929×10^{-4}
Btu/hr	watts	0.2931
Btu/min	foot-lbs/s	12.96
Btu/min	horsepower	0.02356
Btu/min	kilowatts	0.01757
Btu/min	watts	17.57
Btu/sq. ft/min	watts/sq. in	0.1221
bucket (Br. dry)	cubic cm	1.818×10^4
bushels	cubic ft	1.2445
bushels	cubic in	2,150.4
bushels	cubic m	0.03524
bushels	liters	35.24
bushels	pecks	4.0
bushels	pints (dry)	64.0
bushels	quarts (dry)	32.0

C

To Convert	Into	Multiply By
calories, gram (mean)	Btu (mean)	3.9685×10^{-3}
candle/sq. cm	Lamberts	3.142
candle/sq. in	Lamberts	0.4870
centares (centiares)	sq. meters	1.0
Centigrade	Fahrenheit	$(C^{\circ} \times 9/5) + 32$
centigrams	grams	0.01
centiliter	ounce fluid (U.S.)	0.3382
centiliter	cubic inch	0.6103
centiliter	drams	2.705
centiliter	liters	0.01
centimeter	feet	3.281×10^{-2}
centimeter	inches	0.3937
centimeter	kilometers	10^{-5}
centimeter	meters	0.01
centimeter	miles	6.214×10^{-6}
centimeter	millimeters	10.0
centimeter	mils	393.7
centimeter	yards	1.094×10^{-2}
centimeter-dynes	cm-grams	1.020×10^{-3}
centimeter-dynes	meter-kg	1.020×10^{-8}
centimeter-dynes	pound-ft	7.376×10^{-8}
centimeter-grams	cm-dynes	980.7
centimeter-grams	meter-kg	10^{-5}
centimeter-grams	pound-ft	7.233×10^{-5}
centimeters of mercury	atmospheres	0.01316
centimeters of mercury	feet of water	0.4461
centimeters of mercury	kg/sq. meter	136.0
centimeters of mercury	pounds/sq. ft	27.85
centimeters of mercury	pounds/sq. in	0.1934
centimeters/sec	feet/min	1.9686
centimeters/sec	feet/sec	0.03281
centimeters/sec	kilometers/hr	0.036
centimeters/sec	knots	0.1943
centimeters/sec	meters/min	0.6
centimeters/sec	miles/hr	0.02237
centimeters/sec	miles/min	3.728×10^{-4}
centimeters/sec/sec	feet/sec/sec	0.03281
centimeters/sec/sec	km/hr/sec	0.036
centimeters/sec/sec	meters/sec/sec	0.01
centimeters/sec/sec	miles/hr/sec	0.02237
chain	inches	792.00
chain	meters	20.12
chains (surveyors' or Gunter's)	yards	22.00
circular mils	sq. cm	5.067×10^{-6}
circular mils	sq. mils	0.7854

circular mils	sq. inches	7.854×10^{-7}
circumference	Radians	6.283
cord feet	cubic feet	16
cords	cord feet	8
coulomb	statcoulombs	2.998×10^{9}
coulombs	faradays	1.036×10^{-5}
coulombs/sq. cm	coulombs/sq. in	64.52
coulombs/sq. cm	coulombs/sq. meter	10^{4}
coulombs/sq. in	coulombs/sq. cm	0.1550
coulombs/sq. in	coulombs/sq. meter	1,550
coulombs/sq. meter	coulombs/sq. cm	10^{-4}
coulombs/sq. meter	coulombs/sq. in	6.452×10^{-4}
cubic centimeters	cubic feet	3.531×10^{-5}
cubic centimeters	cubic inches	0.06102
cubic centimeters	cubic meters	10^{-6}
cubic centimeters	cubic yards	1.308×10^{-6}
cubic centimeters	gallons (U.S. liq.)	2.642×10^{-4}
cubic centimeters	liters	0.001
cubic centimeters	pints (U.S. liq.)	2.113×10^{-3}
cubic centimeters	quarts (U.S. liq.)	1.057×10^{-3}
cubic feet	bushels (dry)	0.8036
cubic feet	cubic cm	28,320.0
cubic feet	cubic inches	1,728.0
cubic feet	cubic meters	0.02832
cubic feet	cubic yards	0.03704
cubic feet	gallons (U.S. liq.)	7.48052
cubic feet	liters	28.32
cubic feet	pints (U.S. liq.)	59.84
cubic feet	quarts (U.S. liq.)	29.92
cubic feet/min	cubic cm/sec	472.0
cubic feet/min	gallons/sec	0.1247
cubic feet/min	liters/sec	0.4720
cubic feet/min	pounds of water/min	62.43
cubic feet/sec	million gal/day	0.646317
cubic feet/sec	gallons/min	448.831
cubic inches	cubic cm	16.39
cubic inches	cubic feet	5.787×10^{-4}
cubic inches	cubic meters	1.639×10^{-5}
cubic inches	cubic yards	2.143×10^{-5}
cubic inches	gallons	4.329×10^{-3}
cubic inches	liters	0.01639
cubic inches	mil-feet	1.061×10^{5}
cubic inches	pints (U.S. liq.)	0.03463
cubic inches	quarts (U.S. liq.)	0.01732
cubic meters	bushels (dry)	28.38
cubic meters	cubic cm	10^{6}
cubic meters	cubic feet	35.31
cubic meters	cubic inches	61,023.0
cubic meters	cubic yards	1.308
cubic meters	gallons (U.S. liq.)	264.2
cubic meters	liters	1,000.0
cubic meters	pints (U.S. liq.)	2,113.0

cubic meters	quarts (U.S. liq.)	1,057.
cubic yards	cubic cm	7.646×10^{5}
cubic yards	cubic feet	27.0
cubic yards	cubic inches	46,656.0
cubic yards	cubic meters	0.7646
cubic yards	gallons (U.S. liq.)	202.0
cubic yards	liters	764.6
cubic yards	pints (U.S. liq.)	1,615.9
cubic yards	quarts (U.S. liq.)	807.9
cubic yards/min	cubic ft/sec	0.45
cubic yards/min	gallons/sec	3.367
cubic yards/min	liters/sec	12.74

D

To Convert	Into	Multiply By
Dalton	gram	1.650×10^{-24}
days	seconds	86,400.0
decigrams	grams	0.1
deciliters	liters	0.1
decimeters	meters	0.1
degrees (angle)	quadrants	0.01111
degrees (angle)	radians	0.01745
degrees (angle)	seconds	3,600.0
degrees/sec	radians/sec	0.01745
degrees/sec	revolutions/min	0.1667
degrees/sec	revolutions/sec	2.778×10^{-3}
dekagrams	grams	10.0
dekaliters	liters	10.0
dekameters	meters	10.0
Drams (apothecaries' or troy)	ounces (avoirdupois)	0.1371429
Drams (apothecaries' or troy)	ounces (troy)	0.125
Drams (U.S., fluid or apoth.)	cubic cm	3.6967
drams	grams	1.7718
drams	grains	27.3437
drams	ounces	0.0625
dyne/cm	erg/sq. millimeter	0.01
dyne/sq. cm	atmospheres	9.869×10^{-7}
dyne/sq. cm	inch of mercury at 0°C	2.953×10^{-5}
dyne/sq. cm	inch of water at 4°C	4.015×10^{-4}
dynes	grams	1.020×10^{-3}
dynes	joules/cm	10^{-7}
dynes	joules/meter (newtons)	10^{-5}
dynes	kilograms	1.020×10^{-6}
dynes	poundals	7.233×10^{-5}
dynes	pounds	2.248×10^{-6}
dynes/sq. cm	bars	10^{-6}

E

To Convert	Into	Multiply By
ell	cm	114.30
ell	inches	45
em, pica	inch	0.167
em, pica	cm	0.4233
erg/sec	Dyne-cm/sec	1.000
ergs	Btu	9.480×10^{-11}
ergs	dyne-centimeters	1.0
ergs	foot-pounds	7.367×10^{-8}
ergs	gram-calories	0.2389×10^{-7}
ergs	gram-cm	1.020×10^{-3}
ergs	horsepower-hr	3.7250×10^{-14}
ergs	joules	10^{-7}
ergs	kg-calories	2.389×10^{-11}
ergs	kg-meters	1.020×10^{-8}
ergs	kilowatt-hr	0.2778×10^{-13}
ergs	watt-hours	0.2778×10^{-10}
ergs/sec	Btu/min	$5{,}688 \times 10^{-9}$
ergs/sec	ft-lb/min	4.427×10^{-6}
ergs/sec	ft-lb/sec	7.3756×10^{-8}
ergs/sec	horsepower	1.341×10^{-10}
ergs/sec	kg-calories/min	1.433×10^{-9}
ergs/sec	kilowatts	10^{-10}

F

To Convert	Into	Multiply By
farad	microfarads	10^{6}
Faraday/sec	ampere (absolute)	9.6500×10^{4}
faradays	ampere-hours	26.80
faradays	coulombs	9.649×10^{4}
fathom	meter	1.828804
fathoms	feet	6.0
feet	centimeters	30.48
feet	kilometers	3.048×10^{-4}
feet	meters	0.3048
feet	miles (naut.)	1.645×10^{-4}
feet	miles (stat.)	1.894×10^{-4}
feet	millimeters	304.8
feet	mils	1.2×10^{4}
feet of water	atmospheres	0.02950
feet of water	in of mercury	0.8826
feet of water	kg/sq. cm	0.03048
feet of water	kg/sq. meter	304.8

feet of water	pounds/sq. ft	62.43
feet of water	pounds/sq. in	0.4335
feet/min	cm/sec	0.5080
feet/min	feet/sec	0.01667
feet/min	km/hr	0.01829
feet/min	meters/min	0.3048
feet/min	miles/hr	0.01136
feet/sec	cm/sec	30.48
feet/sec	km/hr	1.097
feet/sec	knots	0.5921
feet/sec	meters/min	18.29
feet/sec	miles/hr	0.6818
feet/sec	miles/min	0.01136
feet/sec/sec	cm/sec/sec	30.48
feet/sec/sec	km/hr/sec	1.097
feet/sec/sec	meters/sec/sec	0.3048
feet/sec/sec	miles/hr/sec	0.6818
feet/100 feet	per centigrade	1.0
foot-candle	lumen/sq. meter	10.764
foot-pounds	Btu	1.286×10^{-3}
foot-pounds	ergs	1.356×10^{7}
foot-pounds	gram-calories	0.3238
foot-pounds	hp-hr	5.050×10^{-7}
foot-pounds	joules	1.356
foot-pounds	kg-calories	3.24×10^{-4}
foot-pounds	kg-meters	0.1383
foot-pounds	kilowatt-hr	3.766×10^{-7}
foot-pounds/min	Btu/min	1.286×10^{-3}
foot-pounds/min	foot-pounds/sec	0.01667
foot-pounds/min	horsepower	3.030×10^{-5}
foot-pounds/min	kg-calories/min	3.24×10^{-4}
foot-pounds/min	kilowatts	2.260×10^{-5}
foot-pounds/sec	Btu/hr	4.6263
foot-pounds/sec	Btu/min	0.07717
foot-pounds/sec	horsepower	1.818×10^{-3}
foot-pounds/sec	kg-calories/min	0.01945
foot-pounds/sec	kilowatts	1.356×10^{-3}
Furlongs	miles (U.S.)	0.125
furlongs	rods	40.0
furlongs	feet	660.0

G

To Convert	Into	Multiply By
gallons	cubic cm	3,785.0
gallons	cubic feet	0.1337
gallons	cubic inches	231.0
gallons	cubic meters	3.785×10^{-3}
gallons	cubic yards	4.951×10^{-3}
gallons	liters	3.785

gallons (liq. Br. Imp.)	gallons (U.S. liq.)	1.20095
gallons (U.S.)	gallons (Imp.)	0.83267
gallons of water	pounds of water	8.3453
gallons/min	cubic ft/sec	2.228×10^{-3}
gallons/min	liters/sec	0.06308
gallons/min	cubic ft/hr	8.0208
gausses	lines/sq. in	6.452
gausses	webers/sq. cm	10^{-8}
gausses	webers/sq. in	6.452×10^{-8}
gausses	webers/sq. meter	10^{-4}
gilberts	ampere-turns	0.7958
gilberts/cm	amp-turns/cm	0.7958
gilberts/cm	amp-turns/in	2.021
gilberts/cm	amp-turns/meter	79.58
gills	liters	0.1183
gills	pints (liq.)	0.25
gills (British)	cubic cm	142.07
grade	radian	0.01571
grains	drams (avoirdupois)	0.03657143
grains (troy)	grains (avdp.)	1.0
grains (troy)	grams	0.06480
grains (troy)	ounces (avdp.)	2.0833×10^{-3}
grains (troy)	pennyweight (troy)	0.04167
grains/Imp. gal	parts/million	14.286
grains/U.S. gal	parts/million	17.118
grains/U.S. gal	pounds/million gal	142.86
gram-calories	Btu	3.9683×10^{-3}
gram-calories	ergs	4.1868×10^{7}
gram-calories	foot-pounds	3.0880
gram-calories	horsepower-hr	1.5596×10^{-6}
gram-calories	kilowatt-hr	1.1630×10^{-6}
gram-calories	watt-hr	1.1630×10^{-3}
gram-calories/sec	Btu/hr	14.286
gram-centimeters	Btu	9.297×10^{-8}
gram-centimeters	ergs	980.7
gram-centimeters	joules	9.807×10^{-5}
gram-centimeters	kg-calories	2.343×10^{-8}
gram-centimeters	kg-meters	10^{-5}
grams	dynes	980.7
grams	grains	15.43
grams	joules/cm	9.807×10^{-5}
grams	joules/meter (newtons)	9.807×10^{-3}
grams	kilograms	0.001
grams	milligrams	1,000
grams	ounces (avdp.)	0.03527
grams	ounces (troy)	0.03215
grams	poundals	0.07093
grams	pounds	2.205×10^{-3}
grams/cm	pounds/inch	5.600×10^{-3}
grams/cubic cm	pounds/cubic ft	62.43
grams/cubic cm	pounds/cubic in	0.03613
grams/cubic cm	pounds/mil-foot	3.405×10^{-7}

grams/liter	grains/gal	58.417
grams/liter	pounds/1,000 gal	8.345
grams/liter	pounds/cubic ft	0.062427
grams/liter	parts/million	1,000.0
grams/sq. cm	pounds/sq. ft	2.0481

H

To Convert	Into	Multiply By
hand	cm	10.16
hectares	acres	2.471
hectares	sq. feet	1.076×10^5
hectograms	grams	100.0
hectoliters	liters	100.0
hectometers	meters	100.0
hectowatts	watts	100.0
henries	millihenries	1,000.0
horsepower	Btu/min	42.44
horsepower	foot-lb/min	33,000
horsepower	foot-lb/sec	550.0
horsepower	kg-calories/min	10.68
horsepower	kilowatts	0.7457
horsepower	watts	745.7
horsepower (boiler)	Btu/hr	33.479
horsepower (boiler)	kilowatts	9.803
horsepower (metric) (542.5 ft lb./sec)	horsepower (550 ft lb./sec)	0.9863
horsepower (550 ft lb./sec)	horsepower (metric) (542.5 ft lb./sec)	1.014
horsepower-hr	Btu	2,547
horsepower-hr	ergs	2.6845×10^{13}
horsepower-hr	foot-lb	1.98×10^6
horsepower-hr	gram-calories	641,190
horsepower-hr	joules	2.684×10^6
horsepower-hr	kg-calories	641.1
horsepower-hr	kg-meters	2.737×10^5
horsepower-hr	kilowatt-hr	0.7457
hours	days	4.167×10^{-2}
hours	weeks	5.952×10^{-3}
hundredweights (long)	pounds	112
hundredweights (long)	tons (long)	0.05
hundredweights (short)	ounces (avoirdupois)	1,600
hundredweights (short)	pounds	100
hundredweights (short)	tons (metric)	0.0453592
hundredweights (short)	tons (long)	0.0446429

I

To Convert	Into	Multiply By
inches	centimeters	2.540
inches	meters	2.540×10^{-2}
inches	miles	1.578×10^{-5}
inches	millimeters	25.40
inches	mils	1,000.0
inches	yards	2.778×10^{-2}
inches of mercury	atmospheres	0.03342
inches of mercury	feet of water	1.133
inches of mercury	kg/sq. cm	0.03453
inches of mercury	kg/sq. meter	345.3
inches of mercury	pounds/sq. ft	70.73
inches of mercury	pounds/sq. in	0.4912
inches of water (at 4°C)	atmospheres	2.458×10^{-3}
inches of water (at 4°C)	inches of mercury	0.07355
inches of water (at 4°C)	kg/sq. cm	2.540×10^{-3}
inches of water (at 4°C)	ounces/sq. in	0.5781
inches of water (at 4°C)	pounds/sq. ft	5.204
inches of water (at 4°C)	pounds/sq. in	0.03613
international ampere	ampere (absolute)	0.9998
international Volt	volts (absolute)	1.0003
international volt	joules (absolute)	1.593×10^{-19}
international volt	joules	9.654×10^{4}

J

To Convert	Into	Multiply By
joules	Btu	9.480×10^{-4}
joules	ergs	10^{7}
joules	foot-pounds	0.7376
joules	kg-calories	2.389×10^{-4}
joules	kg-meters	0.1020
joules	watt-hr	2.778×10^{-4}
joules/cm	grams	1.020×10^{4}
joules/cm	dynes	10^{7}
joules/cm	joules/meter (newtons)	100.0
joules/cm	poundals	723.3
joules/cm	pounds	22.48

K

To Convert	Into	Multiply By
kilogram-calories	Btu	3.968
kilogram-calories	foot-pounds	3,088
kilogram-calories	hp-hr	1.560×10^{-3}
kilogram-calories	joules	4,186
kilogram-calories	kg-meters	426.9
kilogram-calories	kilojoules	4.186
kilogram-calories	kilowatt-hr	1.163×10^{-3}
kilogram meters	Btu	9.294×10^{-3}
kilogram meters	ergs	9.804×10^{7}
kilogram meters	foot-pounds	7.233
kilogram meters	joules	9.804
kilogram meters	kg-calories	2.342×10^{-3}
kilogram meters	kilowatt-hr	2.723×10^{-6}
kilograms	dynes	980,665
kilograms	grams	1,000.0
kilograms	joules/cm	0.09807
kilograms	joules/meter (newtons)	9.807
kilograms	poundals	70.93
kilograms	pounds	2.205
kilograms	tons (long)	9.842×10^{-4}
kilograms	tons (short)	1.102×10^{-3}
kilograms/cubic meter	grams/cubic cm	0.001
kilograms/cubic meter	pounds/cubic ft	0.06243
kilograms/cubic meter	pounds/cubic in	3.613×10^{-5}
kilograms/cubic meter	pounds/mil-foot	3.405×10^{-10}
kilograms/meter	pounds/ft	0.6720
kilograms/sq. cm	dynes	980,665
kilograms/sq. cm	atmospheres	0.9678
kilograms/sq. cm	feet of water	32.81
kilograms/sq. cm	inches of mercury	28.96
kilograms/sq. cm	pounds/sq. ft	2,048
kilograms/sq. cm	pounds/sq. in	14.22
kilograms/sq. meter	atmospheres	9.678×10^{-5}
kilograms/sq. meter	bars	98.07×10^{-6}
kilograms/sq. meter	feet of water	3.281×10^{-3}
kilograms/sq. meter	inches of mercury	2.896×10^{-3}
kilograms/sq. meter	pounds/sq. ft	0.2048
kilograms/sq. meter	pounds/sq. in	1.422×10^{-3}
kilograms/sq. mm	kg/sq. meter	10^{6}
kilolines	maxwells	1,000.0
kiloliters	liters	1,000.0
kilometers	centimeters	10^{5}
kilometers	feet	3,281
kilometers	inches	3.937×10^{4}
kilometers	meters	1,000.0
kilometers	miles	0.6214

kilometers	millimeters	10^4
kilometers	yards	1,094
kilometers/hr	cm/sec	27.78
kilometers/hr	feet/min	54.68
kilometers/hr	feet/sec	0.9113
kilometers/hr	knots	0.5396
kilometers/hr	meters/min	16.67
kilometers/hr	miles/hr	0.6214
kilometers/hr/sec	cm/sec/sec	27.78
kilometers/hr/sec	feet/sec/sec	0.9113
kilometers/hr/sec	meters/sec/sec	0.2778
kilometers/hr/sec	miles/hr/sec	0.6214
kilowatt-hr	Btu	3,413
kilowatt-hr	ergs	3.600×10^{13}
kilowatt-hr	foot-lb	2.655×10^6
kilowatt-hr	gram-calories	859,850
kilowatt-hr	horsepower-hr	1.341
kilowatt-hr	joules	3.6×10^6
kilowatt-hr	kg-calories	860.5
kilowatt-hr	kg-meters	3.671×10^5
kilowatt-hr	pounds of water raised from 62° to 212°F	22.75
kilowatts	Btu/min	56.92
kilowatts	foot-lb/min	4.426×10^4
kilowatts	foot-lb/sec	737.6
kilowatts	horsepower	1.341
kilowatts	kg-calories/min	14.34
kilowatts	watts	1,000.0
knots	feet/hr	6,080
knots	kilometers/hr	1.8532
knots	nautical miles/hr	1.0
knots	statute miles/hr	1.151
knots	yards/hr	2,027
knots	feet/sec	1.689

L

To Convert	Into	Multiply By
league	miles (approx.)	3.0
light year	miles	5.9×10^{12}
light year	kilometers	9.4637×10^{12}
lines/sq. cm	gausses	1.0
lines/sq. in	gausses	0.1550
lines/sq. in	webers/sq. cm	1.550×10^{-9}
lines/sq. in	webers/sq. in	10^{-8}
lines/sq. in	webers/sq. meter	1.550×10^{-5}
links (engineer's)	inches	12.0
links (surveyor's)	inches	7.92
liters	bushels (U.S. dry)	0.02838
liters	cubic cm	1,000.0

liters	cubic feet	0.03531
liters	cubic inches	61.02
liters	cubic meters	0.001
liters	cubic yards	1.308×10^{-3}
liters	gallons (U.S. liq.)	0.2642
liters	pints (U.S. liq.)	2.113
liters	quarts (U.S. liq.)	1.057
liters/min	cubic ft/sec	5.886×10^{-4}
liters/min	gal/sec	4.403×10^{-3}
lumen	spherical candle power	0.07958
lumen	watt	0.001496
lumens/sq. ft	foot-candles	1.0
lumens/sq. ft	lumen/sq. meter	10.76
lux	foot-candles	0.0929

M

To Convert	Into	Multiply By
maxwells	kilolines	0.001
maxwells	webers	10^{-8}
megalines	maxwells	10^{6}
megohms	microhms	10^{12}
megohms	ohms	10^{6}
meter-kilograms	cm-dynes	9.807×10^{7}
meter-kilograms	cm-grams	10^{5}
meter-kilograms	pound-feet	7.233
meters	centimeters	100.0
meters	feet	3.281
meters	inches	39.37
meters	kilometers	0.001
meters	miles (naut.)	5.396×10^{-4}
meters	miles (stat.)	6.214×10^{-4}
meters	millimeters	1,000.0
meters	yards	1.094
meters	varas	1.179
meters/min	cm/sec	1,667
meters/min	feet/min	3.281
meters/min	feet/sec	0.05468
meters/min	km/hr	0.06
meters/min	knots	0.03238
meters/min	miles/hr	0.03728
meters/sec	feet/min	196.8
meters/sec	feet/sec	3.281
meters/sec	kilometers/hr	3.6
meters/sec	kilometers/min	0.06
meters/sec	miles/hr	2.237
meters/sec	miles/min	0.03728
meters/sec/sec	cm/sec/sec	100.0
meters/sec/sec	ft/sec/sec	3.281
meters/sec/sec	km/hr/sec	3.6

meters/sec/sec	miles/hr/sec	2.237
microfarad	farads	10^{-6}
micrograms	grams	10^{-6}
microhms	megohms	10^{-12}
microhms	ohms	10^{-6}
microliters	liters	10^{-6}
microns	meters	1×10^{-6}
miles (naut.)	feet	6,080.27
miles (naut.)	kilometers	1.853
miles (naut.)	meters	1,853
miles (naut.)	miles (statute)	1.1516
miles (naut.)	yards	2,027
miles (statute)	centimeters	1.609×10^5
miles (statute)	feet	5,280
miles (statute)	inches	6.336×10^4
miles (statute)	kilometers	1.609
miles (statute)	meters	1,609
miles (statute)	miles (naut.)	0.8684
miles (statute)	yards	1,760
miles/hr	cm/sec	44.70
miles/hr	feet/min	88
miles/hr	feet/sec	1.467
miles/hr	km/hr	1.609
miles/hr	km/min	0.02682
miles/hr	knots	0.8684
miles/hr	meters/min	26.82
miles/hr	miles/min	0.1667
miles/hr/sec	cm/sec/sec	44.70
miles/hr/sec	feet/sec/sec	1.467
miles/hr/sec	km/hr/sec	1.609
miles/hr/sec	meters/sec/sec	0.4470
miles/min	cm/sec	2,682
miles/min	feet/sec	88
miles/min	km/min	1.609
miles/min	knots/min	0.8684
miles/min	miles/hr	60
mil-feet	cubic inches	9.425×10^{-6}
milliers	kilograms	1,000
milligrams	grains	0.01543236
milligrams	grams	0.001
milligrams/liter	parts/million	1.0
millihenries	henries	0.001
milliliters	liters	0.001
millimeters	centimeters	0.1
millimeters	feet	3.281×10^{-3}
millimeters	inches	0.03937
millimeters	kilometers	10^{-6}
millimeters	meters	0.001
millimeters	miles	6.214×10^{-7}
millimeters	mils	39.37
millimeters	yards	1.094×10^{-3}
millimicrons	meters	1×10^{-9}

million gal/day	cubic ft/sec	1.54723
mils	centimeters	2.540×10^{-3}
mils	feet	8.333×10^{-5}
mils	inches	0.001
mils	kilometers	2.540×10^{-8}
mils	yards	2.778×10^{-5}
miner's inches	cubic ft/min	1.5
minims (British)	cubic cm	0.059192
minims (U.S., fluid)	cubic cm	0.061612
minutes (angles)	degrees	0.01667
minutes (angles)	quadrants	1.852×10^{-4}
minutes (angles)	radians	2.909×10^{-4}
minutes (angles)	seconds	60.0
myriagrams	kilograms	10.0
myriameters	kilometers	10.0
myriawatts	kilowatts	10.0

N

To Convert	Into	Multiply By
nepers	decibels	8.686
Newton	dynes	1 x 105

O

To Convert	Into	Multiply By
ohm (international)	ohm (absolute)	1.0005
ohms	megohms	10^{-6}
ohms	microhms	10^{6}
ounces	drams	16.0
ounces	grains	437.5
ounces	grams	28.349527
ounces	pounds	0.0625
ounces	ounces (troy)	0.9115
ounces	tons (long)	2.790×10^{-5}
ounces	tons (metric)	2.835×10^{-5}
ounces (fluid)	cubic inches	1.805
ounces (fluid)	liters	0.02957
ounces (troy)	grains	480.0
ounces (troy)	grams	31.103481
ounces (troy)	ounces (avdp.)	1.09714
ounces (troy)	pennyweights (troy)	20.0
ounces (troy)	pounds (troy)	0.08333
ounces/sq. inch	dynes/sq. cm	4,309
ounces/sq. in	pounds/sq. in	0.0625

P

To Convert	Into	Multiply By
parsec	miles	19×10^{12}
parsec	kilometers	3.084×10^{13}
parts/million	grains/U.S. gal	0.0584
parts/million	grains/Imp. gal	0.07016
parts/million	pounds/million gal	8.345
pecks (British)	cubic inches	554.6
pecks (British)	liters	9.091901
pecks (U.S.)	bushels	0.25
pecks (U.S.)	cubic inches	537.605
pecks (U.S.)	liters	8.809582
pecks (U.S.)	quarts (dry)	8
pennyweights (troy)	grains	24.0
pennyweights (troy)	ounces (troy)	0.05
pennyweights (troy)	grams	1.55517
pennyweights (troy)	pounds (troy)	4.1667×10^{-3}
pints (dry)	cubic inches	33.60
pints (liq.)	cubic cm	473.2
pints (liq.)	cubic feet	0.01671
pints (liq.)	cubic inches	28.87
pints (liq.)	cubic meters	4.732×10^{-4}
pints (liq.)	cubic yards	6.189×10^{-4}
pints (liq.)	gallons	0.125
pints (liq.)	liters	0.4732
pints (liq.)	quarts (liq.)	0.5
Planck's quantum	erg - second	6.624×10^{-27}
poise	gram/cm sec	1.00
poundals	dynes	13,826
poundals	grams	14.10
poundals	joules/cm	1.383×10^{-3}
poundals	joules/meter (newtons)	0.1383
poundals	kilograms	0.01410
poundals	pounds	0.03108
pound-feet	cm-dynes	1.356×10^{7}
pound-feet	cm-grams	13,825
pound-feet	meter-kg	0.1383
pounds	drams	256
pounds	dynes	44.4823×10^{4}
pounds	grains	7,000
pounds	grams	453.5924
pounds	joules/cm	0.04448
pounds	joules/meter (newtons)	4.448
pounds	kilograms	0.4536
pounds	ounces	16.0
pounds	ounces (troy)	14.5833
pounds	poundals	32.17
pounds	pounds (troy)	1.21528

pounds	tons (short)	0.0005
pounds (avoirdupois)	ounces (troy)	14.5833
pounds (troy)	grains	5,760
pounds (troy)	grams	373.24177
pounds (troy)	ounces (avdp.)	13.1657
pounds (troy)	ounces (troy)	12.0
pounds (troy)	pennyweights (troy)	240.0
pounds (troy)	pounds (avdp.)	0.822857
pounds (troy)	tons (long)	3.6735×10^{-4}
pounds (troy)	tons (metric)	3.7324×10^{-4}
pounds (troy)	tons (short)	4.1143×10^{-4}
pounds of water	cubic ft	0.01602
pounds of water	cubic inches	27.68
pounds of water	gallons	0.1198
pounds of water/min	cubic ft/sec	2.670×10^{-4}
pounds/cubic ft	grams/cubic cm	0.01602
pounds/cubic ft	kg/cubic meter	16.02
pounds/cubic ft	pounds/cubic in	5.787×10^{-4}
pounds/cubic ft	pounds/mil-foot	5.456×10^{-9}
pounds/cubic in	gm/cubic cm	27.68
pounds/cubic in	kg/cubic meter	2.768×10^{4}
pounds/cubic in	pounds/cubic ft	1,728
pounds/cubic in	pounds/mil-foot	9.425×10^{-6}
pounds/ft	kg/meter	1.488
pounds/in	gm/cm	178.6
pounds/mil-foot	gm/cubic cm	2.306×10^{6}
pounds/sq. ft	atmospheres	4.725×10^{-4}
pounds/sq. ft	feet of water	0.01602
pounds/sq. ft	inches of mercury	0.01414
pounds/sq. ft	kg/sq. meter	4.882
pounds/sq. ft	pounds/sq. in	6.944×10^{-3}
pounds/sq. in	atmospheres	0.06804
pounds/sq. in	feet of water	2.307
pounds/sq. in	inches of mercury	2.036
pounds/sq. in	kg/sq. meter	703.1
pounds/sq. in	pounds/sq. ft	144.0

Q

To Convert	Into	Multiply By
quadrants (angle)	degrees	90.0
quadrants (angle)	minutes	5,400.0
quadrants (angle)	radians	1.571
quadrants (angle)	seconds	3.24×10^{5}
quarts (dry)	cubic inches	67.20
quarts (liq.)	cubic cm	946.4
quarts (liq.)	cubic feet	0.03342
quarts (liq.)	cubic inches	57.75
quarts (liq.)	cubic meters	9.464×10^{-4}
quarts (liq.)	cubic yards	1.238×10^{-3}

quarts (liq.)	gallons	0.25
quarts (liq.)	liters	0.9463

R

To Convert	Into	Multiply By
radians	degrees	57.30
radians	minutes	3,438
radians	quadrants	0.6366
radians	seconds	2.063×10^5
radians/sec	degrees/sec	57.30
radians/sec	revolutions/min	9.549
radians/sec	revolutions/sec	0.1592
radians/sec/sec	revolutions/min/min	573.0
radians/sec/sec	revolutions/min/sec	9.549
radians/sec/sec	revolutions/sec/sec	0.1592
revolutions	degrees	360.0
revolutions	quadrants	4.0
revolutions	radians	6.283
revolutions/min	degrees/sec	6.0
revolutions/min	radians/sec	0.1047
revolutions/min	revolutions/sec	0.01667
revolutions/min/min	radians/sec/sec	1.745×10^{-3}
revolutions/min/min	revolutions/min/sec	0.01667
revolutions/min/min	revolutions/sec/sec	2.778×10^{-4}
revolutions/sec	degrees/sec	360.0
revolutions/sec	radians/sec	6.283
revolutions/sec	revolutions/min	60.0
revolutions/sec/sec	radians/sec/sec	6.283
revolutions/sec/sec	revolutions/min/min	3,600.0
revolutions/sec/sec	revolutions/min/sec	60.0
rod	chain (Gunters)	0.25
rod	meters	5.029
rods	feet	16.5
rods (surveyors' meas.)	yards	5.5

S

To Convert	Into	Multiply By
scruples	grains	20
seconds (angle)	degrees	2.778×10^{-4}
seconds (angle)	minutes	0.01667
seconds (angle)	quadrants	3.087×10^{-6}
seconds (angle)	radians	4.848×10^{-6}
slug	kilogram	14.59
slug	pounds	32.17
sphere	steradians	12.57

square centimeters	circular mils	1.973×10^{5}
square centimeters	sq. feet	1.076×10^{-3}
square centimeters	sq. inches	0.1550
square centimeters	sq. meters	0.0001
square centimeters	sq. miles	3.861×10^{-11}
square centimeters	sq. millimeters	100.0
square centimeters	sq. yards	1.196×10^{-4}
square feet	acres	2.296×10^{-5}
square feet	circular mils	1.833×10^{8}
square feet	sq. cm	929.0
square feet	sq. inches	144.0
square feet	sq. meters	0.09290
square feet	sq. miles	3.587×10^{-8}
square feet	sq. millimeters	9.290×10^{4}
square feet	sq. yards	0.1111
square inches	circular mils	1.273×10^{6}
square inches	sq. cm	6.452
square inches	sq. feet	6.944×10^{-3}
square inches	sq. millimeters	645.2
square inches	sq. mils	10^{6}
square inches	sq. yards	7.716×10^{-4}
square kilometers	acres	247.1
square kilometers	sq. cm	10^{10}
square kilometers	sq. ft	10.76×10^{6}
square kilometers	sq. inches	1.550×10^{9}
square kilometers	sq. meters	10^{6}
square kilometers	sq. miles	0.3861
square kilometers	sq. yards	1.196×10^{6}
square meters	acres	2.471×10^{-4}
square meters	sq. cm	10^{4}
square meters	sq. feet	10.76
square meters	sq. inches	1,550
square meters	sq. miles	3.861×10^{-7}
square meters	sq. millimeters	10^{6}
square meters	sq. yards	1.196
square miles	acres	640.0
square miles	sq. feet	27.88×10^{6}
square miles	sq. km	2.590
square miles	sq. meters	2.590×10^{6}
square miles	sq. yards	3.098×10^{6}
square millimeters	circular mils	1,973
square millimeters	sq. cm	0.01
square millimeters	sq. feet	1.076×10^{-5}
square millimeters	sq. inches	1.550×10^{-3}
square mils	circular mils	1.273
square mils	sq. cm	6.452×10^{-6}
square mils	sq. inches	10^{-6}
square yards	acres	2.066×10^{-4}
square yards	sq. cm	8,361
square yards	sq. feet	9.0
square yards	sq. inches	1,296
square yards	sq. meters	0.8361

square yards	sq. miles	3.228×10^{-7}
square yards	sq. millimeters	8.361×10^{5}

T

To Convert	Into	Multiply By
temperature (˚C) + 273	absolute temperature (˚C)	1.0
temperature (˚C) + 17.78	temperature (˚F)	1.8
temperature (˚F) + 460	absolute temperature (˚F)	1.0
temperature (˚F) - 32	temperature (˚C)	5/9
tons (long)	kilograms	1,016
tons (long)	pounds	2,240
tons (long)	tons (short)	1.120
tons (metric)	kilograms	1,000
tons (metric)	pounds	2,205
tons (short)	kilograms	907.1848
tons (short)	ounces	32,000
tons (short)	ounces (troy)	29,166.66
tons (short)	pounds	2,000
tons (short)	pounds (troy)	2,430.56
tons (short)	tons (long)	0.89287
tons (short)	tons (metric)	0.9078
tons (short)/sq. ft	kg/sq. meter	9,765
tons (short)/sq. ft	pounds/sq. in	2,000
tons of water/24 hr	pounds of water/hr	83.333
tons of water/24 hr	gallons/min	0.16643
tons of water/24 hr	cubic ft/hr	1.3349

V

To Convert	Into	Multiply By
volt (absolute)	statvolts	0.003336
volt/inch	volt/cm	0.39370

To Convert	Into	Multiply By
watt-hours	Btu	3.413
watt-hours	ergs	3.60×10^{10}
watt-hours	foot-pounds	2,656
watt-hours	gram-calories	859.85
watt-hours	horsepower-hr	1.341×10^{-3}

watt-hours	kilogram-calories	0.8605
watt-hours	kilogram-meters	367.2
watt-hours	kilowatt-hr	0.001
watt (international)	watt (absolute)	1.0002
watts	Btu/hr	3.4129
watts	Btu/min	0.05688
watts	ergs/sec	107
watts	foot-lb/min	44.27
watts	foot-lb/sec	0.7378
watts	horsepower	1.341×10^{-3}
watts	horsepower (metric)	1.360×10^{-3}
watts	kg-calories/min	0.01433
watts	kilowatts	0.001
watts (Abs.)	Btu (mean)/min	0.056884
watts (Abs.)	joules/sec	1
webers	maxwells	10^{8}
webers	kilolines	10^{5}
webers/sq. in	gausses	1.550×10^{7}
webers/sq. in	lines/sq. in	10^{8}
webers/sq. in	webers/sq. cm	0.1550
webers/sq. in	webers/sq. meter	1,550
webers/sq. meter	gausses	10^{4}
webers/sq. meter	lines/sq. in	6.452×10^{4}
webers/sq. meter	webers/sq. cm	10^{-4}
webers/sq. meter	webers/sq. in	6.452×10^{-4}

To Convert	Into	Multiply By
yards	centimeters	91.44
yards	kilometers	9.144×10^{-4}
yards	meters	0.9144
yards	miles (naut.)	4.934×10^{-4}
yards	miles (stat.)	5.682×10^{-4}
yards	millimeters	914.4

11.5 Reference Tables

Table 11.1 Power conversion factors (decibels to watts).

dBm	dBw	Watts	Multiple	Prefix
+150	+120	1,000,000,000,000	10^{12}	1 Terawatt
+140	+110	100,000,000,000	10^{11}	100 Gigawatts
+130	+100	10,000,000,000	10^{10}	10 Gigawatts
+120	+90	1,000,000,000	10^{9}	1 Gigawatt
+110	+80	100,000,000	10^{8}	100 Megawatts
+100	+70	10,000,000	10^{7}	10 Megawatts
+90	+60	1,000,000	10^{6}	1 Megawatt
+80	+50	100,000	10^{5}	100 Kilowatts
+70	+40	10,000	10^{4}	10 Kilowatts
+60	+30	1,000	10^{3}	1 Kilowatt
+50	+20	100	10^{2}	1 Hectrowatt
+40	+10	10	10	1 Decawatt
+30	0	1	1	1 Watt
+20	-10	0.1	10^{-1}	1 Deciwatt
+10	-20	0.01	10^{-2}	1 Centiwatt
0	-30	0.001	10^{-3}	1 Milliwatt
-10	-40	0.0001	10^{-4}	100 Microwatts
-20	-50	0.00001	10^{-5}	10 Microwatts
-30	-60	0.000,001	10^{-6}	1 Microwatt
-40	-70	0.0,000,001	10^{-7}	100 Nanowatts
-50	-80	0.00,000,001	10^{-8}	10 Nanowatts
-60	-90	0.000,000,001	10^{-9}	1 Nanowatt
-70	-100	0.0,000,000,001	10^{-10}	100 Picowatts
-80	-110	0.00,000,000,001	10^{-11}	10 Picowatts
-90	-120	0,000,000,000,001	10^{-12}	1 Picowatt

Table 11.2 Relationships of voltage standing wave ratio and key operating parameters.

VSWR	Reflection coefficient	Return loss	Power ratio	Percent reflected
1.01:1	0.0050	46.1 dB	0.00002	0.002
1.02:1	0.0099	40.1 dB	0.00010	0.010
1.04:1	0.0196	34.2 dB	0.00038	0.038
1.06:1	0.0291	30.7 dB	0.00085	0.085
1.08:1	0.0385	28.3 dB	0.00148	0.148
1.10:1	0.0476	26.4 dB	0.00227	0.227
1.20:1	0.0909	20.8 dB	0.00826	0.826
1.30:1	0.1304	17.7 dB	0.01701	1.7
1.40:1	0.1667	15.6 dB	0.02778	2.8
1.50:1	0.2000	14.0 dB	0.04000	4.0
1.60:1	0.2308	12.7 dB	0.05325	5.3
1.70:1	0.2593	11.7 dB	0.06722	6.7
1.80:1	0.2857	10.9 dB	0.08163	8.2
1.90:1	0.3103	10.2 dB	0.09631	9.6
2.00:1	0.3333	9.5 dB	0.11111	11.1
2.20:1	0.3750	8.5 dB	0.14063	14.1
2.40:1	0.4118	7.7 dB	0.16955	17.0
2.60:1	0.4444	7.0 dB	0.19753	19.8
2.80:1	0.4737	6.5 dB	0.22438	22.4
3.00:1	0.5000	6.0 dB	0.25000	25.0
3.50:1	0.5556	5.1 dB	0.30864	30.9
4.00:1	0.6000	4.4 dB	0.36000	36.0
4.50:1	0.6364	3.9 dB	0.40496	40.5
5.00:1	0.6667	3.5 dB	0.44444	44.4
6.00:1	0.7143	2.9 dB	0.51020	51.0
7.00:1	0.7500	2.5 dB	0.56250	56.3
8.00:1	0.7778	2.2 dB	0.60494	60.5
9.00:1	0.8000	1.9 dB	0.64000	64.0
10.00:1	0.8182	1.7 dB	0.66942	66.9
15.00:1	0.8750	1.2 dB	0.76563	76.6
20.00:1	0.9048	0.9 dB	0.81859	81.9
30.00:1	0.9355	0.6 dB	0.87513	97.5
40.00:1	0.9512	0.4 dB	0.90482	90.5
50.00:1	0.9608	0.3 dB	0.92311	92.3

Table 11.3 Specifications of standard copper wire sizes.

Wire size AWG	Diam. in mils	Circular mil area	Turns per linear inch[1]			Ohms per 100 ft[2]	Current carrying capacity[3]	Diam. in mm
			enam.	SCE	DCC			
1	289.3	83810	-	-	-	0.1239	119.6	7.348
2	257.6	05370	-	-	-	0.1563	94.8	6.544
3	229.4	62640	-	-	-	0.1970	75.2	5.827
4	204.3	41740	-	-	-	0.2485	59.6	5.189
5	181.9	33100	-	-	-	0.3133	47.3	4.621
6	162.0	26250	-	-	-	0.3951	37.5	4.115
7	144.3	20820	-	-	-	0.4982	29.7	3.665
8	128.5	16510	7.6	-	7.1	0.6282	23.6	3.264
9	114.4	13090	8.6	-	7.8	0.7921	18.7	2.906
10	101.9	10380	9.6	9.1	8.9	0.9989	14.8	2.588
11	90.7	8234	10.7	-	9.8	1.26	11.8	2.305
12	80.8	6530	12.0	11.3	10.9	1.588	9.33	2.063
13	72.0	5178	13.5	-	12.8	2.003	7.40	1.828
14	64.1	4107	15.0	14.0	13.8	2.525	5.87	1.628
15	57.1	3257	16.8	-	14.7	3.184	4.65	1.450
16	50.8	2583	18.9	17.3	16.4	4.016	3.69	1.291
17	45.3	2048	21.2	-	18.1	5.064	2.93	1.150
18	40.3	1624	23.6	21.2	19.8	6.386	2.32	1.024
19	35.9	1288	26.4	-	21.8	8.051	1.84	0.912
20	32.0	1022	29.4	25.8	23.8	10.15	1.46	0.812
21	28.5	810	33.1	-	26.0	12.8	1.16	0.723
22	25.3	642	37.0	31.3	30.0	16.14	0.918	0.644
23	22.6	510	41.3	-	37.6	20.36	0.728	0.573
24	20.1	404	46.3	37.6	35.6	25.67	0.577	0.511
25	17.9	320	51.7	-	38.6	32.37	0.458	0.455
26	15.9	254	58.0	46.1	41.8	40.81	0.363	0.406
27	14.2	202	64.9	-	45.0	51.47	0.288	0.361
28	12.6	160	72.7	54.6	48.5	64.9	0.228	0.321
29	11.3	127	81.6	-	51.8	81.83	0.181	0.286
30	10.0	101	90.5	64.1	55.5	103.2	0.144	0.255
31	8.9	50	101	-	59.2	130.1	0.114	0.227
32	8.0	63	113	74.1	61.6	164.1	0.090	0.202
33	7.1	50	127	-	66.3	206.9	0.072	0.180
34	6.3	40	143	86.2	70.0	260.9	0.057	0.160
35	5.6	32	158	-	73.5	329.0	0.045	0.143
36	5.0	25	175	103.1	T7.0	414.8	0.036	0.127
37	4.5	20	198	-	80.3	523.1	0.028	0.113
38	4.0	16	224	116.3	83.6	659.6	0.022	0.101
39	3.5	12	248	-	86.6	831.8	0.018	0.090

1. Based on 25.4 mm.
2. Ohms per 1,000 ft measured at 20°C.
3. Current-carrying capacity at 700 cm/amp.

Table 11.4 Celcius-to-fahrenheit conversion table.

°Celsius	°Fahrenheit	°Celsius	°Fahrenheit
-50	-58	125	257
-45	-49	130	266
-40	-40	135	275
-35	-31	140	284
-30	-22	145	293
-25	-13	150	302
-20	4	155	311
-15	5	160	320
-10	14	165	329
-5	23	170	338
0	32	175	347
5	41	180	356
10	50	185	365
15	59	190	374
20	68	195	383
25	77	200	392
30	86	205	401
35	95	210	410
40	104	215	419
45	113	220	428
50	122	225	437
55	131	230	446
60	140	235	455
65	149	240	464
70	158	245	473
75	167	250	482
80	176	255	491
85	185	260	500
90	194	265	509
95	203	270	518
100	212	275	527
105	221	280	536
110	230	285	545
115	239	290	554
120	248	295	563

Table 11.5 Inch-to-millimeter conversion table.

Inch	0	1/8	1/4	3/8	1/2	5/8	3/4	7/8	Inch
0	0.0	3.18	6.35	9.52	12.70	15.88	19.05	22.22	0
1	25.40	28.58	31.75	34.92	38.10	41.28	44.45	47.62	1
2	50.80	53.98	57.15	60.32	63.50	66.68	69.85	73.02	2
3	76.20	79.38	82.55	85.72	88.90	92.08	95.25	98.42	3
4	101.6	104.8	108.0	111.1	114.3	117.5	120.6	123.8	4
5	127.0	130.2	133.4	136.5	139.7	142.9	146.0	149.2	5
6	152.4	155.6	158.8	161.9	165.1	168.3	171.4	174.6	6
7	177.8	181.0	184.2	187.3	190.5	193.7	196.8	200.0	7
8	203.2	206.4	209.6	212.7	215.9	219.1	222.2	225.4	8
9	228.6	231.8	235.0	238.1	241.3	244.5	247.6	250.8	9
10	254.0	257.2	260.4	263.5	266.7	269.9	273.0	276.2	10
11	279	283	286	289	292	295	298	302	11
12	305	308	311	314	317	321	324	327	12
13	330	333	337	340	343	346	349	352	13
14	356	359	362	365	368	371	375	378	14
15	381	384	387	391	394	397	400	403	15
16	406	410	413	416	419	422	425	429	16
17	432	435	438	441	445	448	451	454	17
18	457	460	464	467	470	473	476	479	18
19	483	486	489	492	495	498	502	505	19
20	508	511	514	518	521	524	527	530	20

Table 11.6 Conversion of millimeters to decimal inches.

mm	Inches	mm	Inches	mm	Inches
1	0.039370	46	1.811020	91	3.582670
2	0.078740	47	1.850390	92	3.622040
3	0.118110	48	1.889760	93	3.661410
4	0.157480	49	1.929130	94	3.700780
5	0.196850	50	1.968500	95	3.740150
6	0.236220	51	2.007870	96	3.779520
7	0.275590	52	2.047240	97	3.818890
8	0.314960	53	2.086610	98	3.858260
9	0.354330	54	2.125980	99	3.897630
10	0.393700	55	2.165350	100	3.937000
11	0.433070	56	2.204720	105	4.133848
12	0.472440	57	2.244090	110	4.330700
13	0.511810	58	2.283460	115	4.527550
14	0.551180	59	2.322830	120	4.724400
15	0.590550	60	2.362200	125	4.921250
16	0.629920	61	2.401570	210	8.267700
17	0.669290	62	2.440940	220	8.661400
18	0.708660	63	2.480310	230	9.055100
19	0.748030	64	2.519680	240	9.448800
20	0.787400	65	2.559050	250	9.842500
21	0.826770	66	2.598420	260	10.236200
22	0.866140	67	2.637790	270	10.629900
23	0.905510	68	2.677160	280	11.032600
24	0.944880	69	2.716530	290	11.417300
25	0.984250	70	2.755900	300	11.811000
26	1.023620	71	2.795270	310	12.204700
27	1.062990	72	2.834640	320	12.598400
28	1.102360	73	2.874010	330	12.992100
29	1.141730	74	2.913380	340	13.385800
30	1.181100	75	2.952750	350	13.779500
31	1.220470	76	2.992120	360	14.173200
32	1.259840	77	3.031490	370	14.566900
33	1.299210	78	3.070860	380	14.960600
34	1.338580	79	3.110230	390	15.354300
35	1.377949	80	3.149600	400	15.748000
36	1.417319	81	3.188970	500	19.685000
37	1.456689	82	3.228340	600	23.622000
38	1.496050	83	3.267710	700	27.559000
39	1.535430	84	3.307080	800	31.496000
40	1.574800	85	3.346450	900	35.433000
41	1.614170	86	3.385820	1000	39.370000
42	1.653540	87	3.425190	2000	78.740000
43	1.692910	88	3.464560	3000	118.110000
44	1.732280	89	3.503903	4000	157.480000
45	1.771650	90	3.543300	5000	196.850000

Table 11.7 Convertion of common fractions to decimal and millimeter units.

Common fractions	Decimal fractions	mm (approx)	Common fractions	Decimal fractions	mm (appox)
1/128	0.008	0.20	1/2	0.500	12.70
1/64	0.016	0.40	33/64	0.516	13.10
1/32	0.031	0.79	17/32	0.531	13.49
3/64	0.047	1.19	35/64	0.547	13.89
1/16	0.063	1.59	9/16	0.563	14.29
5/64	0.078	1.98	37/64	0.578	14.68
3/32	0.094	2.38	19/32	0.594	15.08
7/64	0.109	2.78	39/64	0.609	15.48
1/8	0.125	3.18	5/8	0.625	15.88
9/64	0.141	3.57	41/64	0.641	16.27
5/32	0.156	3.97	21/32	0.656	16.67
11/64	0.172	4.37	43/64	0.672	17.07
3/16	0.188	4.76	11/16	0.688	17.46
13/64	0.203	5.16	45/64	0.703	17.86
7/32	0.219	5.56	23/32	0.719	18.26
15/64	0.234	5.95	47/64	0.734	18.65
1/4	0.250	6.35	3/4	0.750	19.05
17/64	0.266	6.75	49/64	0.766	19.45
9/32	0.281	7.14	25/32	0.781	19.84
19/64	0.297	7.54	51/64	0.797	20.24
5/16	0.313	7.94	13/16	0.813	20.64
21/64	0.328	8.33	53/64	0.828	21.03
11/32	0.344	8.73	27/32	0.844	21.43
23/64	0.359	9.13	55/64	0.859	21.83
3/8	0.375	9.53	7/8	0.875	22.23
25/64	0.391	9.92	57/64	0.891	22.62
13/32	0.406	10.32	29/32	0.906	23.02
27/64	0.422	10.72	59/64	0.922	23.42
7/16	0.438	11.11	15/16	0.938	23.81
29/64	0.453	11.51	61/64	0.953	24.21
15/32	0.469	11.91	31/32	0.969	24.61
31/64	0.484	12.30	63/64	0.984	25.00

Table 11.8 Decimal equivalent size of drill numbers.

Drill no.	Decimal equiv.	Drill no.	Decimal equiv.	Drill no.	Decimal equiv.
80	0.0135	53	0.0595	26	0.1470
79	0.0145	52	0.0635	25	0.1495
78	0.0160	51	0.0670	24	0.1520
77	0.0180	50	0.0700	23	0.1540
76	0.0200	49	0.0730	22	0.1570
75	0.0210	48	0.0760	21	0.1590
74	0.0225	47	0.0785	20	0.1610
73	0.0240	46	0.0810	19	0.1660
72	0.0250	45	0.0820	18	0.1695
71	0.0260	44	0.0860	17	0.1730
70	0.0280	43	0.0890	16	0.1770
69	0.0292	42	0.0935	15	0.1800
68	0.0310	41	0.0960	14	0.1820
67	0.0320	40	0.0980	13	0.1850
66	0.0330	39	0.0995	12	0.1890
65	0.0350	38	0.1015	11	0.1910
64	0.0360	37	0.1040	10	0.1935
63	0.0370	36	0.1065	9	0.1960
62	0.0380	35	0.1100	8	0.1990
61	0.0390	34	0.1110	7	0.2010
60	0.0400	33	0.1130	6	0.2040
59	0.0410	32	0.1160	5	0.2055
58	0.0420	31	0.1200	4	0.2090
57	0.0430	30	0.1285	3	0.2130
56	0.0465	29	0.1360	2	0.2210
55	0.0520	28	0.1405	1	0.2280

Table 11.9 Decimal equivalent size of drill letters.

Letter drill	Decimal equiv.	Letter drill	Decimal equiv.	Letter drill	Decimal equiv.
A	0.234	J	0.277	S	0.348
B	0.238	K	0.281	T	0.358
C	0.242	L	0.290	U	0.368
D	0.246	M	0.295	V	0.377
E	0.250	N	0.302	W	0.386
F	0.257	O	0.316	X	0.397
G	0.261	P	0.323	Y	0.404
H	0.266	Q	0.332	Z	0.413
I	0.272	R	0.339		

Table 11.10 Conversion ratios for length.

Known quantity	Multiply by	Quantity to find
inches (in)	2.54	centimeters (cm)
feet (ft)	30	centimeters (cm)
yards (yd)	0.9	meters (m)
miles (mi)	1.6	kilometers (km)
millimeters (mm)	0.04	inches (in)
centimeters (cm)	0.4	inches (in)
meters (m)	3.3	feet (ft)
meters (m)	1.1	yards (yd)
kilometers (km)	0.6	miles (mi)
centimeters (cm)	10	millimeters (mm)
decimeters (dm)	10	centimeters (cm)
decimeters (dm)	100	millimeters (mm)
meters (m)	10	decimeters (dm)
meters (m)	1000	millimeters (mm)
dekameters (dam)	10	meters (m)
hectometers (hm)	10	dekameters (dam)
hectometers (hm)	100	meters (m)
kilometers (km)	10	hectometers (hm)
kilometers (km)	1000	meters (m)

Table 11.11 Conversion ratios for area.

Known quantity	Multiply by	Quantity to find
square inches (in^2)	6.5	square centimeters (cm^2)
square feet (ft^2)	0.09	square meters (m^2)
square yards (yd^2)	0.8	square meters (m^2)
square miles (mi^2)	2.6	square kilometers (km^2)
acres	0.4	hectares (ha)
square centimeters (cm^2)	0.16	square inches (in^2)
square meters (m^2)	1.2	square yards (yd^2)
square kilometers (km^2)	0.4	square miles (mi^2)
hectares (ha)	2.5	acres
square centimeters (cm^2)	100	square millimeters (mm^2)
square meters (m^2)	10,000	square centimeters (cm^2)
square meters (m^2)	1,000,000	square millimeters (mm^2)
ares (a)	100	square meters (m^2)
hectares (ha)	100	ares (a)
hectares (ha)	10,000	square meters (m^2)
square kilometers (km^2)	100	hectares (ha)
square kilometers (km^2)	1,000	square meters (m^2)

Table 11.12 Conversion ratios for mass.

Known quantity	Multiply by	Quantity to find
ounces (oz)	28	grams (g)
pounds (lb)	0.45	kilograms (kg)
tons	0.9	tonnes (t)
grams (g)	0.035	ounces (oz)
kilograms (kg)	2.2	pounds (lb)
tonnes (t)	100	kilograms (kg)
tonnes (t)	1.1	tons
centigrams (cg)	10	milligrams (mg)
decigrams (dg)	10	centigrams (cg)
decigrams (dg)	100	milligrams (mg)
grams (g)	10	decigrams (dg)
grams (g)	1000	milligrams (mg)
dekagram (dag)	10	grams (g)
hectogram (hg)	10	dekagrams (dag)
hectogram (hg)	100	grams (g)
kilograms (kg)	10	hectograms (hg)
kilograms (kg)	1000	grams (g)

Table 11.13 Conversion ratios for volume.

Known quantity	Multiply by	Quantity to find
milliliters (mL)	0.03	fluid ounces (fl oz)
liters (L)	2.1	pints (pt)
liters (L)	1.06	quarts (qt)
liters (L)	0.26	gallons (gal)
gallons (gal)	3.8	liters (L)
quarts (qt)	0.95	liters (L)
pints (pt)	0.47	liters (L)
cups (c)	0.24	liters (L)
fluid ounces (fl oz)	30	milliliters (mL)
teaspoons (tsp)	5	milliliters (mL)
tablespoons (tbsp)	15	milliliters (mL)
liters (L)	100	milliliters (mL)

Table 11.14 Conversion ratios for cubic measure.

Known quantity	Multiply by	Quantity to find
cubic meters (m^3)	35	cubic feet (ft^3)
cubic meters (m^3)	1.3	cubic yards (yd^3)
cubic yards (yd^3)	0.76	cubic meters (m^3)
cubic feet (ft^3)	0.028	cubic meters (m^3)
cubic centimeters (cm^3)	1000	cubic millimeters (mm^3)
cubic decimeters (dm^3)	1000	cubic centimeters (cm^3)
cubic decimeters (dm^3)	1,000,000	cubic millimeters (mm^3)
cubic meters (m^3)	1000	cubic decimeters (dm^3)
cubic meters (m^3)	1	steres
cubic feet (ft^3)	1728	cubic inches (in^3)
cubic feet (ft^3)	28.32	liters (L)
cubic inches (in^3)	16.39	cubic centimeters (cm^3)
cubic meters (m^3	264	gallons (gal)

Table 11.15 Conversion ratios for electrical quantities.

Known quantity	Multiply by	Quantity to find
Btu per minute	0.024	horsepower (hp)
Btu per minute	17.57	watts (W)
horsepower (hp)	33,000	foot-pounds per min (ft-lb/min)
horsepower (hp)	746	watts (W)
kilowatts (kW)	57	Btu per minute
kilowatts (kW)	1.34	horsepower (hp)

Appendix A

Professional Association Directory

Most professional organizations in the electronics industry have established standards or recommended practices that can serve as a guide to the system engineer planning a new facility or modifying an existing one. The following list includes the primary organizations involved in the disciplines covered in this book.

Advanced Television Systems Committee (ATSC)
1776 K Street, N.W.
Washington, DC 20006
USA
202-828-3130

American National Standards Institute (ANSI)
655 15th Street, N.W.
Suite 300
Washington, DC 20005
USA
202-639-4090

Audio Engineering Society (AES)
60 East 42nd Street
New York, NY 10165
USA
212-661-2355

Cable Television Laboratories (CableLabs)
1050 Walnut Street
Suite 500
Boulder, CO 80302
USA
303-939-8500

Department of Trade and Industry (DoTI)
Radio Regulatory Division
Ashdown House
123 Victoria Street
London, SW1E 6RB
England
+44-71-215-5000

Electronic Industries Association (EIA)
Engineering Department
2001 I Street, N.W.
Washington DC 20006
USA
202-457-4971

European Broadcasting Union (EBU)
Technical Centre
Avenue Albert Lancaster, 32
B-1 180
Brussels, Belgium
+32-2-375-5990

Federal Communications Commission (FCC)
1919 M Street, N.W.
Washington DC 20554
USA
202-653-8247

Illumination Engineering Society of North America (IES)
345 East 47th Street
New York, NY 10017
USA
212-705-7926

Independent Television Commission (ITC)
70, Brompton Road
Knightsbridge
London SW3 1 EY
England
+44-71-584 7011

Institute of Electrical and Electronics Engineers (IEEE)
345 East 47th Street
New York, NY 10017
USA
212-705-7900

Institution of Electrical Engineers (IEE)

P.O. Box 96
Michael Faraday House
Six Hills Way
Stevenage, Herts. SG1 2SD
England
+44-438-313311

International Electrotechnical Commission (IEC)
3, rue de Varembe
P.O. Box 131
1211 Geneva 20
Switzerland
+41-22-34-01-50

International Society for Optical Engineering (SPIE)
P.O. Box 10
Bellingham, WA 98227-0010
USA
206-676-3290

International Telecommunications Union (ITU)
International Radio Consultative Committee
Place des Nations
1211 Geneva 20
Switzerland
+41-22-99-51-11

International Teleproduction Society (ITS)
990 Sixth Avenue
Suite 2E
New York, NY 10018
USA
212-629-3266

International Television Association (ITVA)
6311 N. O'Connor Road
LB-51
Irving, TX 75039
USA
214-869-1112

National Association of Broadcasters (NAB)
1771 N Street, N.W.
Washington, DC 20036
USA
202-429-5300

National Cable Television Association (NCTA)
1724 Massachusetts Avenue, N.W.
Washington, DC 20036
USA
202-775-3550

National Institute of Standards and Technology (NIST)
Department of Commerce
Gaithersburg, MD 20899
USA
202-921-1000

National Telecommunications and Information Administration (NTIA)
14th Street and Constitution Avenue, N.W.
Washington, DC 20203
USA
202-337-1551

National Transcommunications, Ltd. (NTL)
Crawley Court
Winchester
Hampshire S021 2QA
England
+44 962 823434

Society of Broadcast Engineers (SBE)
P.O. Box 20450
Indianapolis, IN 46220
USA
317-253-1640

Society of Motion Picture and Television Engineers (SMPTE)
595 West Heartsease Avenue
White Plains, NY 10607-1824
USA
914-761-1100

Telecommunications Industry Association (TIA)
1722 I Street, N.W.
Suite 440
Washington, DC 20006
USA
202-457-4936

Underwriters Laboratories (UL)
333 Pfingsten Road
Northbrook, IL 60062
USA
312-272-8800

Appendix B

Glossary

A

absolute delay The amount of time a signal is delayed. The delay may be expressed in time or number of pulse events.

absorption The transference of some or all of the energy contained in an electromagnetic wave to the substance or medium in which it is propagating or upon which it is incident.

ac coupling A method of coupling one circuit to another through a capacitor or transformer so as to transmit the varying (ac) characteristics of the signal while blocking the static (dc) characteristics.

ac/dc coupling Coupling between circuits that accommodate the passing of both ac and dc signals (may also be referred to as simply dc coupling).

accelerated life test A special form of reliability testing performed by the manufacturer. The unit under test is subjected to stresses that exceed those typically experienced in normal operation. The goal of an *accelerated life test* is to improve the reliability of products shipped by forcing latent failures in components to become evident before the unit leaves the factory.

acceptable reliability level The maximum number of failures allowed per thousand operating hours of a given component or system.

acceptance test The process of testing newly purchased equipment to ensure that it is fully compliant with contractual specifications.

access The point at which entry is gained to a circuit or facility.

acoustics The physics and study of sound.

acquisition time In a communication system, the amount of time required to attain synchronism.

active Any device or circuit that introduces gain or uses a source of energy other than that inherent in the signal to perform its function.

adapter A fitting or electrical connector which links equipment that cannot be connected directly.

adaptive A device able to adjust or react to a condition or application, as an *adaptive circuit.* This term usually refers to filter circuits.

adaptive system A general name for a system that is capable of reconfiguring itself to meet new requirements.

addressability The capability of one device or process to access the code or data of another.

adjacent channel interference Interference to communications caused by a transmitter operating on an adjacent radio channel. The sidebands of the transmitter mix with the carrier being received on the desired channel, resulting in noise.

admittance A measure of how well alternating current flows in a conductor. It is the reciprocal of *impedance* and is expressed in *siemens*. The real part of admittance is *conductance* and the imaginary part is *susceptance*.

air core An inductor with no magnetic material in its core.

algorithm A prescribed finite set of well-defined rules or processes for the solution of a problem in a finite number of steps.

alignment The adjustment of circuit components so that an entire system meets minimum performance values.

allocation The planned use of certain facilities and equipment to meet current, pending, and/or forecasted circuit- and carrier-system requirements.

alphanumeric A generic term for alphabetic letters, numerical digits, and special characters that may be processed and displayed by a machine. *Alphanumeric displays* provide a character set that includes letters, numbers, and punctuation marks.

alternating current (ac) A continuously variable current, rising to a maximum in one direction, falling to zero, then reversing direction and rising to a maximum in the other direction, then falling to zero and repeating the cycle. Alternating current usually follows a sinusoidal growth and decay curve. Note that the correct usage of the term *ac* is lower case.

alternator A generator that produces alternating current electric power.

ambient electromagnetic environment The radiated or conducted electromagnetic signals and noise at a specific location and time.

ambient level The magnitude of radiated or conducted electromagnetic signals and noise at a specific test location when equipment-under-test is not powered.

ambient noise The normal background noise in a communications environment that is unrelated to noise induced in a transmission medium, for example, acoustic room noise at an operator location.

ambient temperature The temperature of the surrounding medium, typically air, that comes into contact with an apparatus. Ambient temperature may also refer simply to room temperature.

American National Standards Institute (ANSI) A nonprofit organization that coordinates voluntary standards activities in the United States. The institute represents the United States in two major telecommunication organizations: the International Standards Organization (ISO) and the International Electrotechnical Commission (IEC).

American Standard Code for Information Interchange (ASCII) A widely used seven-bit character code for data communications and data processing. An eighth bit is added for parity checking. As a standard, ASCII enables digital communications between computers and similar systems, regardless of manufacturer.

American Wire Gauge (AWG) The standard American method of classifying wire diameter.

ammeter An instrument that measures and records the amount of current in amperes flowing in a circuit.

amp (A) An abbreviation of the term *ampere*.

ampacity A measure of the current-carrying capacity of a power cable. *Ampacity* is determined by the maximum continuous-performance temperature of the insulation, by the heat generated in the cable (as a result of conductor and insulation losses), and by the heat-dissipating properties of the cable and its environment.

ampere (amp) The standard unit of electric current.

ampere-hour The energy that is consumed when a current of one ampere flows for a period of one hour.

amplification The process that results when the output of a circuit is an enlarged reproduction of the input signal. Amplifiers may be designed to provide amplification of voltage, current, or power, or a combination of these quantities.

amplifier A device that receives an input signal and provides as an output a magnified replica of the input waveform.

amplitude The magnitude of a signal in voltage or current, frequently expressed in terms of *peak*, *peak-to-peak*, or *root-mean-square* (RMS). The actual amplitude of a quantity at a particular instant often varies in a sinusoidal manner.

amplitude distortion A distortion mechanism occurring in an amplifier or other device when the output amplitude is not a linear function of the input amplitude under specified conditions.

amplitude equalizer A corrective network that is designed to modify the amplitude characteristics of a circuit or system over a desired frequency range.

amplitude-vs.-frequency distortion The distortion in a transmission system caused by the nonuniform attenuation or gain of the system with respect to frequency under specified conditions.

analog A continuously varying physical variable, such as sound or electrical current.

anchor The device to which an item, such as a cable or junction box, is fastened. In pole route construction and antenna mast erection, anchors of various types are buried in the ground and rods or guy lines are attached to give stability to the above-ground structure.

angle of incidence (or reflection) The angle between a ray and a line perpendicular to a surface at the point the ray is incident to (or reflected from) the surface. The angle of incidence equals the angle of reflection.

anodize The formation of a thin film of oxide on a metallic surface, usually to produce an insulating layer.

antenna A device used to transmit or receive a radio signal. An antenna is usually designed for a specified frequency range and serves to couple electromagnetic energy from a transmission line to and/or from the free space through which it travels. Directional antennas concentrate the energy in a particular horizontal or vertical direction.

apparent power The product of the root-mean-square values of the voltage and current in an alternating-current circuit without a correction for the phase difference between the voltage and current.

arc A sustained luminous discharge between two or more electrodes.

architecture The basic design of a computer or telecommunications system that defines the relationships between the hardware and software components.

arrester A protection device that bypasses high-voltage electrical energy around circuit components and toward a low-impedance path to ground.

artificial line An assembly of resistors, inductors, and capacitors that simulates the electrical characteristics of a transmission line.

aspect ratio In a rectangle, the ratio of the longer dimension to the shorter.

assembly A manufactured part made by combining several other parts or subassemblies. For example, a cable assembly consists of the cable with connectors at each end.

asynchronous A signal lacking synchronization with some other reference signal.

atmospheric noise Radio noise caused by natural atmospheric processes, such as lightning.

attack time The time interval in seconds required for a device to respond to a control stimulus.

attenuation The decrease in amplitude of an electrical signal traveling through a transmission medium caused by dielectric and conductor losses.

attenuation coefficient The rate of decrease in the amplitude of an electrical signal caused by attenuation. The *atten-*

uation coefficient can be expressed in decibels or nepers per unit length. It may also be referred to as the *attenuation constant.*

attenuation distortion The distortion caused by attenuation that varies over the frequency range of a signal.

attenuation-limited operation The condition prevailing when the received signal amplitude (rather than distortion) limits overall system performance.

attenuator A fixed or adjustable component that reduces the amplitude of an electrical signal without causing distortion.

attributes The characteristics of equipment that aid planning and circuit design.

automatic changeover switch A device that monitors the outputs of two devices, such as sync generators in a video system, and automatically switches to the backup device in the event of a failure in the primary device.

automatic test equipment (ATE) A class of instruments that measure performance degradation with minimum reliance on human intervention, and are able to isolate sections or components found to be outside permitted tolerances.

autotransformer A transformer in which both the primary and secondary currents flow through one common part of the coil.

auxiliary power An alternate source of electric power, serving as a backup for the primary utility company ac power.

availability A measure of the degree to which a system, subsystem, or equipment is operable and not in a stage of congestion or failure at any given point in time.

average life The mean value for a normal distribution of product or component lives, generally applied to mechanical failures resulting from *wear-out.*

azimuth The horizontal angle between the north-south line and a particular direction.

azimuth angle The angle of rotation (horizontal) that a ground-based parabolic antenna must be rotated through to point to a specific satellite in a geosynchronous orbit. The azimuth angle for any particular satellite can be determined for any point on the surface of the earth given the latitude and longitude of that point. It is defined with respect to *due north* for convenience.

B

backfill The material used to refill a trench after ducts or cables have been laid therein. *Backfill* is also used as a verb, to refill a trench.

background noise The total system noise in the absence of information transmission, independent of the presence or absence of a signal.

backplane The physical area, usually at the rear of an electronics frame, where modules and cables plug into the system.

backscatter The deflection or reflection of radiant energy through angles greater than 90 degrees with respect to the original angle of travel.

backscatter range The maximum distance from which backscattered radiant energy can be measured.

backscattering A form of wave scattering in which at least one component of the scattered wave is deflected opposite to the direction of propagation of the incident wave.

backup A circuit element or facility used to replace an element that has failed.

backup supply A redundant power supply that takes over if the primary power supply fails.

balance The process of equalizing the voltage, current, or other parameter between two or more circuits or systems.

balanced A circuit having two sides (conductors) carrying voltages that are symmetrical around a common reference point, typically ground.

balanced circuit A circuit whose two sides are electrically equal in all transmission respects.

balanced line A transmission line consisting of two conductors in the presence of ground capable of being operated in such a way that when the voltages of the two conductors at all transverse planes are equal in magnitude and opposite in polarity with respect to ground, the currents in the two conductors are equal in magnitude and opposite in direction.

balanced three-wire system A power distribution system using three conductors, one of which is balanced to have a potential midway between the potentials of the other two.

balanced-to-ground The condition when the impedance to ground on one wire of a two-wire circuit is equal to the impedance to ground on the other wire.

ballistics A term describing the dynamic characteristics of a meter movement, most notably response time, damping, and overshoot.

balun (balanced/unbalanced) A device used to connect balanced circuits with unbalanced circuits, for example, the balanced circuit of twisted-pair wiring with the unbalanced circuit of a coaxial cable.

band A range of frequencies between a specified upper and lower limit.

band elimination filter A filter having a single continuous attenuation band, with neither the upper nor lower cut-off frequencies being zero or infinite. A *band elimination filter* may also be referred to as a *band-stop*, *notch*, or *band reject* filter.

bandpass filter A filter having a single continuous transmission band with neither the upper nor the lower cut-off frequencies being zero or infinite. A bandpass filter permits only a specific band of frequencies to pass; frequencies above or below are attenuated.

bandwidth The range of signal frequencies that can be transmitted by a communications channel with a defined maximum loss or distortion. Bandwidth indicates the information-carrying capacity of a channel.

bandwidth expansion ratio The ratio of the necessary bandwidth to the baseband bandwidth.

bandwidth-limited operation The condition prevailing when the frequency spectrum or bandwidth, rather than the amplitude (or power) of the signal, is the limiting factor in communication capability. This condition is reached when the system distorts the shape of the waveform beyond tolerable limits.

bank A group of similar items connected together in a specified manner and used in conjunction with one another.

bar code label A standard label that contains information required as input for automated inventory-management systems. Typically, a bar code label is read by a laser scanner.

bare A wire conductor that is not enameled or enclosed in an insulating sheath.

baseband The band of frequencies occupied by a signal before it modulates a carrier wave to form a transmitted radio or line signal.

baseband channel A channel that carries a signal without modulation, in contrast to a *passband* channel.

baseband signal The original form of a signal, unchanged by modulation.

bath tub The shape of a typical graph of component failure rates: high during an initial period of operation, falling to an acceptable low level during the normal usage period, and then rising again as the components become time-expired.

battery A group of several cells connected together to furnish current by conversion of chemical, thermal, solar, or nuclear energy into electrical energy. A single cell is itself sometimes also called a battery.

battery eliminator A device consisting of a rectifier and filter capable of supplying dc power at the required voltage without the use of a battery.

baud A unit of signaling speed equal to the number of signal events per second. Baud is equivalent to bits/second in cases where each signal event represents exactly one bit. Often the term *baud rate* is used informally to mean baud, referring to the specified maximum rate of data transmission along an interconnection. Typically, the baud settings of two devices must match if the devices are to communicate with one another.

bay A row or suite of racks on which transmission, switching, and/or processing equipment is mounted.

bel A unit of power measurement, named in honor of Alexander Graham Bell. The commonly used unit is one tenth of a bel, or a decibel (dB). One bel is defined as a tenfold increase in power. If an amplifier increases the power of a signal by 10 times, the power gain of the amplifier is equal to 1 bel or 10 *decibels* (dB). If power is increased by 100 times, the power gain is 2 bels or 20 decibels.

benchmark A specified and standardized task designed so that the performance of different systems can be objectively compared.

bend A transition component between two elements of a transmission waveguide.

bending radius The smallest bend that may be put into a cable under a stated pulling force. The bending radius is typically expressed in inches.

bidirectional An operational qualification which implies that the transmission of information occurs in both directions.

block diagram An overview diagram that uses geometric figures to represent the principal divisions or sections of a circuit, and lines and arrows to show the path of a signal, or to show program functionalities. It is not a *schematic*, which provides greater detail.

blocking capacitor A capacitor included in a circuit to stop the passage of direct current.

BNC An abbreviation for *bayonet Neill-Concelman*. BNC is a type of cable connector used extensively in television and radio frequency equipment (named for its inventor).

board A printed circuit consisting of a flat board of insulating material with conductive circuits etched on its surface. The term *board* may also be used to describe a fully populated printed wiring board that is a component of a larger assembly.

Boltzmann's constant 1.38×10^{-23} joules.

breaking strength The amount of force required to break a wire or optical fiber.

bridge A type of network circuit used to match different circuits to each other, ensuring minimum transmission impairment.

bridging The shunting or paralleling of one circuit with another.

broadband The quality of a communications link having essentially uniform response over a given range of frequencies. A communications link is said to be *broadband* if it offers no perceptible degradation to the signal being transported.

buffer A circuit or component which isolates one electrical circuit from another.

burn-in The operation of a device, sometimes under extreme conditions, to stabilize its characteristics and identify latent component failures before bringing the device into normal service.

bus A central conductor for the primary signal path. The term bus may also refer to a signal path to which a number of inputs may be connected for feed to one or more outputs.

bus driver A circuit that amplifies bus data and/or control signals sufficiently to ensure valid reception at some distant point.

busbar A main dc power bus.

bypass capacitor A capacitor which provides a signal path that effectively shunts or bypasses other components.

bypass relay A switch used to bypass the normal electrical route of a signal or current in the event of power, signal, or equipment failure.

byte An agreed-up group of bits, typically eight. A byte commonly represents one alphabetic or special character, two decimal digits, or eight binary bits of information.

C

C/N (carrier-to-noise ratio) The ratio of the power in a satellite signal carrier to the received noise as measured in decibels. C/N is a measure of the received signal strength for an earth receiving station.

cable An electrically and/or optically conductive interconnecting device.

cable loss Signal loss caused by passing a signal through a coaxial cable. Losses are the result of resistance, capacitance, and inductance in the cable.

cable splice The connection of two pieces of cable by joining them mechanically on a pair-for-pair basis and closing the joint with a weathertight case or sleeve.

cable vault A room below ground level in a central office/telephone exchange or cable terminal/repeater station where cables enter the building, usually in conduit.

cable vault ground bar (CVGB) The copper ground bar located in the cable vault for bonding cable sheaths.

cabling The wiring used to interconnect electronic equipment.

calibrate The process of checking, and adjusting if necessary, a test instrument against one known to be set correctly.

calibration The process of identifying and measuring errors in instruments and/or procedures.

capacitance The property of a device or component that enables it to store energy in an electrostatic field and to release it later. A capacitor consists of two conductors separated by an insulating material. When the conductors have a voltage difference between them, a charge will be stored in the electrostatic field between the conductors.

capacitor A device that stores electrical energy. A capacitor allows the apparent flow of alternating current, while blocking the flow of direct current. The degree to which the device permits ac current flow depends on the frequency of the signal and the size of the capacitor. Capacitors are used in filters, delay-line components, couplers, frequency selectors, timing elements, voltage transient suppression, and other applications.

carrier A single frequency wave which, when transmitted, is modulated by another wave containing information. A carrier may be modulated by manipulating its amplitude and/or frequency in direct relation to one or more applied signals.

carrier frequency The frequency of an unmodulated oscillator or transmitter. Also, the average frequency of a transmitter when a signal is frequency modulated by a symmetrical signal.

cascade connection A tandem arrangement of two or more similar component devices or circuits with the output of one connected to the input of the next.

cascaded An arrangement of two or more circuits in which the output of one circuit is connected to the input of the next circuit.

cell An elementary unit of communication, of power supply, or of equipment.

Celsius A temperature measurement scale, expressed in degrees C, in which water freezs at 0°C and boils at 100°C. To convert to degrees Fahrenheit, multiply by 0.555 and add 32. To convert to Kelvin add 273 (approximately).

center frequency In frequency modulation, the resting frequency or initial frequency of the carrier before modulation.

center tap A connection made at the electrical center of a coil.

central processing unit (CPU) The electronic components of a computer that interpret instructions, perform calculations, move data in main computer storage, and control input/output operations.

channel The smallest subdivision of a circuit that provides a single type of communication service.

channel coding The process by which digital data is converted to a waveform suitable for recording.

channel decoder A device that converts an incoming modulated signal on a given channel back into the source-encoded signal.

channel encoder A device that takes a given signal and converts it into a form suitable for transmission over the communications channel.

channel noise level The ratio of the channel noise at any point in a transmission system to some arbitrary amount of circuit noise chosen as a reference. This ratio is usually expressed in *decibels above reference noise*, abbreviated *dBrn*.

channel reliability The percent of time a channel is available for use in a specific direction during a specified period.

character set A finite set of different characters that are agreed upon and considered to be complete for some purpose.

characteristic The property of a circuit or component.

characteristic impedance The impedance of a transmission line, as measured at the driving point, if the line were of infinite length. In such a line, there would be no standing waves. The *characteristic impedance* may also be referred to as the *surge impedance*.

charge The process of replenishing or replacing the electrical charge in a secondary cell or storage battery.

charger A device used to recharge a battery. Types of charging include: (1) constant voltage charge, (2) equalizing charge, and (3) trickle charge.

chassis ground A connection to the metal frame of an electronic system that holds the components in place. The chassis ground connection serves as the ground return or electrical common for the system.

circuit Any closed path through which an electrical current can flow. In a *parallel circuit*, components are connected between common inputs and outputs such that all paths are parallel to each other. The same voltage appears across all paths. In a *series circuit*, the same current flows through all components.

circuit noise level The ratio of the circuit noise at some given point in a transmission system to an established reference, usually expressed in decibels above the reference.

circuit reliability The percentage of time a circuit is available to the user during a specified period of scheduled availability.

circular mil The measurement unit of the cross-sectional area of a circular conductor. A *circular mil* is the area of a circle whose diameter is one mil, or 0.001 inch.

clear channel A transmission path wherein the full bandwidth is available to the user, with no portions of the channel used for control, framing, or signaling.

clipper A limiting circuit which ensures that a specified output level is not exceeded by restricting the output waveform to a maximum peak amplitude.

clipping The distortion of a signal caused by removing a portion of the waveform through restriction of the amplitude of the signal by a circuit or device.

clock A reference source of timing information for equipment, machines, or systems.

clock frequency The master frequency of periodic pulses that is used to synchronize the operation of equipment.

clock jitter Undesirable random changes in clock phase.

coax A shorthand expression for *coaxial cable*, which is used to transport high frequency signals.

coaxial cable A transmission line consisting of an inner conductor surrounded first by an insulating material and then by an outer conductor, either solid or braided. The mechanical dimensions of the cable determine its *characteristic impedance*.

code A pre-determined set of signal elements representing letters, numbers, and other information characters.

codec An acronym for coder/decoder, a device that codes analog signals into digital signals for transmission and decodes those signals into their analog original at the receiver.

coherence The correlation between the phases of two or more waves.

coherent The condition characterized by a fixed phase relationship among points on an electromagnetic wave.

coherent pulse The condition in which a fixed phase relationship is maintained between consecutive pulses during pulse transmission.

cold joint A soldered connection that was inadequately heated, with the result that the wire is held in place by rosin flux, not solder. A cold joint is sometimes referred to as a *dry joint*.

cold start The process of starting a system by turning the power on.

collate To compare or merge two things into one.

collector One element of a transistor.

comb filter An electrical filter circuit that passes a series of frequencies and rejects the frequencies in between, producing a frequency response similar to the teeth of a comb. A comb filter may be used on encoded video to select the chrominance signal and reject the luminance signal, thereby reducing cross chrominance artifacts, or conversely, to select the luminance signal and reject the chrominance signal, thereby reducing cross luminance artifacts. Comb filtering successfully reduces artifacts but may also cause a certain amount of resolution loss in the picture.

common A point that acts as a reference for circuits, often equal in potential to the local ground.

common carrier A company that carries goods, services, or people from one point to another for the public.

common mode Signals identical with respect to amplitude, frequency, and phase that are applied to both terminals of a cable and/or both the input and reference of an amplifier.

common mode hum Power line interference (usually 60 Hz) that appears on both terminals of a cable with the same phase, amplitude, and frequency.

common mode noise Unwanted signals in the form of voltages appearing between the local ground reference and each of the power conductors, including neutral and the equipment ground.

common mode range The amplitude of the common mode signal that can be applied to the two differential inputs of an amplifier and maintain its performance.

common mode rejection (CMR) A measure of how well a differential amplifier rejects a signal that appears simultaneously and in phase at both input terminals. As a specification, CMR is usually stated as a dB ratio at a given frequency.

common mode rejection ratio (CMRR) The ratio of differential gain to the common mode gain.

common return A return path that is common to two or more circuits, and returns currents to their source or to ground.

common return offset The dc common return potential difference of a line.

communication center A facility responsible for the reception, transmission, and delivery of messages.

communications security equipment Equipment designed to provide security to telecommunications by converting information to a form unintelligible to an unauthorized interceptor, and by reconverting such information to its original form for authorized recipients.

communications system A collection of individual communications networks, transmission systems, relay stations, tributary stations, and terminal equipment capable of interconnection and interoperation to form an integral whole. The individual components must serve a common purpose, be technically compatible, employ common procedures, respond to some form of control, and, in general, operate in unison.

commutation A successive switching process carried out by a commutator.

commutator A circular assembly of contacts, insulated one from another, each leading to a different portion of the circuit or machine.

compact disc read-only memory (CD-ROM) A memory system that stores digital data as a series of microscopic pits on the surface of an optical disc. Light from a low-power laser, reflected off of the disc, is detected by a photodiode to read the data.

compandor An electronic device that improves the apparent signal-to-noise ratio and expands the dynamic range of transmission systems by compressing the amplitude range of a signal at the transmitter and expanding the signal to a normal range at the receiver.

comparator A circuit that is designed to compare two variable signals, or a variable signal against a reference signal.

compatibility The ability of diverse systems to exchange necessary information at appropriate levels of command directly and in usable form. Communications equipment items are compatible if signals can be exchanged between them without the addition of buffering or translation for the specific purpose of achieving workable interface connections, and if the equipment or systems being interconnected possess comparable performance characteristics, including the suppression of undesired radiation.

compensation The process of passing a signal through an element or circuit with characteristics that are the reverse of those in the transmission line, so that the net effect is a received signal with an acceptable level/frequency characteristic.

complex wave A waveform consisting of two or more sinewave components. At any instant of time, a complex wave is the algebraic sum of all its sinewave components. Examples include voice and music signals.

compliance For mechanical systems, a property which is the reciprocal of stiffness.

component An assembly, or part thereof, that is essential to the operation of some larger circuit or system. A *component* is an immediate subdivision of the assembly to which it belongs.

composite two-tone test signal A test signal used for intermodulation distortion measurements.

compound error Two or more errors of any type.

computer A device, circuit, or system that processes information according to a set of stored instructions. Typically, an electronic computer consists of data input and output circuits, a central processing unit, memory storage, and facilities for interaction with a user, such as a keyboard and a visual monitor or readout.

concentricity A measure of the deviation of the center conductor position relative to its ideal location in the exact center of the dielectric cross-section of a coaxial cable.

conditioning The adjustment of a channel in order to provide the appropriate transmission characteristics needed for data or other special services.

conditioning equipment The equipment used to match transmission levels and impedances, and to provide equalization between facilities.

conductance A measure of the capability of a material to conduct electricity. It is the reciprocal of *resistance* and is expressed in *siemens*. (Formerly expressed as *mho*, or *ohm* spelled backwards.)

conducted emission An electromagnetic energy propagated along a conductor.

conduction The transfer of energy through a medium, such as the conduction of electricity by a wire, or of heat by a metallic frame, or of sound by air.

conduction band A partially filled or empty atomic energy band in which electrons are free to move easily, allowing the material to carry an electric current.

conductivity The conductance per unit length.

conductor Any material that is capable of carrying an electric current.

conduit A pipe, tube, or compartmentalized structure placed underground to form ducts through which cables can pass. Conduit is also used inside building walls or floors for the same purpose.

conduit run The route to be followed by a nest of ducts.

cone of protection In reference to lightning, the space enclosed by a cone formed with its apex at the highest point of a lightning rod or protecting tower, the diameter of the base of the cone having a definite relationship to the height of the rod or tower. When overhead ground wires are used, the space protected is referred to as a *protected zone*.

configuration A relative arrangement of parts.

connect time The amount of time that a given circuit is in use.

connection A point at which a junction of two or more conductors is made.

connector A device mounted on the end of a wire or fiber-optic cable that mates to a similar device on a specific piece of equipment or another cable.

constant-current source A source with infinitely high output impedance so that output current is independent of voltage, for a specified range of output voltages.

constant-voltage charge A method of charging a secondary cell or storage battery during which the terminal voltage is kept at a constant value.

constant-voltage source A source with low, ideally zero, internal impedance, so that voltage will remain constant, independent of current supplied.

contact The points that are brought together or separated to complete or break an electric circuit.

contact bounce The rebound of a contact, which temporarily opens the circuit after its initial *make*.

contact form The configuration of a contact assembly on a relay. Many different configurations are possible from simple single-*make* contacts to complex arrangements involving *breaks* and *makes*.

contact noise A noise resulting from current flow through an electrical contact that has a rapidly varying resistance, as when the contacts are corroded or dirty.

contact resistance The resistance at the surface when two conductors make contact.

contingency planning Planning of a course of action such that the plan will only be invoked if the contingency materializes.

continuity A continuous path for the flow of current in an electrical circuit.

continuous wave An electromagnetic signal in which successive oscillations of the waves are identical.

control The supervision that an operator or device exercises over a circuit or system.

control panel A device used for entering operational commands to a device or system.

control processor Circuits used to generate or alter control signals.

convention A generally acceptable symbol, sign, or practice in a given industry.

copper loss The loss resulting from the heating effect of current.

copper wire counterpoise ground (CWCG) A length of bare copper wire buried at least 18 inches below ground to provide an earth interface.

copper-clad aluminum Aluminum wire with welded copper cladding, somewhat less expensive than pure hard drawn copper but similar in electrical characteristics.

copper-clad steel A type of steel wire with welded copper cladding. Copper-clad steel wire can be used in areas where long cable spans are necessary.

cord The movable portion of a jack-and-plug connection. A cord usually has a plug at each end that is connected by two or more conductors.

corona A bluish luminous discharge resulting from ionization of the air near a conductor carrying a voltage gradient above a certain *critical level*.

corrective maintenance The necessary tests, measurements, and adjustments required to remove or correct a fault.

coulomb The standard unit of electric quantity or charge. One *coulomb* is equal to the quantity of electricity transported in 1 second by a current of 1 ampere.

Coulomb's Law The attraction and repulsion of electric charges act on a line between them. The charges are inversely proportional to the square of the distance between them, and proportional to the product of their magnitudes. (Named for the French physicist Charles-Augustine de Coulomb, 1736 - 1806.)

couple The process of linking two circuits by inductance, so that energy is transferred from one circuit to another.

coupled mode The selection of either ac or dc coupling.

coupling The relationship between two components that enables the transfer of energy between them. Included are *direct coupling* through a direct electrical connection, such as a wire; *capacitive coupling* through the capacitance formed by two adjacent conductors; and *inductive coupling* in which energy is transferred through a magnetic field. Capacitive coupling is also called *electrostatic coupling*. Inductive coupling is often referred to as *electromagnetic coupling*.

coupling coefficient A measure of the electrical coupling that exists between two circuits. The *coupling coefficient* is equal to the ratio of the mutual impedance to the square root of the product of the self impedances of the coupled circuits, all impedances being of the same.

crimp termination A terminal connected to a conductor by high-pressure crimping of a lug onto a wire.

critical path method A project management system. All those activities that build up to the complete project are drawn out on a diagram showing which items must be completed before others can start. The particular chain of events that is the longest determines the completion date of the project, and is referred to as the *critical path*. Any actions or delays that affect this longest chain of events will directly affect completion of the project.

cross coupling The coupling of a signal from one channel, circuit, or conductor to another, where it becomes an undesired signal.

crossover distortion A distortion that results in an amplifier when an irregularity is introduced into the signal as it crosses through a zero reference point. If an amplifier is properly designed and biased, the upper half cycle and lower half cycle of the signal coincide at the zero crossover reference.

crossover frequency The frequency at which output signals pass from one channel to the other in a *crossover network*. At the *crossover frequency* itself the outputs to each side are equal.

crossover network A type of filter that divides an incoming signal into two or more outputs, with higher frequencies directed to one output, and lower frequencies to another.

crossover point The break-even point at which the costs of two competing technologies are equal.

crosspoint A single element in an array of elements that comprise a switch. A *crosspoint* is a set of physical or logical contacts that operate together to perform some switching action.

crosstalk Undesired transmission of signals from one circuit into another circuit in the same system. Crosstalk is usually caused by unintentional capacitive (ac) coupling.

crosstalk coupling The ratio of the power in a disturbing circuit to the induced power in the disturbed circuit, observed at a particular point under specified conditions. Crosstalk coupling is typically expressed in dB.

crowbar A short-circuit or low resistance path placed across the input to a circuit.

CRT (cathode ray tube) A vacuum tube device that produces light when energized by the electron beam generated inside the tube. A CRT includes an electron gun, deflection mechanism, and phosphor-covered faceplate.

crystal A solidified form of a substance that has atoms and molecules arranged in a symmetrical pattern.

crystal-controlled oscillator An oscillator in which a piezoelectric-effect crystal is coupled to a tuned oscillator circuit in such a way that the crystal pulls the oscillator frequency to its own natural frequency and does not allow frequency drift.

crystal filter A filter that uses piezoelectric crystals to create resonant or antiresonant circuits.

crystal oscillator An oscillator using a piezoelectric crystal as the tuned circuit that controls the resonant frequency.

current A general term for the transfer of electricity, or the movement of electrons or *holes*.

current amplifier A low output impedance amplifier capable of providing high current output.

current-carrying capacity A measure of the maximum current that can be carried continuously without damage to components or devices in a circuit.

current probe A sensor, clamped around an electrical conductor, in which an induced current is developed from the magnetic field surrounding the conductor. For measurements, the current probe is connected to a suitable test instrument.

current transformer A transformer-type of instrument in which the primary carries the current to be measured and the secondary is in series with a low current ammeter. A current transformer is used to measure high values of alternating current.

cut-off frequency The frequency above or below which the output current in a circuit is reduced to a specified level.

cycle The interval of time or space required for a periodic signal to complete one period.

cycles per second The standard unit of frequency, expressed in Hertz (one cycle per second).

D

D connector A type of connector that has a trapezoidal shell resembling the letter *D*.

damped oscillation An oscillation exhibiting a progressive diminution of amplitude with time.

damping The dissipation and resultant reduction of any type of energy, such as sound or electromagnetic waves. The use of sound-absorbing material for soundproofing or the use of electronic components to reduce the amplitude of an oscillator signal are examples of damping.

dart leader stroke The initial discharge that largely determines the path taken by a lightning flash. A *dart leader* develops continuously, a *stepped leader* develops in short steps. Both are followed by the high-current discharge flash, usually in the reverse direction to the leader stroke.

data bank A comprehensive collection of data.

daughter board A small printed circuit board mounted on a standard-sized (or *mother*) board.

dB (decibel) A measure of voltage, current or power gain equal to 0.1 bel. Decibels are given by the equations 20 log Vout/Vin, 20 log Iout/Iin, or 10 log Pout/Pin.

dBk A measure of power relative to 1 kilowatt. 0 dBk equals 1 kW.

dBm (decibels above 1 milliwatt) A logarithmic measure of power with respect to a reference power of one milliwatt.

dBmv A measure of voltage gain relative to 1 millivolt at 75 ohms.

dBr The power difference expressed in dB between any point and a reference point selected as the *zero relative transmission level* point. Any power expressed in dBr does not specify the absolute power; it is a relative measurement only.

dBrn (decibels above reference noise) The noise power relative to one picowatt, or -90 dBm.

dBu A term that reflects comparison between a measured value of voltage and a reference value of 0.775 V, expressed under conditions in which the impedance at the point of measurement (and of the reference source) is not considered.

dbV A measure of voltage gain relative to 1 V.

dBW A measure of power relative to 1 watt. 0 dBW equals 1 W.

dc An abbreviation for *direct current*. Note that the preferred usage of the term *dc* is lower case.

dc amplifier A circuit capable of amplifying dc and slowly varying alternating current signals.

dc component The portion of a signal that consists of direct current. This term may also refer to the average value of a signal.

dc coupled A connection configured so that both the signal (ac component) and the constant voltage on which it is riding (dc component) are passed from one stage to the next.

dc coupling A method of coupling one circuit to another so as to transmit the static (dc) characteristics of the signal, as well as the varying (ac) characteristics. Any dc offset present on the input signal is maintained and will be present in the output.

dc offset The amount that the dc component of a given signal has shifted from its correct level.

dc signal bounce Overshoot of the proper dc voltage level resulting from multiple ac couplings in a signal path.

dc/dc converter A unit that converts one dc voltage value into one or more different values.

debugging The process of detecting, diagnosing, and correcting errors (*bugs*) that may occur in a software program.

deca A prefix meaning *ten*.

decay The reduction in amplitude of a signal on an exponential basis.

decay time The time required for a signal to fall to a certain fraction of its original value.

decibel (dB) One tenth of a bel. The decibel is a logarithmic measure of the ratio between two powers. In telecommunications, it is commonly used to express transmission gain, loss, level, and similar quantities. The number of decibels is equal to 10 times the logarithm to the base 10 of the power ratio.

deck A tape recording or playback device.

decode The process of recovering information from a signal into which the information has been encoded.

decoder A device capable of deciphering encoded signals. A decoder interprets input instructions and initiates the appropriate control operations as a result.

decoding In telecommunications, the process of converting encoded words into the original signal, possibly via an intermediate step of pulse-amplitude modulation (PAM). In pulse-code modulation (PCM), this is the conversion of 7-bit (D1 type) or 8-bit (including D2, D3, D4, and D5 type) pulse-code modulation (PCM) words to analog signals. Decoding is the inverse of encoding.

decoupling The reduction or removal of undesired coupling between two circuits or stages.

de-emphasis The reduction of the high-frequency components of a received signal to reverse the pre-emphasis that was placed on them to overcome attenuation and noise in the transmission process.

de-energized A system from which sources of power have been disconnected.

defect An error made during initial planning that is normally detected and corrected during the development phase. Note that a *fault* is an error that occurs in an in-service system.

degauss The process of demagnetizing (erasing) all recorded material on a magnetic medium, such as video or audio tape.

degradation In susceptibility testing, any undesirable change in the operational performance of a test specimen.

This term does not necessarily mean malfunction or catastrophic failure.

degradation failure A failure that results from a gradual change in performance characteristics of a system or part with time.

delay The amount of time by which a signal is delayed or an event retarded.

delay circuit A circuit designed to delay a signal passing through it by a specified amount.

delay distortion The distortion resulting from the difference in phase delays at two frequencies of interest.

delay equalizer A network that adjusts the velocity of propagation of the frequency components of a complex signal to counteract the delay distortion characteristics of a transmission channel.

delay line A transmission network that increases the propagation time of a signal traveling through it.

delta-connected system A three-phase power distribution system where a single-phase output can be derived from each of the adjacent pairs of an equilateral triangle formed by the service drop transformer secondary windings.

delta connection A common method of joining together a three-phase power supply, with each phase across a different pair of the three wires used.

demand The number of circuits required on a route.

demand factor The ratio of the maximum demand on a power system to the total connected load of the system.

demand load The total electrical power required by a facility.

demand meter A measuring device used to monitor the power demand of a system; it compares the peak power of the system with the average power.

demodulator Any device that recovers the original signal after it has modulated a high frequency carrier. The output from the unit may be in baseband composite form.

demultiplexer (demux) A device used to separate two or more signals that were previously combined by a compatible multiplexer and are transmitted over a single channel.

derating factor An operating safety margin provided for a component or system to ensure reliable performance. A *derating allowance* also is typically provided for operation under extreme environmental conditions, or under stringent reliability requirements.

desiccant A drying agent used for drying out cable splices or sensitive equipment.

design A layout of all the necessary equipment and facilities required to make a trunk, special circuit, piece of equipment, or system work.

design objective The desired electrical or mechanical performance characteristic for electronic circuits and equipment.

detection The rectification process that results in the modulating signal being separated out from a modulated wave.

detector A device that converts one type of energy into another.

device A functional circuit, component, or network unit, such as a transistor or an integrated circuit.

dewpoint The temperature at which moisture will condense.

diagnosis The process of locating errors in software, or equipment faults in hardware.

diagnostic routine A software program designed to trace errors in software, locate hardware faults, or identify the cause of a breakdown.

diaphragm A thin material that vibrates in response to an incident sound wave, such as in a telephone transmitter, or is electrically vibrated to produce sound waves, such as in a telephone receiver.

dielectric An insulating material that separates the elements of various components, including capacitors and transmission lines. Dielectric materials include air, plastic, mica, ceramic, and Teflon. A dielectric material must be an insulator. *Teflon* is a registered trademark of DuPont.

dielectric constant The ratio of the capacitance of a capacitor with a certain dielectric material to the capacitance with a vacuum as the dielectric. The *dielectric constant* is considered a measure of the capability of a dielectric material to store an electrostatic charge.

dielectric strength The potential gradient at which electric breakdown occurs.

differential amplifier An input circuit which rejects voltages that are the same at both input terminals but amplifies any voltage difference between the inputs. Use of a differential amplifier causes any signal present on both terminals, such as common mode hum, to cancel itself.

differential dc The maximum dc voltage that can be applied between the differential inputs of an amplifier while maintaining linear operation.

differential gain The difference in output amplitude (expressed in percent or dB) of a small high frequency sinewave signal at two stated levels of a low frequency signal on which it is superimposed.

differential-mode interference An interference source that causes a change in potential of one side of a signal transmission path relative to the other side.

differential phase The difference in output phase of a small high frequency sinewave signal at two stated levels of a low frequency signal on which it is superimposed.

diffraction The deviation of a wavefront from the path predicted by geometric optics when the wavefront is restricted by an opening or the edge of an object. Diffraction occurs whenever a light or radio beam is restricted in any way, and may still be important when the opening is many orders of magnitude larger than the wavelength.

diffuse reflection The scattering effect that occurs when light, radio, or sound waves strike a rough surface.

diffusion The spreading or scattering of a wave, such as an optical wave or sound wave.

digit One of the elements that combine to form numbers in a numbering system. For example, the decimal system uses arabic numerals 0 through 9 and the binary system uses numerals 0 and 1. Within the public switched telephone network, a digit is a signaling element that is combined with other digits in specified formats to form addresses, destination codes, routing, and other information.

digital coding A method of representing information by discrete bits.

diode A semiconductor or vacuum tube with two electrodes that passes electric current in one direction only. Diodes are used in rectifiers, gates, modulators, and detectors.

dip switch (DIP switch) An abbreviation for *dual in-line package* switch. A DIP switch is a printed circuit board-mounted device package of dual in-line style, typically consisting of two to eight switches.

direct coupling A coupling method between stages that permits dc current to flow between the stages.

direct current An electrical signal in which the direction of current flow and the value remain constant.

discharge The conversion of stored energy, as in a battery or capacitor, into an electric current.

discontinuity An abrupt nonuniform point of change in a transmission circuit that causes a disruption of normal operation.

discrete An individual circuit component. A discrete circuit is one that uses individual transistors and other components rather than integrated circuits.

discrete component A separately contained circuit element with its own external connections.

dish An antenna system consisting of a parabolic shaped reflector with a signal feed element at the focal point. Dish antennas commonly are used for transmission and reception from microwave stations and communications satellites.

dispersion The chromatic or wavelength dependence of a parameter.

display The representation of text and images on a cathode-ray tube, an array of light-emitting diodes, or a liquid-crystal readout.

display device An output unit that provides a visual representation of data.

distortion The difference between the waveshape of an original signal and the signal after it has traversed the transmission circuit.

distortion-limited operation The condition prevailing when the shape of the signal, rather than the amplitude (or power), is the limiting factor in communication capability. This condition is reached when the system distorts the shape of the waveform beyond tolerable limits. For linear systems, *distortion-limited* operation is equivalent to *bandwidth-limited* operation.

distribution substation A utility company installation that includes step-down transformers, switching hardware, circuit protection devices, and voltage control circuits to process incoming ac energy so that it can be efficiently utilized by local industrial and residential customers.

distribution voltage drop The voltage drop between any two defined points of interest in a power distribution system.

disturbance The interference with normal conditions and communications by some external energy source.

disturbance current The unwanted current of any irregular phenomenon associated with transmission that tends to limit or interfere with the interchange of information.

disturbance power The unwanted power of any irregular phenomenon associated with transmission which tends to limit or interfere with the interchange of information.

disturbance voltage The unwanted voltage of any irregular phenomenon associated with transmission which tends to limit or interfere with the interchange of information.

diversity factor The ratio of the sum of the individual maximum demands of the various parts of a power distribution system to the maximum demand of the whole system. The *diversity factor* is always greater than unity.

diversity receiver A receiver using two antennas connected through circuitry that senses which antenna is receiving the stronger signal. Electronic gating permits the stronger source to be routed to the receiving system.

documentation A written description of a program. *Documentation* can be considered as any record that has permanence and can be read by humans or machines.

down-lead A leading-in wire from an antenna to a receiver.

downlink The portion of a communication link used for transmission of signals from a satellite or airborne platform to a surface terminal.

downstream A specified signal modification occurring after other given devices in a signal path.

downtime The time during which equipment is not capable of doing useful work because of malfunction. This does not include preventive maintenance time. In other words, *downtime* is measured from the occurrence of a malfunction to the correction of that malfunction.

drift A slow change in a nominally constant signal characteristic, such as frequency.

drilled well ground (DWG) A well drilled for the sole purpose of providing an earth interface for the ground conductor.

drive The input signal to a circuit, particularly to an amplifier.

drive pulse A term commonly used to describe a set of signals needed by source equipment, such as a video camera. This signal set may be composed of any of the following: sync, blanking, subcarrier, horizontal drive, vertical drive, PAL pulse, and burst flag.

driver An electronic circuit that supplies an isolated output to drive the input of another circuit.

dropout The momentary loss of a signal.

drop-out value The value of current or voltage at which a relay will cease to be operated.

dropping resistor A resistor designed to carry current that will make a required dropped voltage available.

dry circuit A circuit that carries ac signals, but no dc current.

dry contacts A set of contacts that carry ac signals, but no dc current.

dual in-line package A standard method of packaging integrated circuits with input/output pins bent at right angles and in lines along the two long sides of the unit so that they go straight into holes in a printed circuit board.

duct A pipe or conduit installed underground or in a building whose purpose is to protect the cables installed therein.

duplex separation The frequency spacing required in a communications system between the *forward* and *return* channels to maintain interference at an acceptably low level.

duplex signaling A configuration permitting signaling in both transmission directions simultaneously.

duty cycle The ratio of operating time to total elapsed time of a device which operates intermittently, expressed in percent.

dynamic A situation in which the operating parameters and/or requirements of a given system are continually changing.

dynamic range The maximum range or extremes in amplitude, from the lowest to the highest (noise floor to system clipping), that a system is capable of reproducing. The dynamic range is expressed in dB against a reference level.

dynamo A rotating machine, normally a dc generator.

dynamotor A rotating machine used to convert dc into ac.

E

earth A large conducting body with no electrical potential, also called *ground.*

earth capacitance The capacitance between a given circuit or component and a point at ground potential.

earth current A current that flows to earth/ground, especially one that follows from a fault in the system. *Earth current* may also refer to a current that flows in the earth, resulting from ionospheric disturbances, lightning, or faults on power lines.

earth fault A fault that occurs when a conductor is accidentally grounded/earthed, or when the resistance to earth of an insulator falls below a specified value.

earth ground A large conducting body that represents *zero level* in the scale of electrical potential. An *earth ground* is a connection made either accidentally or by design between a conductor and earth.

earth potential The potential taken to be the arbitrary zero in a scale of electric potential.

effective ground A connection to ground through a medium of sufficiently low impedance and adequate current-carrying capacity to prevent the buildup of voltages that might be hazardous to equipment or personnel.

effective resistance The increased resistance of a conductor to an alternating current resulting from the skin effect, relative to the direct-current resistance of the conductor. Higher frequencies tend to travel only on the outer skin of the conductor, whereas dc flows uniformly through the entire area.

efficiency The useful power output of an electrical device or circuit divided by the total power input, expressed in percent.

elapsed time The total time during which a circuit or device is in use.

electret A substance exhibiting a permanent electric charge. It is the electric equivalent of a permanent magnet.

electric charge An excess of either electrons or protons within a given space or material.

electric field strength The magnitude, measured in volts per meter, of the electric field in an electromagnetic wave.

electric flux The amount of electric charge, measured in coulombs, across a dielectric of specified area. *Electric flux* may also refer simply to electric lines of force.

electricity An energy force derived from the movement of negative and positive electric charges.

electrode An electrical terminal that emits, collects, or controls an electric current.

electrolysis A chemical change induced in a substance resulting from the passage of electric current through an electrolyte.

electrolyte A nonmetallic conductor of electricity in which current is carried by the physical movement of ions.

electromagnet An iron or steel core surrounded by a wire coil. The core becomes magnetized when current flows through the coil but loses its magnetism when the current flow is stopped. Electromagnets are the motivating component of most mechanical relays and solenoids.

electromagnetic compatibility The capability of electronic equipment or systems to operate in a specific electromagnetic environment, at designated levels of efficiency and within a defined margin of safety, without interfering with itself or other systems.

electromagnetic field The electric and magnetic fields associated with radio and light waves.

electromagnetic induction An electromotive force created with a conductor by the relative motion between the conductor and a nearby magnetic field.

electromagnetic pulse (EMP) An intense pulse of electromagnetic energy resulting from a nuclear explosion. The potentially destructive effects of EMP are considered in the design of telecommunications systems.

electromagnetism The study of phenomena associated with varying magnetic fields, electromagnetic radiation, and moving electric charges.

electromotive force (EMF) An electrical potential, measured in volts, that can produce movement of electrical charges.

electron A stable elementary particle with a negative charge that is mainly responsible for electrical conduction. Electrons move when under the influence of an electric field. This movement constitutes an *electric current.*

electron beam A stream of emitted electrons, usually in a vacuum.

electron gun A hot cathode that produces a finely focused stream of fast electrons, which are necessary for the operation of a cathode ray tube. The gun is made up of a hot cathode electron source, a control grid, accelerating anodes, and (usually) focusing electrodes.

electron lens A device used for focusing an electron beam in a cathode ray tube. Such focusing can be accomplished by either magnetic forces, in which external coils are used to create the proper magnetic field within the tube, or electrostatic forces, where metallic plates within the tube are charged electrically in such a way as to control the movement of electrons in the beam.

electron volt The energy acquired by an electron in passing through a potential difference of one volt, in vacuum.

electronic A description of devices (or systems) that are dependent on the flow of electrons in electron tubes, semiconductors, and other devices, and not solely on electron flow in ordinary wires, inductors, capacitors, and similar passive components.

electronic switch A transistor, semiconductor diode, or a vacuum tube used as an on/off switch in an electrical circuit. Electronic switches can be controlled manually, by other circuits, or by computers for fast, flexible switching.

electronics The field of science and engineering that deals with electron devices and their utilization.

electro-optic device An electronic device whose operation depends on changes made to the refractive index of its semiconductor material by electric fields.

electroplate The process of coating a given material with a deposit of metal by electrolytic action.

electrostatic The condition pertaining to electric charges that are at rest.

electrostatic discharge (ESD) The flow of electrical current between two objects that are at different electrical potentials or voltages. This flow removes the voltage difference. The currents that flow during discharge can destroy electronic components used in telecommunications equipment. Electrostatic discharge can be controlled through proper equipment design, and by protective measures, including proper equipment and personnel grounding techniques.

electrostatic field The space in which there is electric stress produced by static electric charges.

electrostatic induction The process of inducing static electric charges on a body by bringing it near other bodies that carry high electrostatic charges.

electrostatic printing A type of printing commonly used in photocopying machines. A pattern of electrostatic charges is produced on the surface of paper, and a fine powder is then applied to the paper, which sticks to the charged lines and letters. The image is made permanent by applying heat.

element A substance that consists of atoms of the same atomic number. Elements are the basic units in all chemical changes other than those in which *atomic changes*, such as fusion and fission, are involved.

EMI (electromagnetic interference) Undesirable electromagnetic waves that are radiated unintentionally from an electronic circuit or device into other circuits or devices, disrupting their operation.

emphasis The intentional alteration of the frequency-amplitude characteristics of a signal to reduce the adverse effects of noise in a communication system.

empirical A conclusion not based on pure theory, but on practical and experimental work.

emulation The use of one system to imitate the capabilities of another system.

enable To prepare a circuit for operation or to allow an item to function.

enabling signal A signal that permits the occurrence of a specified event.

encapsulation The process of providing a protective outer casing (usually plastic) for a circuit or device.

encode The conversion of information from one form into another to obtain characteristics required by a transmission or storage system.

encoder A device that processes one or more input signals into a specified form for transmission and/or storage.

energized The condition when a circuit is switched on, or powered up.

energy spectral density A frequency-domain description of the energy in each of the frequency components of a pulse.

engineered circuit A circuit that is designed and designated to meet a specific requirement.

envelope The boundary of the family of curves obtained by varying a parameter of a wave.

envelope delay The difference in absolute delay between the fastest and slowest propagating frequencies within a specified bandwidth.

envelope delay distortion The maximum difference or deviation (in microseconds) of the envelope-delay characteristic between any two specified frequencies.

envelope detection A demodulation process that senses the shape of the modulated RF envelope. A diode detector is one type of envelope detection device.

environmental An equipment specification category relating to temperature and humidity.

EQ (equalization) network A network connected to a circuit to correct or control its transmission frequency characteristics.

equalization (EQ) The reduction of frequency distortion and/or phase distortion of a circuit through the introduction of one or more networks to compensate for the difference in attenuation, time delay, or both, at the various frequencies in the transmission band.

equalize The process of inserting in a line a network with complementary transmission characteristics to those of the line, so that when the loss or delay in the line and that in the equalizer are combined, the overall loss or delay is approximately equal at all frequencies.

equalizer A network that corrects the transmission-frequency characteristics of a circuit to allow it to transmit selected frequencies in a uniform manner.

equalizing charge A charge given a battery until all cells are as fully charged as possible. An *equalizing charge* ensures complete restoration of the active material in the plates of all cells.

equipment A general term for electrical apparatus and hardware, switching systems, and transmission components.

equipment failure The condition when a hardware fault stops the successful completion of a task.

equipment ground A protective ground consisting of a conducting path to ground of non-current carrying metal parts.

equivalent circuit A simplified network that emulates the characteristics of the real circuit it replaces. An equivalent circuit is typically used for mathematical analysis.

equivalent network A network that may replace another network without altering the performance of that portion of the system external to the network.

equivalent noise resistance A quantitative representation in resistance units of the spectral density of a noise voltage generator at a specified frequency.

ergonomic A study relating to biotechnology focusing on the design of equipment to suit human habits and dimensions.

error A collective term that includes all types of edit rejects, inconsistencies, transmission deviations, and control failures.

error checking A procedure performed by computer-based communications programs to ensure complete and accurate data transfer.

estimate A written statement of proposed costs, procedures, and timetables for a given project.

excitation The current that energizes field coils in a generator.

expandor A device with a nonlinear gain characteristic that acts to increase the gain more on larger input signals than it does on smaller input signals.

extender board An adapter board that extends a module outside of its frame to allow easier access to the module's components for troubleshooting and alignment.

external clock An external circuit or device that provides clocking or timing information.

extract A body of data derived from one data source to be used for a different application.

extremely high frequency (EHF) The band of microwave frequencies between the limits of 30 GHz and 300 GHz (wavelengths between 1 cm and 1 mm).

extremely low frequency The radio signals with operating frequencies below 300 Hz (wavelengths longer than 1000 km).

F

fail-safe operation A type of control architecture for a system that prevents improper functioning in the event of circuit or operator failure.

failure A detected cessation of ability to perform a specified function or functions within previously established limits. A *failure* is beyond adjustment by the operator by means of controls normally accessible during routine operation of the system. (This requires that measurable limits be established to define "satisfactory performance.")

failure effect The result of the malfunction or failure of a device or component.

failure in time (FIT) A unit value that indicates the reliability of a component or device. One failure in time corresponds to a failure rate of 10^{-9} per hour.

failure mode and effects analysis (FMEA) An iterative documented process performed to identify basic faults at the component level and determine their effects at higher levels of assembly.

failure rate The ratio of the number of actual failures to the number of times each item has been subjected to a set of specified stress conditions.

fall time The length of time during which a pulse decreases from 90 percent to 10 percent of its maximum amplitude.

fan in The greatest number of separate inputs acceptable to a single specified logic circuit without adversely affecting performance.

far end The distant end to that being considered; not the end where testing is being carried out.

far-end crosstalk Crosstalk that is propagated in a disturbed channel in the same direction as the propagation of the signal in the disturbing channel. The terminals of the disturbed channel where the far-end crosstalk is present and the energized terminals of the disturbing channel are usually remote from each other.

farad The standard unit of capacitance equal to the value of a capacitor with a potential of one volt between its plates when the charge on one plate is one coulomb and there is an equal and opposite charge on the other plate. The farad is a large value and is more commonly expressed in *microfarads* or *picofarads*. The *farad* is named for the English chemist and physicist Michael Faraday (1791 - 1867).

fatigue The reduction in strength of a metal caused by the formation of crystals resulting from repeated flexing of the part in question.

fault A condition that causes a device, a component, or an element to fail to perform in a required manner. Examples include a short circuit, broken wire, or intermittent connection.

fault finder A test set or other device that enables faults to be identified and localized.

fault location A procedure of electrical tests made from a terminal station to determine the location (and sometimes the cause) of system malfunction.

fault to ground A fault caused by the failure of insulation and the consequent establishment of a direct path to ground from a part of the circuit which should not normally be grounded.

fault-tolerant network A network designed or engineered to remain in operation in the event of system or component failures. For example, a fault-tolerant network might use alternate routing.

fault tree analysis (FTA) An iterative documented process of a systematic nature performed to identify basic faults, determine their causes and effects, and establish their probabilities of occurrence.

feature A distinctive characteristic or part of a system or piece of equipment, usually visible to end users and designed for their convenience.

feedback The return of a portion of the output of a device to the input. *Positive feedback* adds to the input, *negative feedback* subtracts from the input.

feedback amplifier An amplifier with the components required to feed a portion of the output back into the input to alter the characteristics of the output signal.

feedline A transmission line, typically coaxial cable, that connects a high frequency energy source to its load.

feedthrough A method of connecting a circuit on one layer on a printed circuit board with one on another layer. Complex boards often involve ten or more layers, so feedthrough alignment is critical.

ferrite A ceramic material made of powdered and compressed ferric oxide, plus other oxides (mainly cobalt, nickel, zinc, yttrium-iron, and manganese). These materials have low eddy current losses at high frequencies.

ferromagnetic material A material with low relative permeability and high coercive force so that it is difficult to magnetize and demagnetize. Hard ferromagnetic materials retain magnetism well, and are commonly used in permanent magnets.

fidelity The degree to which a system, or a portion of a system, accurately reproduces at its output the essential characteristics of the signal impressed upon its input.

field strength The strength of an electric, magnetic, or electromagnetic field.

field upgrade A product upgrade that takes place at the customer's site.

filler A material used to fill in gaps or voids, such as spaces between wires in a multi-pair cable.

filter A network that passes desired frequencies but greatly attenuates other frequencies.

filtered noise White noise that has been passed through a filter. The power spectral density of filtered white noise has the same shape as the transfer function of the filter.

finite impulse response (FIR) filter A type of digital filter that uses over-sampling techniques to enable implementation of anti-aliasing without the undesirable side effects found with analog filters.

fitting A coupling or other mechanical device that joins one component with another.

fixed A system or device that is not changeable or movable.

flashover An arc or spark between two conductors.

flashover voltage The voltage between conductors at which flashover just occurs.

flat face tube The design of CRT tube with an almost flat face, giving improved legibility of text and reduced reflection of ambient light.

flat level A signal that has an equal amplitude response for all frequencies within a stated range.

flat loss A circuit, device, or channel that attenuates all frequencies of interest by the same amount, also called *flat slope*.

flat noise A noise whose power per unit of frequency is essentially independent of frequency over a specified frequency range.

flat response A performance parameter of a system in which the output signal amplitude of the system is a faithful reproduction of the input amplitude over some range of specified input frequencies.

flicker A brightness fluctuation slower than the persistence of vision, typically less than about 30 fluctuations per second.

float To operate a power load such as a telephone central office on a mains-driven rectifier in parallel with a low impedance storage battery that is kept fully charged by the rectifier and is itself only called upon to provide power during temporary and short-duration peaks for which the rectifier output is insufficient.

floating A circuit or device that is not connected to any source of potential or to ground.

flow chart A graphic portrayal that shows the sequence in which functions are performed, from the beginning of a job to the end.

fluorescence The characteristic of a material to produce light when excited by an external energy source. Minimal or no heat results from the process.

flutter Deviations from the correct pitch in a recording as a result of changes in the mean velocity of the reproducing medium exceeding 0.1 percent. The term *flutter* is generally applied to cyclic deviations occurring at a relatively high rate (10 Hz). Cyclic variation at lower rates (once per revolution of a disk) is called *wow*.

flux The electric or magnetic lines of force resulting from an applied energy source.

flywheel effect The characteristic of an oscillator that enables it to sustain oscillations after removal of the control stimulus. This characteristic may be desirable, as in the case of a phase-locked loop employed in a synchronous system, or undesirable, as in the case of a voltage-controlled oscillator.

focusing A method of making beams of radiation converge on a target, such as the face of a CRT.

forced air reclamation method A technique for removing water from the air core of polyethylene insulated cable by pumping through large amounts of nitrogen or dry air.

frame A segment of an analog or digital signal that has a repetitive characteristic, in that corresponding elements of successive *frames* represent the same things.

free electron An electron that is not attached to an atom and is thus mobile when an electromotive force is applied.

free-running oscillator An oscillator that is not synchronized with an external timing source.

frequency The number of complete cycles of a periodic waveform that occur within a given length of time. Frequency is usually specified in cycles per second (*Hertz*). Frequency is the reciprocal of wavelength. The higher the frequency, the shorter the wavelength. In general, the higher the frequency of a signal, the more capacity it has to carry information, the smaller an antenna is required, and the more susceptible the signal is to absorption by the atmosphere and physical structures. At microwave frequencies, radio signals take on a *line-of-sight* characteristic and require highly directional and focused antennas to be used successfully.

frequency accuracy The degree of conformity of a given signal to the specified value of a frequency.

frequency allocation The designation of radio-frequency bands for use by specific radio services.

frequency changer An electro-mechanical or electronic device used for changing the operating frequency of a power supply.

frequency converter A circuit or device used to change a signal of one frequency into another of a different frequency.

frequency coordination The process of analyzing frequencies in use in various bands of the spectrum to achieve reliable performance for current and new services.

frequency counter An instrument or test set used to measure the frequency of a radio signal or any other alternating waveform.

frequency departure An unintentional deviation from the nominal frequency value.

frequency difference The algebraic difference between two frequencies. The two frequencies can be of identical or different nominal values.

frequency displacement The end-to-end shift in frequency that may result from independent frequency translation errors in a circuit.

frequency distortion The distortion of a multi-frequency signal caused by unequal attenuation or amplification at the different frequencies of the signal. This term may also be referred to as *amplitude distortion*.

frequency-division multiple access (FDMA) The provision of multiple access to a transmission facility, such as an earth satellite, by assigning each transmitter its own frequency band.

frequency-division multiplexing (FDM) The process of transmitting multiple analog signals by an orderly assignment of frequency slots, that is, by dividing transmission bandwidth into several narrow bands, each of which carries a single communication and is sent simultaneously with others over a common transmission path. In fiber optics, the term is *wavelength-division multiplexing*.

frequency domain A representation of signals as a function of frequency, rather than of time.

frequency modulation (FM) The modulation of a carrier signal so that its instantaneous frequency is proportional to the instantaneous value of the modulating wave.

frequency multiplier A circuit that provides as an output an exact multiple of the input frequency.

frequency offset A frequency shift that occurs when a signal is sent over an analog transmission facility in which the modulating and demodulating frequencies are not identical. A channel with frequency offset does not preserve the waveform of a transmitted signal.

frequency response The measure of system linearity in reproducing signals across a specified bandwidth. Frequency response is expressed as a frequency range with a specified amplitude tolerance in dB. Frequency response in digital audio systems is limited to one half the sampling frequency (Nyquist limit).

frequency response characteristic The variation in the transmission performance (gain or loss) of a system with respect to variations in frequency.

frequency reuse A technique used to expand the capacity of a given set of frequencies or channels by separating the signals either geographically or through the use of different polarization techniques. Frequency reuse is a common element of the *frequency coordination* process.

frequency selectivity The ability of equipment to separate or differentiate between signals at different frequencies.

frequency shift The difference between the frequency of a signal applied at the input of a circuit and the frequency of that signal at the output.

frequency shift keying (FSK) A commonly used method of digital modulation in which a one and a zero (the two possible states) are each transmitted as separate frequencies.

frequency stability A measure of the variations of the frequency of an oscillator from its mean frequency over a specified period of time.

frequency standard An oscillator with an output frequency sufficiently stable and accurate that it is used as a reference.

fuel cell A chemical cell that produces electrical energy from a chemical reaction.

full duplex A communications system capable of transmission simultaneously in two directions.

full-wave rectifier A circuit configuration in which both positive and negative half-cycles of the incoming ac signal are rectified to produce a unidirectional (dc) current through the load.

functional block In specification description language, an object of manageable size and relevant internal relationship, containing one or more processes.

functional block diagram A diagram illustrating the definition of a problem on a logical and functional basis.

functional description In specification description language, the actual behavior of the implementation of the functional requirements of a system in terms of the internal structure and logic processes within the system.

functional unit An entity of hardware and/or software capable of accomplishing a given purpose.

fundamental frequency The lowest frequency of a complex signal; higher components are often harmonics or multiples of the fundamental.

fundamental mode The lowest-order mode of an optical waveguide, the only mode capable of propagation in a single-mode (mono-mode) fiber.

fuse A protective device used to limit current flow in a circuit to a specified level. The fuse consists of a metallic link that melts and opens the circuit at a specified current level.

fuse wire A fine-gauge wire made of an alloy that overheats and melts at the relatively low temperatures produced when the wire carries overload currents. When used in a fuse, the wire is called a *fuse link*.

fused fiber splice A splice accomplished by the application of localized heat sufficient to fuse or melt the ends of two lengths of optical fiber, forming, in effect, a continuous, single fiber.

fused quartz The precise term for glass made by melting natural quartz crystals.

G

gain An increase or decrease in the level of an electrical signal. Gain is measured in terms of decibels or number of times of magnification. Strictly speaking, *gain* refers to an increase in level. Negative numbers, however, are commonly used to denote a decrease in level.

gain bandwidth The gain times the frequency of measurement when a device is biased for maximum obtainable gain.

gain/frequency characteristic The gain-vs.-frequency characteristic of a channel over the bandwidth provided, also referred to as *frequency response*.

gain/frequency distortion A circuit defect in which a change in frequency causes a change in signal amplitude. When this type of distortion occurs to a television signal, it can cause adverse changes in color saturation, and a lack of vertical line resolution as a result of luminance pulse ringing.

galvanic A device that produces direct current by chemical action.

gang The mechanical connection of two or more circuit devices so that they can all be adjusted simultaneously. An example is the ganged capacitor in a superheterodyne radio receiver that adjusts the input and oscillator stages at the same time.

gang capacitor A variable capacitor with more than one set of moving plates linked together.

gang tuning The simultaneous tuning of several different circuits by turning a single shaft on which ganged capacitors are mounted.

ganged A device that is mechanically coupled, normally through the use of a shared shaft.

gas breakdown The ionization of a gas between two electrodes caused by the application of a voltage that exceeds a threshold value. The ionized path has a low impedance. Certain types of circuit and line protectors rely on gas breakdown to divert hazardous currents away from protected equipment.

gas-discharge tube A gas-filled tube designed to carry current during gas breakdown. The gas-discharge tube is commonly used as a protective device, preventing high voltages from damaging sensitive equipment.

gauge A measure of wire diameter. In measuring wire gauge, the lower the number, the thicker the wire.

Gaussian beam A beam of light whose radial intensity distribution is Gaussian.

Gaussian distribution A statistical distribution, also called the *normal* distribution. The graph of a Gaussian distribution is a bell-shaped curve.

Gaussian noise Noise in which the distribution of amplitude follows a Gaussian model; that is, the noise is random but distributed about a reference voltage of zero.

Gaussian pulse A pulse that has the same form as its own Fourier transform.

general parameters In specification description language, the basic operating parameters of a unit or system.

general purpose interface (GPI) A parallel interconnection scheme that allows remote control of certain functions of a device. One wire is dedicated to each function.

generator A machine that converts mechanical energy into electrical energy.

geosynchronous The attribute of a satellite in which the relative position of the satellite as viewed from the surface of a given planet is stationary. For earth, the geosynchronous position is 22,300 miles above the planet.

germanium A metallic semiconductor material used in making transistors.

giga A prefix meaning *one billion*.

gigahertz (GHz) A measure of frequency equal to one billion cycles per second. Signals operating above 1 gigahertz are commonly known as *microwaves*, and begin to take on the characteristics of visible light.

glitch A general term used to describe a wide variety of momentary signal discontinuities.

graceful degradation An equipment failure mode in which the system suffers reduced capability, but does not fail altogether.

graticule A fixed pattern of reference marking used with oscilloscope CRTs to simplify measurements. The graticule may be etched on a transparent plate covering the front of the CRT or, for greater accuracy in readings, may be electrically generated within the CRT itself.

grid A mesh electrode within an electron tube that controls the flow of electrons between the cathode and plate of the tube.

ground An electrical connection to earth or to a common conductor usually connected to earth.

ground clamp A clamp used to connect a ground wire to a ground rod or system.

ground field/earth interface A configuration of metallic conductors, rods, and pipes used to establish a low resistance contact with the earth.

ground loop An undesirable circulating ground current in a circuit grounded via multiple connections or at multiple points.

ground plane A conducting material at ground potential, physically close to other equipment, so that connections may be made readily to ground the equipment at the required points.

ground potential The point at zero electric potential.

ground return A conductor used as a path for one or more circuits back to the ground plane or central facility ground point.

ground rod A metal rod driven into the earth and connected into a mesh of interconnected rods so as to provide a low resistance link to ground.

ground window A single-point interface between the integrated ground plane of a building and an isolated ground plane.

ground wire A copper conductor used to extend a good low-resistance earth ground to protective devices in a facility.

grounded The connection of a piece of equipment to earth via a low resistance path.

grounding The act of connecting a device or circuit to ground or to a conductor that is grounded.

group delay A condition where the different frequency elements of a given signal suffer differing propagation delays through a circuit or a system. The delay at a lower frequency is different than the delay at a higher frequency, resulting in a time-related distortion of the signal at the receiving point.

group delay time The rate of change of the total phase shift of a waveform with angular frequency through a device or transmission facility.

group velocity The speed of a pulse on a transmission line.

guard band A narrow bandwidth between adjacent channels intended to reduce interference or crosstalk.

guy-anchor A rigid mechanical point to which a guy wire is attached.

guy wire A steel wire or rope used to hold a pole or tower upright. The top end of the guy wire is fixed to the pole, and the lower end is fixed to an anchor buried securely in the ground.

H

half-wave rectifier A circuit or device that changes only positive or negative half-cycle inputs of alternating current into direct current.

hard disk A digital data storage device using a rigid, magnetic disk as the storage medium.

harden The process of constructing military telecommunications facilities so as to protect them from damage by enemy action, especially *electromagnetic pulse* (EMP) radiation.

hardware Physical equipment, such as mechanical, magnetic, electrical, or electronic devices or components.

hardwire The process of installing wire or cable directly between equipment, without passing through test jacks or cross-connect points.

hardwired Electrical devices connected through physical wiring. This term may also refer to an electronic programming technique using physical connections and, therefore, essentially unalterable.

harmonic A periodic wave having a frequency that is an integral multiple of the fundamental frequency. For example, a wave with twice the frequency of the fundamental is called the *second harmonic*.

harmonic analyzer A test set capable of identifying the frequencies of the individual signals that make up a complex wave.

harmonic distortion The production of harmonics at the output of a circuit when a periodic wave is applied to its input. The level of the distortion is usually expressed as a percentage of the level of the input.

hazard A condition that could lead to danger for operating personnel.

head A device that erases, records, or reads information from a magnetic tape passed under it.

headroom The difference, in decibels, between the typical operating signal level and a peak overload level.

heads-up display An optical system that transfers operational or navigational information into a pilot's line of sight so that the pilot need not look down at instruments, but may keep looking ahead through the windshield.

heat detector A temperature-sensitive device that provides an indication when a specified temperature has been reached.

heat loss The loss of useful electrical energy resulting from conversion into unwanted heat.

heat sink A device that conducts heat away from a heat-producing component so that it stays within a safe working temperature range.

heater In an electron tube, the filament that heats the cathode to enable it to emit electrons.

hecto A prefix meaning *100*.

henry The standard unit of electrical inductance, equal to the self-inductance of a circuit or the mutual inductance of two circuits when there is an induced electromotive force of one volt and a current change of one ampere per second. The symbol for inductance is *H*, named for the American physicist Joseph Henry (1797-1878).

hertz (Hz) The unit of frequency that is equal to one cycle per second. Hertz is the reciprocal of *period*, the interval after which the same portion of a periodic waveform recurs. Hertz was named for the German physicist Heinrich R. Hertz (1857 - 1894).

heterodyne The mixing of two signals in a nonlinear device in order to produce two additional signals at frequencies that are the sum and difference of the original frequencies.

heterodyne frequency The sum of, or the difference between, two frequencies, produced by combining the two signals together in a modulator or similar device.

hexadecimal numeral A numeral in the hexadecimal (base 16) numbering system, represented by the characters: 0, 1, 2, 3, 4, 5, 6, 7, 8, 9, A, B, C, D, E, and F, optionally preceded by H.

high frequency loss Loss of signal amplitude at higher frequencies through a given circuit or medium. For example, high frequency loss could be caused by passing a signal through a coaxial cable.

high Q An inductance or capacitance whose ratio of reactance to resistance is high.

high tension A high voltage circuit.

high Z looping input A high impedance input circuit that also includes an output to enable routing the signal to another piece of equipment.

high-capacity digital service A high-capacity digital facility that operates at line rates faster than 44.736 Mbit/s.

high-pass filter A network that passes signals of higher than a specified frequency but attenuates signals of all lower frequencies.

hookup A connection or circuit made for a special purpose.

horn gap A lightning arrester utilizing a gap between two horns. When lightning causes a discharge between the horns, the heat produced lengthens the arc and breaks it.

horsepower The basic unit of mechanical power. One horsepower (hp) equals 550 foot-pounds per second or 746 watts.

hot A charged electrical circuit or device.

hot dip galvanized The process of galvanizing steel by dipping it into a bath of molten zinc.

hot standby System equipment that is fully powered but not in service. A *hot standby* can rapidly replace a primary system in the event of a failure.

hum Undesirable coupling of the 60 Hz power sine wave into other electrical signals and/or circuits.

hum rejection The ability of a circuit to cancel interference in a video or audio signal, usually at the 50 Hz or 60 Hz power line frequency.

human factors engineering Engineering that applies the study of ergonomics to the design of equipment and software to create safe, easy-to-use systems.

HVAC An abreviation for *heating, ventilation, and air conditioning* system.

hybrid system A communication system that accommodates both digital and analog signals.

hydrometer A testing device used to measure specific gravity, particularly the specific gravity of the dilute sulfuric acid in a lead-acid storage battery, to learn the state of charge of the battery.

hysteresis The property of an element evidenced by the dependence of the value of the output, for a given excursion of the input, upon the history of prior excursions and direction of the input. Originally, *hysteresis* was the name for magnetic phenomenon only—the lagging of flux density behind the change in value of the magnetizing flux—but now, the term is also used to describe other nonelastic behavior.

hysteresis loop The plot of magnetizing current against magnetic flux density (or of other similarly related pairs of parameters), which appears as a loop. The area within the loop is proportional to the power loss resulting from hysteresis.

hysteresis loss The loss in a magnetic core resulting from hysteresis.

I

I^2R loss The power lost as a result of the heating effect of current passing through resistance.

idling current The current drawn by a circuit, such as an amplifier, when no signal is present at its input.

image converter An electron tube that produces a visible image from an image usually produced by infrared radiation.

image frequency A frequency on which a carrier signal, when heterodyned with the local oscillator in a superheterodyne receiver, will cause a sum or difference frequency that is the same as the intermediate frequency of the receiver. Thus, a signal on an image frequency will be demodulated along with the desired signal and will interfere with it.

imagery Collectively, the representations of objects reproduced electronically or by optical means on film, electronic display devices, or other media.

impact ionization The ionization of an atom or molecule as a result of a high energy collision.

impairment The loss of quality of service provided by an individual circuit when its transmission limits are exceeded or signaling functions, such as seizure, disconnect, and automatic number identification, are experiencing intermittent failures.

impedance The total passive opposition offered to the flow of an alternating current. *Impedance* consists of a combination of resistance, inductive reactance, and capacitive reactance. It is the vector sum of resistance and reactance ($R + jX$), or the vector of magnitude Z at an angle.

impedance characteristic A graph of the impedance of a circuit showing how it varies with frequency.

impedance irregularity A discontinuity in an impedance characteristic, caused for example by the use of different coaxial cable types.

impedance matching The adjustment of the impedances of adjoining circuit components to a common value so as to minimize reflected energy from the junction and to maximize energy transfer across it. Incorrect adjustment results in an *impedance mismatch.*

impedance-matching transformer A transformer used between two circuits of different impedances with a turns ratio that provides for maximum power transfer and minimum loss by reflection.

impulse A short, high energy surge of electrical current in a circuit or on a line.

impulse current A current that rises rapidly to a peak then decays to zero without oscillating.

impulse excitation The production of an oscillatory current in a circuit by impressing a voltage for a relatively short period compared with the duration of the current produced.

impulse noise A noise signal consisting of random occurrences of energy spikes, having random amplitude and bandwidth.

impulse response The amplitude-vs.-time output of a transmission facility or device in response to an impulse.

impulse voltage A unidirectional voltage that rises rapidly to a peak and then falls to zero, without any appreciable oscillation.

in-band signaling A signaling method in which control data is sent over the same transmission channel or circuit as the user communications, and in the same frequency band as that provided for the user.

incidence angle The angle between the perpendicular to a surface and the direction of arrival of a signal.

increment A small change in the value of a quantity.

induce To produce an electrical or magnetic effect in one conductor by changing the condition or position of another conductor.

induced current The current that flows in a conductor because a voltage has been induced across two points in, or connected to, the conductor.

induced voltage A voltage developed in a conductor when the conductor passes through magnetic lines of force.

inductance The property of an inductor that opposes any change in a current that flows through it. The standard unit of inductance is the *henry.*

induction The electrical and magnetic interaction process by which a changing current in one circuit produces a voltage change not only in its own circuit (*self-inductance*) but also in other circuits to which it is linked magnetically.

induction motor An alternating current motor whose primary winding (usually static) is connected to the external power supply, and induced currents flow through a secondary winding (usually on the rotor).

inductive A circuit element exhibiting inductive reactance.

inductive kick A voltage surge produced when a current flowing through an inductance is interrupted.

inductive load A load that possesses a net inductive reactance.

inductive reactance The reactance of a circuit resulting from the presence of inductance and the phenomenon of induction.

inductor A coil of wire, usually wound on a core of high permeability, that provides high inductance without necessarily exhibiting high resistance.

inert An inactive unit, or a unit that has no power requirements.

inert cell A type of primary power cell that contains all needed chemicals in dry form, and begins to operate only when water is added.

infinite line A transmission line that appears to be of infinite length. There are no reflections back from the far end because it is terminated in its characteristic impedance.

information A collection of facts, data, numbers, letters, and symbols that have meaning.

infrared The band of electromagnetic energy between the extreme low end of the visible part of the spectrum, or about 0.75 mm (micrometers), and the highest microwave frequencies of about 1,000 mm.

infrared radiation An electromagnetic radiation next in frequency to visible red light and extending down into the microwave frequencies. Wavelengths between 740 nm and 1 mm usually cover this band. The *IR region* is sometimes subdivided into *near infrared*, *middle infrared*, and *far infrared* segments.

inhibit A control signal that prevents a device or circuit from operating.

injection The application of a signal to an electronic device.

in-phase The property of alternating current signals of the same frequency that achieve their peak positive, peak negative, and zero amplitude values simultaneously.

input The waveform fed into a circuit, or the terminals that receive the input waveform.

input transformer A transformer placed at the input of a device or circuit to match the impedance of the device to that of the preceding stage, or to isolate the preceding stage from the device.

input-looping A circuit arrangement that permits the input of a device to be connected to another system downstream. The looping connector may or may not be buffered from the input signal.

input/output unit (I/O) A collection of peripheral devices, such as teleprinters or tape drives, used to provide input or output to a computer system.

insertion gain The gain resulting from the insertion of a transducer in a transmission system, expressed as the ratio of the power delivered to that part of the system following the transducer to the power delivered to that same part before insertion. If more than one component is involved in the input or output, the particular component used must be specified. This ratio is usually expressed in decibels. If the resulting number is negative, an *insertion loss* is indicated.

insertion loss The signal loss within a circuit, usually expressed in decibels as the ratio of input power to output power.

insertion loss-vs.-frequency characteristic The amplitude transfer characteristic of a system or component as a function of frequency. The amplitude response may be stated as actual gain, loss, amplification, or attenuation, or as a ratio of any one of these quantities at a particular frequency, with respect to that at a specified reference frequency.

inspection lot A collection of units of product from which a sample is drawn and inspected to determine conformance with acceptability criteria.

instantaneous value The value of a varying waveform at a given instant of time. The value can be in volts, amperes, or phase angle.

instrument multiplier A measuring device that enables a high voltage to be measured using a meter with only a low voltage range.

instrument rating The range within which an instrument has been designed to operate without damage.

insulate The process of separating one conducting body from another conductor.

insulation The material that surrounds and insulates an electrical wire from other wires or circuits. *Insulation* may also refer to any material that does not ionize easily and thus presents a large impedance to the flow of electrical current.

insulator A material or device used to separate one conducting body from another.

integrated circuit An array of electronic components, both active and passive, integrated into a semiconductor substrate or deposited on it, and capable of performing circuit functions.

integration The production of complete and complex circuits on a single chip, usually of silicon. The generally accepted ranges of integrated circuits are: *small scale integration* (SSI), simple circuits of up to about 10 circuit elements per chip; *medium scale integration* (MSI), more complex circuits of up to about 1,000 elements per chip; *large scale integration* (LSI), complex circuits with up to 16 kbits of memory capacity (some makers say up to 64 kbits) or circuit elements; and *very large scale integration* (VLSI), circuits with 32 kbits, 64 kbits, or up to 1 Mbits of memory capacity or circuit elements.

intelligence signal A signal containing information.

intensity The strength of a given signal under specified conditions.

interconnect cable A short-distance cable intended for use between equipment (generally less than 3 m in length).

interface A device or circuit used to interconnect two pieces of electronic equipment.

interface device A unit that joins two interconnecting systems.

interference emission An emission that results in an electrical signal being propagated into and interfering with the proper operation of electrical or electronic equipment.

interlock A protection device or system designed to remove all dangerous voltages from a machine or piece of equipment when access doors or panels are opened or removed.

intermediate frequency A frequency that results from combining a signal of interest with a signal generated within a radio receiver. In superheterodyne receivers, all incoming signals are converted to a single intermediate frequency for which the amplifiers and filters of the receiver have been optimized.

intermittent A noncontinuous recurring event, often used to denote a problem that is difficult to find because of its unpredictable nature.

intermodulation The production, in a nonlinear transducer element, of frequencies corresponding to the sums and differences of the fundamentals and harmonics of two or more frequencies that are transmitted through the transducer.

intermodulation distortion (IMD) The distortion that results from the mixing of two input signals in a nonlinear system. The resulting output contains new frequencies that represent the sum and difference of the input signals and the sums and differences of their harmonics. IMD is also called *intermodulation noise*.

intermodulation noise In a transmission path or device, the noise signal that is contingent upon modulation and demodulation, resulting from nonlinear characteristics in the path or device.

internal resistance The actual resistance of a source of electric power. The total electromotive force produced by a power source is not available for external use; some of the energy is used in driving current through the source itself.

interoperability The condition achieved among communications and electronics systems or equipment when information or services can be exchanged directly between them or their users, or both.

interpolate The process of estimating unknown values based on a knowledge of comparable data that falls on both sides of the point in question.

interrupting capacity The rating of a circuit breaker or fuse that specifies the maximum current the device is designed to interrupt at its rated voltage.

interval The points or numbers lying between two specified endpoints.

inverse voltage The effective value of voltage across a rectifying device, which conducts a current in one direction during one half cycle of the alternating input, during the half cycle when current is not flowing.

inversion The change in the polarity of a pulse, such as from positive to negative.

inverter A circuit or device that converts a direct current into an alternating current.

ionizing radiation The form of electromagnetic radiation that can turn an atom into an ion by knocking one or more of its electrons loose. Examples of ionizing radiation include x-rays, gamma rays, and cosmic rays

IR drop A drop in voltage because of the flow of current (I) through a resistance (R), also called *resistance drop*.

IR loss The conversion of electrical power to heat caused by the flow of electrical current through a resistance.

isochronous A signal in which the time interval separating any two significant instants is theoretically equal to the unit interval or to an integral multiple of the unit interval.

isolated ground A ground circuit that is isolated from all equipment framework and any other grounds, except for a single-point external connection.

isolated ground plane A set of connected frames that are grounded through a single connection to a ground reference point. That point and all parts of the frames are insulated from any other ground system in a building.

isolated pulse A pulse uninfluenced by other pulses in the same signal.

isophasing amplifier A timing device that corrects for small timing errors.

isotropic A quantity exhibiting the same properties in all planes and directions.

J

jack A receptacle or connector that makes electrical contact with the mating contacts of a plug. In combination, the plug and jack provide a ready means for making connections in electrical circuits.

jack field The location of jacks or jack strips on a piece of equipment, or on a panel or groups of panels in a central office.

jacket An insulating layer of material surrounding a wire or fiber in a cable.

jitter Small, rapid variations in a waveform resulting from fluctuations in supply voltage or other causes.

joint The joining of one complete cable to another, including the interconnection of conductors and the closure of the overall sheath.

joule The standard unit of work that is equal to the work done by one newton of force when the point at which the force is applied is displaced a distance of one meter in the direction of the force. The *joule* is named for the English physicist James Prescott Joule (1818 - 1889).

K

Kelvin (K) The standard unit of thermodynamic temperature. Zero degrees Kelvin represents *absolute zero*. Water freezes at 273°K and water boils at 373°K under standard pressure conditions.

key A short pin or other projection that slides into a mating slot or groove to guide two parts being assembled. The *key* is used to prevent a connector interface from rotating.

kilo A prefix meaning *one thousand*.

kilobaud A unit of measurement of data transmission speed equal to one thousand baud.

kilohertz (kHz) A unit of measure of frequency equal to 1,000 Hz.

kilovar A unit equal to one thousand volt-amperes.

kilovolt (kV) A unit of measure of electrical voltage equal to 1,000 V.

kilowatt A unit equal to one thousand watts.

knee In a response curve, the region of maximum curvature.

ku band Radio frequencies in the range of 15.35 GHz to 17.25 GHz, typically used for satellite telecommunications.

L

lacquer A resin applied to the ends of a cable with fiber insulants to prevent the fibers from fraying or absorbing moisture.

ladder network A type of filter with components alternately across the line and in the line.

lag The difference in phase between a current and the voltage that produced it, expressed in electrical degrees.

lagging current A current that lags behind the alternating electromotive force that produced it. A circuit that produces a *lagging current* is one containing inductance alone, or whose effective impedance is inductive.

lagging load A load whose combined inductive reactance exceeds its capacitive reactance. When an alternating voltage is applied, the current lags behind the voltage.

laminate A material consisting of layers of the same or different materials bonded together and built up to the required thickness.

language A set of symbols, characters, conventions, and rules used for conveying information.

large-scale integration (LSI) The fabrication, through a single manufacturing process, of numerous complex electronic circuits on a single semiconductor chip.

lash To fasten together by lashing, as with an aerial cable to its supporting messenger strand.

lashed cable An aerial cable lashed to a messenger strand, usually by a long-pitch helical lashing wire.

latch An electronic circuit that holds a digital signal after it has been selected. To *latch* a signal means to hold it.

latitude An angular measurement of a point on the earth above or below the equator. The equator represents 0, the north pole +90, and the south pole -90.

lay The length, direction, or angle of twist of a pair of wires in a telephone cable. Pairs in the same cable have different twists to prevent crosstalk.

layout A proposed or actual arrangement or allocation of equipment.

LC circuit An electrical circuit with both inductance (*L*) and capacitance (*C*) that is resonant at a particular frequency.

LC ratio The ratio of inductance to capacitance in a given circuit.

lead An electrical wire, usually insulated.

leader stroke In lightning, the first stroke, which usually determines the path to be followed by the *return stroke*, where most of the energy is carried.

leading edge The initial portion of a pulse or wave in which voltage or current rise rapidly from zero to a final value.

leading load A reactive load in which the reactance of capacitance is greater than that of inductance. Current through such a load *leads* the applied voltage causing the current.

leakage The loss of energy resulting from the flow of electricity past an insulating material, the escape of electromagnetic radiation beyond its shielding, or the extension of magnetic lines of force beyond their intended working area.

leakage resistance The resistance of a path through which leakage current flows.

learn The ability of a system to store control panel data, including keystrokes, into memory registers for recall at a later time.

least significant bit (LSB) In a digital word, the bit that defines the smallest increment of resolution and is usually the last bit in the word sequence.

level The strength or intensity of a given signal.

level alignment The adjustment of transmission levels of single links and links in tandem to prevent overloading of transmission subsystems.

life cycle The predicted useful life of a class of equipment, operating under normal (specified) working conditions.

life safety system A system designed to protect life and property, such as emergency lighting, fire alarms, smoke exhaust and ventilating fans, and site security.

life test A test in which random samples of a product are checked to see how long they can continue to perform their functions satisfactorily. A form of *stress testing* is used, including temperature, current, voltage, and/or vibration effects, cycled at many times the rate that would apply in normal usage.

lightning An electric discharge resulting from the flow of current from part of a charged cloud to the ground.

lightning flash An electrostatic atmospheric discharge. The typical duration of a lightning flash is approximately 0.5 s. A single flash is made up of various discharge components, usually including three or four high-current pulses called *strokes*.

limiter An electronic device in which some characteristic of the output is automatically prevented from exceeding a pre-determined value.

limiter circuit A circuit of nonlinear elements that restricts the electrical excursion of a variable in accordance with some specified criteria.

limiting A process by which some characteristic at the output of a device is prevented from exceeding a pre-determined value.

line cord The ac power input cable of an electrical device or system.

line level The signal level, usually expressed in dB, at a particular point.

line loss The total end-to-end loss in decibels in a transmission line.

line-up The process of adjusting transmission parameters to bring a circuit to its specified values.

line voltage The voltage of public utility power supplies, normally 117 Vac in the United States and 240 Vac in England.

linear A circuit, device, or channel whose output is directly proportional to its input.

linear distortion A distortion mechanism that is independent of signal amplitude.

linearity A constant relationship, over a designated range, between the input and output characteristics of a circuit or device.

liquid crystal display (LCD) A display panel of low-viscosity material that reflects or blocks light depending on the electric field applied to it. If the material is arranged in pixel patterns, text and images can be displayed.

lissajous pattern The looping patterns generated by a CRT spot when the horizontal (X) and vertical (Y) deflection signals are sinusoids. The lissajous pattern is useful for evaluating the delay or phase of two sinusoids of the same frequency.

live A device or system connected to a source of electric potential.

load The work required of an electrical or mechanical system.

load factor The ratio of the average load over a designated period of time to the peak load occurring during the same period.

load line A straight line drawn across a grouping of plate current/plate voltage characteristic curves showing the relationship between grid voltage and plate current for a particular plate load resistance of an electron tube.

loading The addition of electrical inductance to a metallic transmission line to improve the frequency characteristics of the line. Loading a line increases the distance over which a quality signal can be sent.

logarithm The power to which a base must be raised to produce a given number. Common logarithms are to base 10.

logarithmic scale A meter scale with displacement proportional to the logarithm of the quantity represented.

logic The formal principles of reasoning.

longitude The angular measurement of a point on the surface of the earth in relation to the meridian of Greenwich (London). The earth is divided into 360° of longitude, beginning at the Greenwich mean. As one travels west around the globe, the longitude increases.

longitudinal current A current that travels in the same direction on both wires of a pair. The return current either flows in another pair or via a ground return path.

loop wire A wire that links several terminals or adjacent components.

loopback A test of the transmission capability of a system in which a signal is transmitted through a loop that returns the signal to the source. The test verifies the ability of the source to transmit and receive signals.

loss The power dissipated in a circuit, usually expressed in decibels, that performs no useful work.

loss deviation The change of actual loss in a circuit or system from a designed value.

loss variation The change in actual measured loss over time.

lossy The condition when the line loss per unit length is significantly greater than some defined normal parameter.

lossy cable A coaxial cable constructed to have high transmission loss so it can be used as an artificial load or as an attenuator.

lot size A specific quantity of similar material or a collection of similar units from a common source; in inspection work, the quantity offered for inspection and acceptance at any one time. The *lot size* may be a collection of raw material, parts, subassemblies inspected during production, or a consignment of finished products to be sent out for service.

low-pass filter A filter network that passes all frequencies below a specified frequency with little or no loss, but that significantly attenuates higher frequencies.

lug A tag or projecting terminal onto which a wire may be connected by wrapping, soldering, or crimping.

lumped constant A resistance, inductance, or capacitance connected at a point, and not distributed uniformly throughout the length of a route or circuit.

M

mA An abbreviation for *milliamperes* (0.001 A).

magnetic field An energy field that exists around magnetic materials and current-carrying conductors. Magnetic fields combine with electric fields in light and radio waves.

magnetic flux The field produced in the area surrounding a magnet or electric current. The standard unit of flux is the *weber*.

magnetic flux density A vector quantity measured by a standard unit called the *tesla*. The *magnetic flux density* is the number of magnetic lines of force per unit area, at right angles to the lines.

magnetic leakage The magnetic flux that does not follow a useful path.

magnetomotive force The force that tends to produce lines of force in a magnetic circuit. The *magnetomotive force* bears the same relationship to a magnetic circuit that voltage does to an electrical circuit.

maintainability The probability that a failure will be repaired within a specified time after the failure occurs.

maintenance Any activity intended to keep a functional unit in satisfactory working condition. The term includes the tests, measurements, replacements, adjustments, and repairs necessary to keep a device or system operating properly.

make contacts A contact set that is open when the device is at rest and *make* (close) when the device is operated.

malfunction An equipment failure or a fault.

manometer A test device for measuring gas pressure.

margin The difference between the value of an operating parameter and the value that would result in unsatisfactory operation. Typical *margin* parameters include signal level, signal-to-noise ratio, distortion, crosstalk coupling, and/or undesired emission level.

Markov Model A statistical model of the behavior of a complex system over time in which the probabilities of the occurrence of various future states depend only on the present state of the system, and not on the path by which the present state was achieved. This term was named for the Russian mathematician Andrei Andreevich Markov (1856 - 1922).

mask A device used in the production of thin-film circuits and other components as a means of restricting patterns or deposits.

masking The process of covering selected areas of a semiconductor prior to depositing materials on its surface or etching them away.

mass media services Information services that include television and radio broadcasting, pay television, and printed or electronic publications, such as newspapers and magazines.

mass splicing The simultaneous splicing of multiple conductors or fibers in a cable.

master clock An accurate timing device that generates a synchronous signal to control other clocks or equipment.

master oscillator A stable oscillator that provides a standard frequency signal for other hardware and/or systems.

master-slave timing A system wherein one station or node supplies the timing reference for all other interconnected stations or nodes.

matched termination A termination that absorbs all the incident power and so produces no reflected waves or mismatch loss.

matching The connection of channels, circuits, or devices in a manner that results in minimal reflected energy.

matrix A logical network configured in a rectangular array of intersections of input/output signals.

Mbit/s (megabits per second) A digital transmission speed in millions of bits per second.

mean An arithmetic average in which values are added and divided by the number of such values.

mean time between failures (MTBF) For a particular interval, the total functioning life of a population of an item divided by the total number of failures within the population during the measurement interval.

mean time to failure (MTTF) The measured operating time of a single piece of equipment divided by the total number of failures during the measured period of time. This measurement is normally made during that period between early life and wear-out failures.

mean time to repair (MTTR) The total corrective maintenance time on a component or system divided by the total number of corrective maintenance actions during a given period of time.

measurement A procedure for determining the amount of a quantity.

median A value in a series that has as many readings or values above it as below.

medium An electronic pathway or mechanism for passing information from one point to another.

megabyte One million bytes (actually 1,048,576); one thousand kilobytes.

megahertz (MHz) A quantity equal to one million hertz (cycles per second).

megohm A quantity equal to one million ohms.

message Any idea expressed briefly in a plain or secret language and prepared in a form suitable for transmission by any means of communication.

metal-clad switchgear An indoor or outdoor metal structure containing switching equipment and other associated hardware such as instrument transformers, buses, and connections. The devices are insulated as required and placed in separate grounded metal compartments.

metal-oxide semiconductor (MOS) A device whose operating characteristics are determined by conditions at the interface between a semiconductor layer, usually silicon, and an insulator layer, usually silicon dioxide.

metal-oxide varistor A solid-state voltage clamping device used for transient suppression applications.

metal tape A generic term used to describe a magnetic media formulation of fine, densely packed iron, cobalt, and nickel metal particles layered onto a thin base film. This construction results in improved performance, relative to conventional tape formulations. There are two types of metal tape: *metal powder tape* (a metal powder alloy consisting of cobalt and nickel mixed with binders and lubricants, and then dispursed onto a base film), and *metal evaporated tape* (a metal powder alloy consisting of iron, nickel, and cobalt evaporated onto the base film).

meter The standard unit of length.

metric system A decimal system of measurement based on the meter, the kilogram, and the second.

microelectronics A technology used to build integrated circuits and other small electronic devices and components, sometimes referred to as *microminiaturization*.

micron A unit of length equal to one millionth of a meter (1/25,000 of an inch).

microphonic(s) Unintended noise introduced into an audio or video system by mechanical vibration of electrical components.

micropositioner A device used to hold and align small parts, such as integrated circuits or optical fibers.

microsecond (s) One-millionth of a second (0.000001 s).

microvolt A quantity equal to one-millionth of a volt.

military time Time represented on a 24-hour rather than a 12-hour clock. For example, *1600* is the same as *4:00 p.m.*

milliammeter A test instrument for measuring electrical current, often part of a *multimeter*.

millihenry A quantity equal to one-thousandth of a henry.

milliwatt A quantity equal to one-thousandth of a watt.

minimum discernible signal The smallest input that will produce a discernible change in the output of a circuit or device.

mixer A circuit used to combine two or more signals to produce a third signal that is a function of the input waveforms.

mixing ratio The ratio of the mass of water vapor to the mass of dry air in a given volume of air. The *mixing ratio* affects radio propagation.

mnemonic A memory aid in which an abbreviation or arrangement of symbols has an easily remembered relationship to the subject.

mnemonic address A simple address code with some easily remembered relationship to the actual name of the destination, often using initials or other letters from the name to make up a pronounceable word.

mode An electromagnetic field distribution that satisfies theoretical requirements for propagation in a waveguide or oscillation in a cavity.

modem A device or circuit that converts digital signals to and from analog signals for transmission over conventional analog telephone lines. The term *modem* may also refer to a device or circuit that converts analog signals from one frequency band to another. The term is a contraction of modulator/demodulator.

modified refractive index The sum of the refractive index of the air at a given height above sea level, and the ratio of this height to the radius of the earth.

modular An equipment design in which major elements are readily separable, and which the user may replace, reducing the need for service calls.

modulation The process whereby the amplitude, frequency, or phase of a single-frequency wave (the *carrier*) is varied in step with the instantaneous value of, or samples of, a complex wave (the *modulating* wave).

modulator A device that enables the intelligence in an information carrying modulating wave to be conveyed by a signal at a higher frequency. A *modulator* modifies a carrier wave by amplitude, phase, and/or frequency as a function of a control signal that carries intelligence. Signals are *modulated* in this way to permit more efficient and/or reliable transmission over any of several media.

module An assembly replaceable as an entity, often as an interchangeable plug-in item. A *module* is not normally capable of being disassembled.

molded case circuit breaker A circuit breaker assembled as an integral unit and enclosed in an insulated housing. Molded case circuit breakers are used on systems rated at 600 V and less.

monitoring The process of listening to or viewing a communication service for the purpose of determining its quality or whether or not it is free from trouble or interference.

monolithic integrated circuit A microcircuit fabricated as a single component consisting of elements formed in or on a single semi-conducting substrate by diffusion, implantation, or deposition.

monostable A device that is stable in one state only. An input pulse causes the device to change state, but it reverts immediately to its stable state.

motherboard A circuit board that accommodates plug-in cards or *daughterboards* and provides for interconnections between them. A motherboard may also provide input/output connections.

motor A machine that converts electrical energy into mechanical energy.

mounting A support, such as a mounting plate, that holds and integrates a plug-in unit to other circuitry.

moving coil Any device that utilizes a coil of wire in a magnetic field in such a way that the coil is made to move by varying the applied current, or itself produces a varying voltage because of its movement.

multi-conductor A cable containing more than one pair of conductors in the same cable sheath.

multi-layer A type of printed circuit board that has several layers of pattern interconnected by electroplated holes from one plane to another.

multimeter A test instrument fitted with several ranges for measuring voltage, resistance, and current and equipped with an analog meter or digital display readout. The *multimeter* is also known as a *volt-ohm-milliammeter*, or *VOM*.

multiplex (MUX) The use of a common channel to make two or more channels. This is done either by splitting of the common channel frequency band into narrower bands, each of which is used to constitute a distinct channel (*frequency division multiplex*), or by allotting this common channel to multiple users in turn, to constitute different intermittent channels (*time division multiplex*).

multiplexer (MUX) A device or circuit that combines several signals onto a single line.

multiplexing A technique that uses a single transmission path to carry multiple voice and data channels. In time division multiplexing (TDM), path time is shared. For frequency division multiplexing (FDM) or wavelength division multiplexing (WDM), signals are divided into individual channels sent along the same path but at different frequencies.

multiplication Signal mixing that occurs within a multiplier circuit.

multiplier A circuit in which one or more input signals are mixed under the control of one or more control signals. The resulting output is a composite of the input signals, the characteristics of which are determined by the scaling specified for the circuit.

mV An abbreviation for *millivolt* (0.001 V).

mW An abbreviation for *milliwatt* (0.001 W).

N

nanosecond (ns) One-billionth of a second (1×10^{-9} s).

narrowband A communications channel of restricted bandwidth, often resulting in degradation of the transmitted signal.

narrowband emission An emission having a spectrum exhibiting one or more sharp peaks that are narrow in width compared to the nominal bandwidth of the measuring instrument, and are far enough apart in frequency to be resolvable by the instrument.

National Electrical Code (NEC) A document providing rules for the installation of electric wiring and equipment in public and private buildings, published by the National Fire Protection Association. The NEC has been adopted as law by many states and municipalities in the United States.

negative In a conductor or semiconductor material, an excess of electrons or a deficiency of positive charge.

negative feedback The return of a portion of the output signal from a circuit to the input but 180° out of phase. This type of feedback decreases signal amplitude but stabilizes the amplifier and reduces distortion and noise.

negative impedance An impedance characterized by a decrease in voltage drop across a device as the current through the device is increased, or a decrease in current through the device as the voltage across it is increased.

neutral A device or object having no electrical charge.

neutral conductor A conductor in a power distribution system connected to a point in the system that is designed to be at neutral potential. In a balanced system, the neutral conductor carries no current.

neutral ground An intentional ground applied to the neutral conductor or neutral point of a circuit, transformer, machine, apparatus, or system.

nickel-cadmium cell A type of dry rechargeable power cell commonly used in mobile communications applications.

nitrogen A gas widely used to pressurize communications cables and radio frequency transmission lines. If a small puncture occurs in the cable sheath, the nitrogen keeps moisture out so that service is not adversely affected.

node The points at which the current is at minimum in a transmission system in which standing waves are present.

noise Any random disturbance or unwanted signal in a communication system that tends to obscure the clarity of a signal in relation to its intended use.

noise factor (NF) The ratio of the noise power measured at the output of a receiver to the noise power that would be present at the output if the thermal noise resulting from the resistive component of the source impedance were the only source of noise in the system.

noise figure A measure of the noise in dB generated at the input of an amplifier, compared with the noise generated by an impedance-method resistor at a specified temperature.

noise filter A network that attenuates noise frequencies.

noise immunity The ability of a device to discern valid data in the presence of noise.

noise power ratio (NPR) The ratio, expressed in decibels, of signal power to intermodulation product power plus residual noise power, measured at the baseband level.

noise suppressor A filter or digital signal processing circuit in a receiver or transmitter that automatically reduces or eliminates noise.

noise temperature The temperature, expressed in Kelvins, at which a resistor will develop a particular noise voltage. The noise temperature of a radio receiver is the value by which the temperature of the resistive component of the source impedance should be increased—if it were the only source of noise in the system—to cause the noise power at the output of the receiver to be the same as in the real system.

noise to ground As checked on a noise measuring set, longitudinal noise current flowing to ground on one or more conductors.

noise weighting The assignment of a specified amplitude-vs.-frequency characteristic to a noise signal prior to measurement so that the measured value closely approximates the relative effect on a customer using the circuit.

nominal The most common value for a component or parameter that falls between the maximum and minimum limits of a tolerance range.

nominal value A specified or intended value independent of any uncertainty in its realization.

nomogram A chart showing three or more scales across which a straight edge may be held in order to read off a graphical solution to a three-variable equation.

nonconductor A material that does not conduct energy, such as electricity, heat, or sound.

noncritical technical load That part of the technical power load for a facility not required for minimum acceptable operation.

noninductive A device or circuit without significant inductance.

nonionizing radiation Electromagnetic radiation that does not turn an atom into an ion. Examples of non-ionizing radiation include visible light and radio waves

nonlinearity A distortion in which the output of a circuit or system does not rise or fall in direct proportion to the input.

nontechnical load The part of the total operational load of a facility used for such purposes as general lighting, air conditioning, and ventilating equipment during normal operation.

nonvolatile A memory device or system whose stored data is unaffected by the removal of operating power.

normal A line perpendicular to another line or to a surface.

normalized frequency The ratio between the actual frequency and its nominal value.

normally closed Switch contacts that are closed in their nonoperated state, or relay contacts that are closed when the relay is deenergized.

normally open Switch contacts that are open in their nonoperated state, or relay contacts that are open when the relay is deenergized.

normal-mode noise Unwanted signals in the form of voltages appearing in line-to-line and line-to-neutral signals.

north pole The pole of a magnet that seeks the north magnetic pole of the earth.

notch filter A circuit designed to attenuate a specific frequency band; also known as a *band stop filter*.

notched noise A noise signal in which a narrow band of frequencies has been removed.

ns An abbreviation for *nanosecond*.

null A zero or minimum amount or position.

Nyquist interval The maximum time interval between regularly spaced instantaneous samples of a wave of bandwidth *W* for complete determination of the waveform of the signal. The *Nyquist interval* is numerically equal to W/2 s.

Nyquist rate The maximum rate at which data can be transmitted over a limited bandwidth channel without intersymbol interference. The Nyquist rate, in *baud*, is twice the channel bandwidth in *hertz*. The term was named for the American physicist who determined the rate, Harry Nyquist (1889 - 1976).

O

octal The base-8 numbering system.

octave Any frequency band in which the highest frequency is twice the lowest frequency.

off-line A condition wherein devices or subsystems are not connected into, do not form a part of, and are not subject to the same controls as an operational system.

offset An intentional difference between the realized value and the nominal value.

ohm The unit of electric resistance through which one ampere of current will flow when there is a difference of one volt. The quantity is named for the German physicist Georg Simon Ohm (1787 - 1854).

ohmic loss The power dissipation in a line or circuit caused by electrical resistance.

ohmmeter A test instrument used for measuring resistance, often part of a *multimeter*.

Ohm's Law A law that sets forth the relationship between voltage (E), current (I), and resistance (R). The law states that $E = IR$, $I = E/R$, and $R = E/I$.

on-line A device or system that is energized and operational, and ready to perform useful work.

open An interruption in the flow of electrical current, as caused by a broken wire or connection.

open circuit A defined loop or path that closes on itself and contains an infinite impedance.

open circuit impedance The input impedance of a circuit when its output terminals are open, that is, not terminated.

open-circuit voltage The voltage measured at the terminals of a circuit when there is no load and, hence, no current flowing.

operating lifetime The period of time during which the principal parameters of a component or system remain within a prescribed range.

optical disc A form of data storage utilizing a laser to optically record the data bits on a disc which is read with a low power laser pickup. There are three primary types of optical discs: *read only* (RO), *write once read many* (WORM), and erasable/recordable (*thermo magneto optical*, TMO, and *phase change*, PC).

optimize The process of adjusting for the best output or maximum response from a circuit or system.

optoisolator A coupling device consisting of a light emitter and a photodetector used to couple signals without any electrical connection. Optoisolators provide voltage and/or noise isolation between input and output, while transferring the desired signal.

orbit The path, relative to a specified frame of reference, described by the center of mass of a satellite or other object in space, subjected solely to natural forces (mainly gravitational attraction).

original equipment manufacturer (OEM) A manufacturer of equipment that is used in systems assembled and sold by others.

oscillation A variation with time of the magnitude of a quantity with respect to a specified reference when the magnitude is alternately greater than and smaller than the reference.

oscillator A nonrotating device for producing alternating current, the output frequency of which is determined by the characteristics of the circuit.

oscilloscope A test instrument that uses a display, usually a cathode-ray tube, to show the instantaneous values and waveforms of a signal that varies with time or some other parameter.

outage duration The average elapsed time between the start and the end of an outage period.

outage probability The probability that the outage state will occur within a specified time period. In the absence of specific known causes of outages, the *outage probability* is the sum of all outage durations divided by the time period of measurement.

outage threshold A defined value for a supported performance parameter that establishes the minimum operational service performance level for that parameter.

out-of-band energy Energy emitted by a transmission system that falls outside the frequency spectrum of the intended transmission.

output impedance The impedance presented at the output terminals of a circuit, device, or channel.

output stage The final driving circuit in a piece of electronic equipment.

ovenized crystal oscillator (OXO) A crystal oscillator that is enclosed within a temperature regulated heater (oven) to maintain a stable frequency despite external temperature variations.

overload In a transmission system, a power greater than the amount the system was designed to carry. In a power system, an overload could cause excessive heating. In a communications system, distortion of a signal could result.

overshoot The first maximum excursion of a pulse beyond the 100 percent level. Overshoot is the portion of the pulse that exceeds its defined level temporarily before settling to the correct level. Overshoot amplitude is expressed as a percentage of the defined level.

P

propagation time delay The time required for a signal to travel from one point to another.

proximity effect A nonuniform current distribution in a conductor, caused by current flow in a nearby conductor.

pseudo-noise In a spread-spectrum system, a seemingly random series of pulses whose frequency spectrum resembles that of continuous noise.

pseudo-random A sequence of signals that appears to be completely random but have, in fact, been carefully drawn up and repeat after a significant time interval.

pseudo-random noise A noise signal that satisfies one or more of the standard tests for statistical randomness. Although it seems to lack any definite pattern, there is a sequence of pulses that repeats after a long time interval.

pseudo-random number sequence A sequence of numbers that satisfies one or more of the standard tests for statistical randomness. Although it seems to lack any definite pattern, there is a sequence that repeats after a long time interval.

psophometer A test instrument for measuring the noise voltage in a circuit. The instrument includes a weighting network that can be varied to assist the testing of a variety of circuit types.

pull box A small box with above-ground access inserted in a long run conduit to facilitate pulling a cable through the duct.

pulsating direct current A current changing in value at regular or irregular intervals but which has the same direction at all times.

pulse One of the elements of a repetitive signal characterized by the rise and decay in time of its magnitude. A *pulse* is usually short in relation to the time span of interest.

pulse decay time The time required for the trailing edge of a pulse to decrease from 90 percent to 10 percent of its peak amplitude.

pulse duration The time interval between the points on the leading and trailing edges of a pulse at which the instantaneous value bears a specified relation to the peak pulse amplitude.

pulse duration modulation (PDM) The modulation of a pulse carrier by varying the width of the pulses according to the instantaneous values of the voltage samples of the modulating signal (also called *pulse width modulation*).

pulse edge The leading or trailing edge of a pulse, defined as the 50 percent point of the pulse rise or fall time.

pulse fall time The interval of time required for the edge of a pulse to fall from 90 percent to 10 percent of its peak amplitude.

pulse interval The time between the start of one pulse and the start of the next.

pulse length The duration of a pulse (also called *pulse width*).

pulse level The voltage amplitude of a pulse.

pulse period The time between the start of one pulse and the start of the next.

pulse ratio The ratio of the length of any pulse to the total pulse period.

pulse repetition period The time interval from the beginning of one pulse to the beginning of the next pulse.

pulse repetition rate The number of times each second that pulses are transmitted.

pulse return pattern A method of locating faults in an aerial or buried wiring system whereby pulses sent out from the test set are reflected back by any impedance irregularity or discontinuities. This provides accurate localization and determination of the nature of the fault.

pulse rise time The time required for the leading edge of a pulse to rise from 10 percent to 90 percent of its peak amplitude.

pulse train A series of pulses having similar characteristics.

pulse width The measured interval between the 50 percent amplitude points of the leading and trailing edges of a pulse.

puncture A breakdown of insulation or of a dielectric, such as in a cable sheath or in the insulant around a conductor.

pW An abbreviation for picowatt, a unit of power equal to 10^{-12} W (-90 dBm).

Q

Q (quality factor) A figure of merit that defines how close a coil comes to functioning as a pure inductor. *High Q* describes an inductor with little energy loss resulting from resistance. *Q* is determined by dividing the inductive reactance of a device by its resistance.

quad-in-line (QUIL) A method of packaging a large scale integrated circuit with two rows of staggered pins on each side of the device, providing for 48 or more pins on one packaged chip.

quadrature A state of alternating current signals separated by one quarter of a cycle (90°).

quadrature amplitude modulation (AM) A process that allows two different signals to modulate a single carrier frequency. The two signals of interest amplitude modulate two samples of the carrier that are of the same frequency, but differ in phase by 90°. The two resultant signals can be added together and both signals recovered at a decoder when then they are demodulated 90° apart.

quadrature component The component of a voltage or current at an angle of 90° to a reference signal, resulting from inductive or capacitive reactance.

quadrature phase shift keying (QPSK) A type of phase shift keying using four phase states.

quality The absence of objectionable distortion.

quality assurance (QA) All those activities, including surveillance, inspection, control, and documentation, aimed at ensuring that a given product will meet its performance specifications.

quality control (QC) A function whereby management exercises control over the quality of raw material or intermediate products in order to prevent the production of defective devices or systems.

quantum-limited operation An operation wherein the minimum detectable signal is limited by quantum noise.

quantum noise Any noise attributable to the discrete nature of electromagnetic radiation. Examples include shot noise, photon noise, and recombination noise.

quartz A crystalline mineral that when electrically excited vibrates with a stable period. Quartz is typically used as the frequency determining element in oscillators and filters.

quasi-peak detector A detector that delivers an output voltage that is some fraction of the peak value of the regularly repeated pulses applied to it. The fraction increases toward unity as the pulse repetition rate increases.

queueing The ordering of requests on a waiting list.

quick-break fuse A fuse in which the fusible link is under tension, providing for rapid operation.

quiescent An inactive device, signal, or system.

quiescent current The current that flows in a device in the absence of an applied signal.

R

raceway A covered trough or channel for internal wiring and cabling.

rack An equipment rack, usually measuring 19 in (48.26 cm) wide at the front mounting rails.

rack unit (RU) A unit of measure of vertical space in an equipment enclosure. One rack unit is equal to 1.75 in (4.45 cm).

radiate The process of emitting electromagnetic energy.

radiation The emission and propagation of electromagnetic energy in the form of waves. *Radiation* is also called *radiant energy.*

radio The transmission of signals over a distance by means of electromagnetic waves in the approximate frequency range of 150 kHz to 300 GHz. The term may also be used to describe the equipment used to transmit or receive electromagnetic waves.

radio frequency interference (RFI) The intrusion of unwanted signals or electromagnetic noise into various types of equipment resulting from radio frequency transmission equipment or other devices using radio frequencies.

radio frequency spectrum Those frequency bands in the electromagnetic spectrum that range from several hundred thousand cycles per second (*very low frequency*) to several billion cycles per second (*microwave frequencies*).

random noise Electromagnetic signals originating in transient electrical disturbances and which have random time and amplitude patterns. Random noise is generally undesirable; however, it may also be generated for testing purposes.

random number A number formed by a set of digits in which each successive digit is equally likely to be any of the digits in a specified set.

rated output power The power available from an amplifier or other device under specified conditions of operation.

RC constant The time constant of a resistor-capacitor circuit. The *RC constant* is the time in seconds required for current in an RC circuit to rise to 63 percent of its final steady value or fall to 37 percent of its original steady value, obtained by multiplying the resistance value in ohms by the capacitance value in farads.

RC network A circuit that contains resistors and capacitors, normally connected in series.

reactance The part of the impedance of a network resulting from inductance or capacitance. The *reactance* of a component varies with the frequency of the applied signal.

reactive power The power circulating in an ac circuit. It is delivered to the circuit during part of the cycle and is returned during the other half of the cycle. The *reactive power* is obtained by multiplying the voltage, current, and the sine of the phase angle between them.

reactor A component with inductive reactance.

readout A visual display of the output of a device or system.

real time clock (RTC) A hardware device that keeps track of the date, hours, minutes, and seconds. The RTC may also provide a set of interval timer functions for generating statistics or activating control circuits.

received signal level (RSL) The value of a specified bandwidth of signals at the receiver input terminals relative to an established reference.

receiver Any device for receiving electrical signals and converting them to audible sound, visible light, or both.

receptacle An electrical socket designed to receive a mating plug.

reception The act of receiving, listening to, or watching information-carrying signals.

rectification The conversion of alternating current into direct current.

rectifier A device for converting alternating current into direct current. A *rectifier* normally includes filters so that the output is, within specified limits, smooth and free of ac components.

rectify The process of converting alternating current into direct current.

redundancy A system design that provides a backup for key circuits or components in the event of a failure. Redundancy improves the overall reliability of a system.

redundant A configuration when two complete systems are available at one time. If the on-line system fails, the backup will take over with no loss of service.

reference voltage A voltage used for control or comparison purposes.

reflectance The ratio of reflected power to incident power.

reflection An abrupt change, resulting from an impedance mismatch, in the direction of propagation of an electromagnetic wave. For light, at the interface of two dissimilar materials, the incident wave is returned to its medium of origin.

reflection coefficient The ratio between the amplitude of a reflected wave and the amplitude of the incident wave. For large, smooth surfaces, the reflection coefficient may be near unity.

reflection gain The increase in signal strength that results when a reflected wave combines, in phase, with an incident wave.

reflection loss The apparent loss of signal strength caused by an impedance mismatch in a transmission line or circuit. The loss results from the reflection of part of the signal back toward the source from the point of the impedance discontinuity. The greater the mismatch, the greater the loss.

reflectometer A device that measures energy traveling in each direction in a waveguide, used in determining the standing wave ratio.

refraction The bending of a sound, radio, or light wave as it passes obliquely from a medium of one density to a medium of another density that varies its speed.

regenerative repeater A repeater in which digital pulse signals are amplified, reshaped, retimed, and retransmitted.

regulation The process of adjusting the level of some quantity, such as circuit gain, by means of an electronic system that monitors an output and feeds back a controlling signal to constantly maintain a desired level.

regulator A device that maintains its output voltage at a constant level.

relative envelope delay The difference in envelope delay at various frequencies when compared with a reference frequency that is chosen as having zero delay.

relative humidity The ratio of the quantity of water vapor in the atmosphere to the quantity that would cause saturation at the ambient temperature.

relative transmission level The ratio of the signal power in a transmission system to the signal power at some point chosen as a reference. The ratio is usually determined by applying a standard test signal at the input to the system and measuring the gain or loss at the location of interest.

relay A device by which current flowing in one circuit causes contacts to operate that control the flow of current in another circuit.

relay bypass A device that, in the event of a loss of power or other failure, routes a critical signal around the equipment that has failed.

release current The value to which relay current must fall in order for a previously operated relay to release.

release time The time required for a tone or pulse to drop from steady-state level to zero, also referred to as the *decay time*.

reliability The ability of a system or subsystem to perform within the prescribed parameters of quality of service. *Reliability* is often expressed as the probability that a system or subsystem will perform its intended function for a specified interval under stated conditions.

reliability growth The action taken to move a hardware item toward its reliability potential, during development or subsequent manufacturing or operation.

reliability predictions The compiled failure rates for parts, components, subassemblies, assemblies, and systems. These generic failure rates are used as basic data to predict the reliability of a given device or system.

remote control A system used to control a device from a distance.

remote station A station or terminal that is physically remote from a main station or computer but can gain access through a communication channel.

repeater The equipment between two circuits that receives a signal degraded by normal factors during transmission and amplifies the signal to its original level for retransmission. For pulses, the repeater can amplify, reshape, or retime the input signal prior to retransmission.

repetition rate The rate at which regularly recurring pulses are repeated.

reply A transmitted message that is a direct response to an original message.

reset The act of restoring a device to its default or original state. *Reset* may also refer to restoring a counter or logic device to a known state, often a zero output.

residual voltage The vector sum of the voltages in all the phase wires of an unbalanced polyphase power system.

resistance The opposition of a material to the flow of electrical current. Resistance is equal to the voltage drop through a given material divided by the current flow through it.

resistance drop The fall in potential (volts) between two points, the product of the current and resistance.

resistance-grounded A circuit or system grounded for safety through a resistance, which limits the value of the current flowing through the circuit in the event of a fault.

resistive load A load in which the voltage is in phase with the current.

resistivity The resistance per unit volume or per unit area.

resistor A device whose primary function is to introduce resistance into an electrical circuit.

resistor color code The colored markings on a resistor that indicate the value and tolerance of the device.

resonance A tuned condition conducive to oscillation, when the reactance resulting from capacitance in a circuit is equal in value to the reactance resulting from inductance.

resonant frequency The frequency at which the inductive reactance and capacitive reactance of a circuit are equal.

resonator A resonant cavity.

responsivity A measure of the sensitivity of a photosensor. *Responsivity* is the ratio of the output current or voltage to the input flux in watts or lumens. When responsivity is indicated at a particular wavelength (in amperes/watt), it denotes the *spectral response* of the device.

retiming An adjustment of the intervals between corresponding significant instants of a digital signal, using a timing signal as the reference.

return A return path for current, sometimes through ground.

reversal A change in magnetic polarity, in the direction of current flow.

reverse current A small current that flows through a diode when the voltage across it is such that normal forward current does not flow.

reverse voltage A voltage in the reverse direction from that normally applied.

RG-58 A common type of 50 Ω coaxial cable.

ribbon cable Flat cable with multiple parallel conductors that have been individually insulated.

ripple An ac voltage superimposed on the output of a dc power supply, usually resulting from imperfect filtering.

rise time The time required for a pulse to rise from 10 percent to 90 percent of its peak value.

riser A duct, conduit, or cable that runs vertically in a building from floor to floor.

riser cable A high-strength cable intended for use in vertical shafts between floors in a building

roll-off A gradual attenuation of gain-frequency response at either or both ends of a transmission pass band.

root-mean-square (RMS) The square root of the average value of the squares of all the instantaneous values of current or voltage during one half-cycle of an alternating current. For an alternating current, the RMS voltage or current is equal to the amount of direct current or voltage that would produce the same heating effect in a purely resistive circuit. For a sinewave, the root-mean-square value is equal to 0.707 times the peak value. RMS is also called the *effective value.*

rosin joint A dry joint in which the wire is held in place by dry flux and there is little, or no, electrical contact.

rotor The rotating part of an electric generator or motor.

RU An abbreviation for *rack unit.*

run The route followed by cables, conduit, and wires in a communications system.

S

slope The rate of change, with respect to frequency, of transmission line attenuation over a given frequency spectrum.

slope equalizer A device or circuit used to achieve a specified slope in a transmission line.

slot A narrow band of frequencies.

smoothing circuit A filter designed to reduce the amount of ripple in a circuit, usually a dc power supply.

SMT (surface mount technology) A method of constructing a printed circuit board where the active components (integrated circuits, transistors, resistors, and other devices) are bonded directly to the board using the smallest amount of component packaging possible for reliable operation. SMT boards are smaller and, in many instances, more reliable than conventional printed circuit boards.

snubber An electronic circuit used to suppress high frequency noise.

solder A lead or tin alloy that melts readily and is used in a wide variety of wire, terminal, and component connecting applications.

soldering The process of joining metals by fusing them by means of a molten metal with a relatively low melting point.

soldering iron A wedge-shaped piece of copper fitted with a long shank and a heat-isolated handle and used to melt solder. Usually heated electrically by a resistive element, the soldering iron can also be heated by an external source, such as a torch, especially when used for heavy soldering jobs, such as the repair of lead cable sheaths. When fitted with a pistol grip, the device is called a *soldering gun.*

solid A single wire conductor, as contrasted with a stranded, braided, or rope-type wire.

solid-state The use of semiconductors rather than electromechanical relays and electron tubes in a circuit or system.

solid-state device A device that depends on the movement of charged particles rather than on mechanical movement for operation.

sort The process of arranging and grouping items according to a system of classification.

source The part of a system from which signals or messages are considered to originate.

source terminated A circuit whose output is terminated for correct impedance matching with standard cable.

span A digital or analog cable that carries signals from one central office or customer to another to provide end-to-end connectivity.

spare A system that is available but not presently in use.

spare wire A spare pair placed in a cable for use when a regular pair develops a fault.

specification A document intended primarily for use in procurement, which clearly describes the essential technical requirements for items, materials, or services, including the procedures by which it will be determined that the requirements have been met.

spectrum A continuous band of frequencies within which waves have some common characteristics.

spectrum analyzer A test instrument that presents a graphic display of signals over a selected frequency bandwidth. A cathode-ray tube is often used for the display.

spectrum designation of frequency A method of referring to a range of communication frequencies. In American practice, the designation is a two- or three-letter acronym for the name. The ranges are: below 300 Hz, ELF (extremely low frequency); 300 Hz - 3000 Hz, ILF (infra low frequency); 3 kHz - 30 kHz, VLF (very low frequency); 30 kHz - 300 kHz, LF (low frequency); 300 kHz - 3000 kHz, MF (medium frequency); 3 MHz - 30 MHz, HF (high frequency); 30 MHz - 300 MHz, VHF (very high frequency); 300 MHz - 3000 MHz, UHF (ultra high frequency); 3 GHz - 30 GHz, SHF (super high frequency); 30 GHz - 300 GHz, EHF (extremely high frequency); 300 GHz - 3000 GHz, THF (tremendously high frequency).

spherical antenna A type of satellite receiving antenna that permits more than one satellite to be accessed at any given time. A spherical antenna has a broader angle of acceptance than a parabolic antenna.

spike A high-amplitude, short-duration pulse superimposed on an otherwise regular waveform.

splice The process of joining together two entities permanently, to provide an electric or optic path from one wire or waveguide to another.

splice case A metal or plastic housing used to enclose and protect a cable splice.

split-phase A device that derives a second phase from a single-phase power supply by passing it through a capacitive or inductive reactor.

splitter A circuit or device that accepts one input signal and distributes it to several outputs.

splitting ratio The ratio of the power emerging from the output ports of a coupler.

sporadic An event occurring at random and infrequent intervals.

spread spectrum A communications technique in which all the frequency components of a narrowband signal are spread over a wide band. The resulting signal resembles white noise. The technique is used to achieve signal security and privacy, and to enable the use of a common band by many users. In telephony, a low bit-rate signal can be spread to resemble white noise and then multiplexed with a conventional speech signal.

spurious signal Any portion of a given signal that is not part of the fundamental waveform. Spurious signals include transients, noise, and hum.

square-wave A square or rectangular-shaped periodic wave that alternately assumes two fixed values for equal lengths of time, the transition being negligible in comparison with the duration of each fixed value.

square wave testing The use of a square wave containing many odd harmonics of the fundamental frequency as an input signal to a device. Visual examination of the output signal on an oscilloscope indicates the amount of distortion introduced.

stability The ability of a device or circuit to remain stable in frequency, power level, and/or other specified parameters.

standard The specific signal configuration, reference pulses, voltage levels, and other parameters that describe the input/output requirements for a particular type of equipment. Some standards have been established by professional groups or government bodies (such as SMPTE or EBU). Others are determined by equipment vendors and/or users.

standard time and frequency signal A time-controlled radio signal broadcast at scheduled intervals on a number of different frequencies by government-operated radio stations to provide a method for calibrating instruments.

standing wave ratio (SWR) The ratio of the maximum to the minimum value of a component of a wave in a transmission line or waveguide, such as the maximum voltage to the minimum voltage.

state The condition or value of a signal element at a point in time. For example, in a binary circuit, the state can be one or zero.

static A nonmoving electric charge, such as the charge on a capacitor plate.

static charge An electric charge on the surface of an object, particularly a dielectric.

station One of the input or output points in a communications system.

status The present condition of a device.

steady-state A condition in which circuit values remain essentially constant, occurring after all initial transients or fluctuating conditions have passed.

steady-state condition A condition occurring after all initial transient or fluctuating conditions have damped out in which currents, voltages, or fields remain essentially constant or oscillate uniformly without changes in characteristics such as amplitude, frequency, or waveshape.

steep wavefront A rapid rise in voltage of a given signal, indicating the presence of high frequency odd harmonics of a fundamental wave frequency.

step up (or down) The process of increasing (or decreasing) the voltage of an electrical signal, as in a step-up (or step-down) transformer.

storm loading The characteristics of a particular geographical area, such as ice buildup, wind speed, and ambient temperature, that affect the design of aerial cable installations.

strand A group of wires twisted together to form a strong and flexible cable.

stray capacitance An unintended—and usually undesired—capacitance between wires and components in a circuit or system.

stray current A current through a path other than the intended one.

strength member A steel, fiberglass epoxy rod, or other material used to increase the tensile strength of a cable.

stress The force per unit of cross-sectional area on a given object or structure.

stripping The removal of the outer sheath from a cable prior to splicing or termination.

stub A short length of cable spliced at a point on a main cable where branch feeder, distribution, or other main cable is, or is expected to be, connected.

subassembly A functional unit of a system.

subcarrier (SC) A carrier applied as modulation on another carrier, or on an intermediate subcarrier.

subharmonic A frequency equal to the fundamental frequency of a given signal divided by a whole number.

submodule A small circuit board that mounts on a larger module.

subrefraction A refraction for which the refractivity gradient is greater than standard.

substrate The support substance of an integrated circuit, either a semiconductor or an insulator.

subsystem A functional unit of a system.

superheterodyne receiver A radio receiver in which all signals are first converted to a common frequency for which the intermediate stages of the receiver have been optimized, both for tuning and filtering. Signals are converted by mixing them with the output of a local oscillator whose output is varied in accord with the frequency of the received signals so as to maintain the desired intermediate frequency.

surface leakage A leakage current from line to ground over the face of an insulator supporting an open wire route.

surface mount A method of mounting subminiature integrated circuits and other components directly on the surface of a printed circuit board. Surface mount construction permits greater component density on boards, making the electronic equipment smaller.

surface-mounted assembly A method of mounting components and devices onto the surface of a printed circuit board by direct solder attachment to landing traces etched on the board, rather than through-board component leads.

surge A rapid rise in current or voltage, usually followed by a fall back to the normal value.

survivability The ability of a communications network to continue to provide service after major damage to any part of the system.

sweep The process of varying the frequency of a signal over a specified bandwidth.

sweep generator A test oscillator, the frequency of which is constantly varied over a specified bandwidth.

switch A mechanical or solid state-device that opens or closes circuits, changes operating parameters, or selects paths or circuits.

switching The process of making and breaking (connecting and disconnecting) two or more electrical circuits.

synchronization The process of adjusting the corresponding significant instants of signals—for example, the zero-crossings—to make them synchronous. The term *synchronization* is often abbreviated as *sync*.

synchronize The process of causing two systems to operate at the same speed.

synchronous In step or in phase, as applied to two or more devices; a system in which all events occur in a pre-determined timed sequence.

synchronous detection A demodulation process in which the original signal is recovered by multiplying the modulated signal by the output of a synchronous oscillator locked to the carrier.

synchronous system A system in which the transmitter and receiver are operating in a fixed-time relationship.

syntax The relationships among characters or groups of characters, independent of their meanings or the manner of their interpretation and use.

syntax diagram A method of depicting the syntax of an input and output language by pictorial representation.

system standards The minimum required electrical performance characteristics of a specific collection of hardware and/or software.

systems analysis An analysis of a given activity to determine precisely what must be accomplished and how it is to be done.

T

tolerance The permissible variation from a standard.

tone A single-frequency audio signal used as a level-setting reference.

torque A moment of force acting on a body and tending to produce rotation about an axis.

total harmonic distortion (THD) The ratio of the sum of the amplitudes of all signals harmonically related to the fundamental vs. the amplitude of the fundamental signal. THD is expressed in percent.

trace The pattern on an oscilloscope screen when displaying a signal.

track The portion of a moving-type storage medium that is accessible to a given reading station.

tracking The locking of tuned stages in a radio receiver so that all stages are changed appropriately as the receiver tuning is changed.

tradeoff The process of weighing conflicting requirements and reaching a compromise decision in the design of a component or a subsystem.

transceiver Any circuit or device that receives and transmits signals.

transducer A device that converts energy from one form to another. Examples of a transducer include the microphone, which converts sound energy to electrical impulses, or the loudspeaker, which converts electrical signals into sound.

transfer characteristics The intrinsic parameters of a system, subsystem, or unit of equipment which, when applied to the input of the system, subsystem, or unit of equipment, will fully describe its output.

transformer A device consisting of two or more windings wrapped around a single core or linked by a common magnetic circuit.

transformer ratio The ratio of the number of turns in the secondary winding of a transformer to the number of turns in the primary winding, also known as the *turns ratio*.

transient A sudden variance of current or voltage from a steady-state value. A transient normally results from changes in load or effects related to switching action.

transient disturbance A voltage pulse of high energy and short duration impressed upon the ac waveform. The overvoltage pulse may be one to 100 times the normal ac potential and may last up to 15 ms. Rise times measure in the nanosecond range.

transient response The time response of a system under test to a stated input stimulus.

transmission The transfer of electrical power, signals, or intelligence from one location to another by wire, fiber optic, or radio means.

transmission facility A transmission medium and all the associated equipment required to transmit information.

transmission level point A specification, in decibels, of the relative power at a particular point in a transmission system compared to the level at an identified zero-level transmission point.

transmission line Any transmission medium, including free space.

transmission loss The ratio, in decibels, of the power of a signal at a point along a transmission path to the power of the same signal at a more distant point along the same path. This value is often used as a measure of the quality of the transmission medium for conveying signals. Changes in power level are normally expressed in decibels by calculating ten times the logarithm (base 10) of the ratio of the two powers.

transmission mode One of the field patterns in a waveguide in a plane transverse to the direction of propagation.

transmission system The set of equipment that provides single or multi-channel telecommunications facilities capable of carrying audio, video, or data signal.

transmitter The device or circuit that launches a signal into a passive medium, such as a copper-wire pair or an optical fiber. For example, a laser diode can be used as an optical transmitter.

transparency The property of a communications system that enables it to carry a signal without altering or otherwise affecting the electrical characteristics of the signal.

transportable A class of equipment that can be divided up or broken down into units which are readily moved, and upon arrival at the new site, rapidly reassembled into a working terminal.

tray The metal cabinet that holds circuit boards.

triac A gated switching device designed to conduct in either direction.

triangular wave An oscillation, the values of which rise and fall linearly, and immediately change upon reaching their peak maximum and minimum. A graphical representation of a triangular wave resembles a triangle.

triax A special form of coaxial cable containing three conductors.

trigger point An event, incident, or development in an outside-plant network that initiates a step or stimulates a reaction. An example of a trigger point is the exhaustion of existing facilities or structure.

trim The process of making fine adjustments to a circuit or a circuit element.

trimmer A small mechanically-adjustable component connected in parallel or series with a major component so that the net value of the two can be finely adjusted for tuning purposes.

trip free A power circuit breaker that will trip and break a faulty circuit even if the operating handle is held closed.

trouble A failure or fault affecting the service provided by a system.

troubleshoot The process of investigating, localizing, and (if possible) correcting a fault.

tune The process of adjusting the frequency of a device or circuit, such as for resonance or for maximum response to an input signal.

tuned trap A series resonant network bridged across a circuit that eliminates ("traps") the frequency of the resonant network.

tuner The radio frequency and intermediate frequency parts of a radio receiver that produce a low level audio output signal.

tuning The process of adjusting a given frequency—in particular, to adjust for resonance or for maximum response to a particular incoming signal.

turns ratio In a transformer, the ratio of the number of turns on the secondary to the number of turns on the primary.

tweaking The process of adjusting an electronic circuit to optimize its performance.

twin axial cable A shielded coaxial cable with two central conducting leads.

twin-line A feeder cable with two parallel, insulated conductors.

twisted pair A pair of insulated copper wires used in transmission circuits to provide bidirectional communications. The wires are twisted about one another to minimize electrical coupling with other circuits. Paired cable is made up of a few to several thousand twisted pairs.

two-phase A source of alternating current circuit with two sinusoidal voltages that are 90° apart.

U

unattended operation A system that permits a station to receive and transmit messages without the presence of an attendant or operator.

unavailability A measure of the degree to which a system, subsystem, or piece of equipment is not operable and not in a committable state at the start of a mission, when the mission is called for at a random point in time.

unbalanced circuit A two-wire circuit with legs that differ from one another in resistance, capacity to earth or to other conductors, leakage, or inductance.

unbalanced line A transmission line in which the magnitudes of the voltages on the two conductors are not equal with respect to ground. A coaxial cable is an example of an unbalanced line.

unbalanced modulator A modulator whose output includes the carrier signal.

unbalanced output An output with one leg at ground potential.

unbalanced wire circuit A circuit whose two sides are inherently electrically unlike.

uncertainty An expression of the magnitude of a possible deviation of a measured value from the true value. Frequently it is possible to distinguish two components: the *systematic uncertainty* and the *random uncertainty*. The random uncertainty is expressed by the standard deviation or by a multiple of the standard deviation. The systematic uncertainty is generally estimated on the basis of the parameter characteristics.

undervoltage protection The automatic disconnection by a circuit breaker of loads from a power source when the incoming voltage is too low for safe and reliable operation.

ungrounded A circuit or line not connected to ground.

unicoupler A device used to couple a balanced circuit to an unbalanced circuit.

unidirectional A signal or current flowing in one direction only.

uniform transmission line A transmission line with electrical characteristics that are identical, per unit length, over its entire length.

unit An assembly of equipment and associated wiring that together forms a complete system or independent subsystem.

unity coupling In a theoretically perfect transformer, complete electromagnetic coupling between the primary and secondary windings with no loss of power.

unity gain An amplifier or active circuit in which the output amplitude is the same as the input amplitude.

unity power factor A power factor of 1, which means that the load is—in effect—a pure resistance, with ac voltage and current completely in phase.

unshielded Wiring not protected from electromagnetic and radio frequency interference by a conductive braid or foil.

unterminated A device or system that is not terminated.

up-converter A frequency translation device in which the frequency of the output signal is greater than that of the input signal. Such devices are commonly found in microwave radio and satellite systems.

upgrade An action to improve service by offering better facilities.

uplink A transmission system for sending radio signals from the ground to a satellite or aircraft.

upstream A device or system placed ahead of other devices or systems in a signal path.

up-time The uninterrupted period of time that network or computer resources are accessible and available to a user.

useful life The period during which a low, constant failure rate can be expected for a given device or system. The *useful life* is the portion of a product life cycle between break-in and wear out.

user A person, organization, or group that employs the services of a system for the transfer of information, data processing, or other purposes.

user authentication A security strategy that verifies the identity of a person requesting access to a network, system, and/or computer.

V

VA An abbreviation for *volt-amperes*, volts times amperes.

vacuum fluorescent display (VFD) A light-emitting triode utilizing fluorescent phosphors that can be used in alphanumeric display panels.

vacuum relay A relay whose contacts are enclosed in an evacuated space, usually to provide reliable long-term operation.

vacuum switch A switch whose contacts are enclosed in an evacuated container so that spark formation is discouraged.

validity check A test designed to ensure that the quality of transmission is maintained over a given system.

varactor diode A semiconductor device whose capacitance is a function of the applied voltage. A varactor diode, also called a *variable reactance diode* or simply a *varactor*, is often used to tune the operating frequency of a radio circuit.

variable frequency oscillator (VFO) An oscillator whose frequency can be set to any required value in a given range of frequencies.

variable-gain amplifier An amplifier whose gain can be controlled by an external signal source.

variable impedance A capacitor, inductor, or resistor that is adjustable in value.

varistor A semiconductor device whose resistance is a function of the applied voltage. The resistance of a varistor decreases nonlinearly as the applied voltage is increased.

VCXO (voltage controlled crystal oscillator) A device whose output frequency is determined by an input control voltage.

vector A quantity having both magnitude and direction.

vector diagram A diagram using vectors to indicate the relationship between voltage and current in a circuit.

vector sum The sum of two vectors which, when they are at right angles to each other, equal the length of the hypotenuse of the right triangle so formed. In the general case, the vector sum of the two vectors equals the diagonal of the parallelogram formed on the two vectors.

velocity of light The speed of propagation of electromagnetic waves in a vacuum: equal to 299,792,458 m/s, or approximately 186,000 mi/s. For rough calculations, the figure of 300,000 km/s is used.

velocity of propagation The velocity of signal transmission. In free space electromagnetic waves travel at the speed of light. In a cable the velocity is substantially lower.

vernier A device that enables precision reading of a measuring set or gauge, or the setting of a dial with precision.

vestigial sideband A form of transmission in which one sideband is significantly attenuated. The carrier and the other sideband are transmitted without attenuation.

vibration testing A testing procedure whereby subsystems are mounted on a test base that vibrates, thereby revealing any faults resulting from badly soldered joints or other poor mechanical design features.

volt The standard unit of electromotive force equal to the potential difference between two points in a conductor that is carrying a constant current of one ampere when the power dissipated between the two points is equal to one watt. One *volt* is equivalent to the potential difference across a resistance of one ohm when one ampere is flowing through it. The volt is named for the Italian physicist Alessandro Volta (1745 - 1827).

voltage The potential difference between two points.

voltage drop A decrease in electrical potential resulting from current flow through a resistance.

voltage gradient The continuous drop in electrical potential, per unit length, along a uniform conductor or thickness of a uniform dielectric.

voltage level The ratio of the voltage at a given point to the voltage at an arbitrary reference point.

voltage regulation The deviation from a nominal voltage, expressed as a percentage of the nominal voltage.

voltage regulator A circuit used for controlling and maintaining a voltage at a constant level.

voltage stabilizer A device that produces a constant or substantially constant output voltage despite variations in input voltage or output load current.

voltage to ground The voltage between any given portion of a piece of equipment and the ground potential.

voltaic cell A primary cell that produces electricity by chemical changes. An ordinary dry battery may be referred to as a *voltaic cell.*

volt-ampere (VA) The apparent power in an ac circuit (volts times amperes).

voltmeter An instrument used to measure differences in electrical potential.

volt-ohm-milliammeter (VOM) A general purpose multi-range test meter used to measure voltage, resistance, and current.

vox A voice-operated relay circuit that permits the equivalent of push-to-talk operation of a transmitter by the operator.

VSAT (very small aperture terminal) A satellite Ku-band earth station intended for fixed or portable use. The antenna diameter of a VSAT is on the order of 1.5 m or less.

W

warm start The process of rebooting a computer system without turning the power off.

waterproof cable A cable containing a filling compound in all available spaces in the core to resist the entrance of moisture.

watt The unit of power equal to the work done at one joule per second, or the rate of work measured as a current of one ampere under an electric potential of one volt. Designated by the symbol *W*, the watt is named after the Scottish inventor James Watt (1736 - 1819).

watt-hour The work performed by one watt over a one-hour period.

watt meter A meter indicating in watts the rate of consumption of electrical energy.

wave A disturbance that is a function of time or space, or both, and is propagated in a medium or through space.

waveband A band of wavelengths defined for some given purpose.

waveform The characteristic shape of a periodic wave, determined by the frequencies present and their amplitudes and relative phases.

wavefront A continuous surface that is a locus of points having the same phase at a given instant. A *wavefront* is a surface at right angles to rays that proceed from the wave source. The surface passes through those parts of the wave that are in the same phase and travel in the same direction. For parallel rays the wavefront is a plane; for rays that radiate from a point, the wavefront is spherical.

waveguide Generally, a rectangular or circular pipe that constrains the propagation of an acoustic or electromagnetic wave along a path between two locations. The dimensions of a waveguide determine the frequencies for optimum transmission.

wavelength For a sinusoidal wave, the distance between points of corresponding phase of two consecutive cycles.

weber The unit of magnetic flux equal to the flux that, when linked to a circuit of one turn, produces an electromotive force of one volt as the flux is reduced at a uniform rate to zero in one second. The *weber* is named for the German physicist Wilhelm Eduard Weber (1804 - 1891).

weighted The condition when a correction factor is applied.

weighting The adjustment of a measured value to account for conditions that would otherwise be different during a measurement.

weighting network A circuit, used with a test instrument, that has a specified amplitude-vs.-frequency characteristic.

wideband The passing or processing of a wide range of frequencies. The meaning varies with the context. In an audio system, *wideband* can mean a band of up to 20 kHz wide, but in a television system the band can be many megahertz wide.

wire A single metallic conductor, usually solid-drawn and circular in cross section.

wire line A telecommunications service using copper wire technology, not radio links.

wire mile The resistance of one conductor that is one mile long.

wire stripper A hand tool that enables the removal of insulation from a wire without damaging the conductor itself.

wire-wrapping The termination of wires on tags by firmly wrapping the wire around a sharp-cornered tag that bites through the insulator to the conductor.

working range The permitted range of values of an analog signal over which transmitting or other processing equipment can operate.

working voltage The rated voltage that may safely be applied continuously to a given circuit or device.

wow and flutter Common expression relating to the stability of a tape transport. *Wow* refers to low-frequency variations in pitch; *flutter* refers to high frequency variations in pitch caused by variations in the tape-to-head speed of the machine.

wrap The process of making a connection between a wire and a tag by tightly wrapping the wire around the tag with a special tool.

X

x-cut A method of cutting a quartz plate for an oscillator, with the x-axis of the crystal perpendicular to the faces of the plate.

Y

y-cut A method of cutting a quartz plate for an oscillator, with the y-axis of the crystal perpendicular to the faces of the plate.

yield strength The magnitude of mechanical stress at which a material will begin to deform. Beyond the *yield strength* point, extension is no longer proportional to stress and rupture is possible.

yoke A material that interconnects magnetic cores.

Z

zener A diode, usually fabricated of silicon, in which reverse-voltage breakdown results from the *zener effect.* In this mode, the diode provides a constant voltage drop that is useful as a reference, or in a voltage regulator circuit.

zener breakdown In a semiconductor, a sudden nondestructive breakdown that results when the electric field in the barrier region is sufficiently high to cause a field emission that greatly increases the number of mobile carriers.

Index

D

E

I

J

L

M

N

O

P

Q

R

S

About the Authors

Jerry Whitaker

Jerry Whitaker is a technical writer based in Beaverton, OR. He is a Fellow of the Society of Broadcast Engineers and an SBE-certified senior AM-FM engineer. He is also a member of the Society of Motion Picture and Television Engineers, Audio Engineering Society, International Television Association, and Institute of Electrical and Electronics Engineers (Broadcast Society, Communications Society, Power Electronics Society, and Reliability and Maintainability Society). He has written and lectured extensively on the topic of electronic system installation and maintenance.

Mr. Whitaker is a former radio station chief engineer and television news producer. He is the author of the CRC Press publications *Maintaining Electronic Systems* and *AC Power Systems Handbook*. Mr. Whitaker is also author of the McGraw-Hill publication *Radio Frequency Transmission Systems: Design and Operation*, and co-author of McGraw-Hill's *Television and Audio Handbook for Technicians and Engineers*. Mr. Whitaker is co-editor of the McGraw-Hill *Television Engineering Handbook*, Revised Edition, and a contributor to the McGraw-Hill *Audio Engineering Handbook*. He is co-editor of the Intertec Publishing *Information Age Dictionary*. Mr. Whitaker is also a contributor to the National Association of Broadcaster's *NAB Engineering Handbook*, 7th and 8th Editions.

Mr. Whitaker has twice received a Jesse H. Neal Award *Certificate of Merit* from the Association of Business Publishers for editorial excellence.

Gene De Santis

Gene De Santis is an electrical engineer with more than 25 years of engineering, design, and project management experience. He has been involved with the television industry since 1971. An experienced video systems engineer, Mr. De Santis has designed complete facilities for many companies, including IBM and Price Waterhouse.

After spending two years in the military doing radio frequency interference research on guided missiles, he began work as a project engineer with the Video Systems Engineering group at Sony (now the Sony Broadcast Division). In that position, he was intimately involved with the birth and development of the videocassette industry, which Sony initiated in 1970. In 1973, he became director of engineering at S/T Videocassette Duplicating Corp., a joint venture between Sony and Teletronics International (now VCA Teletronics). During his eight years with S/T, Mr. De Santis developed most of the technical systems, standards, and practices used in the videocassette duplicating industry as we know it today. In 1981, he formed De Santis Associates, an independent consulting firm specializing in professional television systems engineering and design.

Mr. De Santis received his B.S. degree in electrical engineering from New Jersey Institute of Technology in 1969. He is a member of the Society of Motion Picture and Television Engineers and the Institute of Electrical and Electronics Engineers.

C. Robert Paulson

Bob Paulson is managing partner of AVP Communications, a consulting practice he founded in 1974 to provide business/product planning and creative services to manufacturers and end users of electronic communications products, systems, and services. During this period, he has published more than 200 articles and technical papers in a score of United States and British trade publications and professional journals. He authored the *ENG/Field Production Handbook* in 1976 and the *ENG/EFP/EPP Handbook* in 1981. Mr. Paulson is a contributor to the *NAB Engineering Handbook*, 8th Edition. He is a regular columnist for *Television Broadcast* magazine and the *RTNDA Communicator* magazine.

Mr. Paulson received B.A. and M.S. degrees in electrical engineering from Dartmouth College in communications arts and sciences, and later pursued graduate studies in managerial economics and organizational management at the University of California at Berkeley.

Mr. Paulson is a Fellow and past Governor of the Society of Motion Picture and Television Engineers, past Chairman of the SMPTE Public Relations Advisory Committee, a member of the SMPTE *Journal* Board of Editors, a long-time member of SMPTE technology committees, and has chaired SMPTE study groups on the videodisc and fiber optics. He is a life member of the Institute of Electrical and Electronics Engineers, a life fellow of the Armed Forces Communications and Electronics Association, a Society of Broadcast Engineers-certified senior broadcast engineer, and an active member of the Audio Engineering Society, International Television Association, and the Distance Learning Association.